CONTENTS

D1081303

BMA LIBRARY
WITHDRAWN FROM LIBRARY
BRITISH MEDICAL ASSOCIATION

CONTENTS

WITHDRAWN BMA LIBRARY
BRITISH MEDICAL ASSOCIATION

Tortora

PRINCIPLES OF ANATOMY & PHYSIOLOGY

Study Guide

Global Edition

Gerard J. Tortora

Bergen Community College

Bryan Derrickson

Valencia Community College

WILEY

BMA LIBRARY
WITHDRAWN FROM
LIBRARY
BRITISH MEDICAL ASSOCIATION

BRITISH MEDICAL ASSOCIATION
0826333

Copyright © The content provided in this textbook is based on Tortora and Derrickson's Principles of Anatomy and Physiology 15th edition [2017]. John Wiley & Sons Singapore Pte. Ltd.

Cover image: ©Mopic/Shutterstock.

Founded in 1807, John Wiley & Sons, Inc. has been a valued source of knowledge and understanding for more than 200 years, helping people around the world meet their needs and fulfill their aspirations. Our company is built on a foundation of principles that include responsibility to the communities we serve and where we live and work. In 2008, we launched a Corporate Citizenship Initiative, a global effort to address the environmental, social, economic, and ethical challenges we face in our business. Among the issues we are addressing are carbon impact, paper specifications and procurement, ethical conduct within our business and among our vendors, and community and charitable support. For more information, please visit our website: www.wiley.com/go/citizenship.

All rights reserved. This book is authorized for sale in Europe, Asia, Africa, and the Middle East only and may not be exported outside of these territories. Exportation from or importation of this book to another region without the Publisher's authorization is illegal and is a violation of the Publisher's rights. The Publisher may take legal action to enforce its rights. The Publisher may recover damages and costs, including but not limited to lost profits and attorney's fees, in the event legal action is required.

No part of this publication may be reproduced, stored in a retrieval system, or transmitted in any form or by any means, electronic, mechanical, photocopying, recording, scanning, or otherwise, except as permitted under Section 107 or 108 of the 1976 United States Copyright Act, without either the prior written permission of the Publisher or authorization through payment of the appropriate per-copy fee to the Copyright Clearance Center, Inc., 222 Rosewood Drive, Danvers, MA 01923, website www.copyright.com. Requests to the Publisher for permission should be addressed to the Permissions Department, John Wiley & Sons, Inc., 111 River Street, Hoboken, NJ 07030, (201) 748-6011, fax (201) 748-6008, website http://www.wiley.com/go/permissions.

ISBN:

Tortora PAP SET 978-1-119-40006-6

PAP Textbook 978-1-119-38292-8

PAP Study Guide 978-1-119-39993-3

Printed in Markono Print Media Pte Ltd

10 9 8 7 6 5 4 3 2

1 AN INTRODUCTION TO THE HUMAN BODY

1.1 Anatomy and Physiology Defined

1. Anatomy is the science of body structures and the relationships among structures; physiology is the science of body functions.
2. Dissection is the careful cutting apart of body structures to study their relationships.
3. Some branches of anatomy are embryology, developmental biology, cell biology, histology, gross anatomy, systemic anatomy, regional anatomy, surface anatomy, radiographic anatomy, and pathological anatomy (see **Table 1.1**).
4. Some branches of physiology are molecular physiology, neurophysiology, endocrinology, cardiovascular physiology, immunology, respiratory physiology, renal physiology, exercise physiology, and pathophysiology (see **Table 1.1**).

1.2 Levels of Structural Organization and Body Systems

1. The human body consists of six levels of structural organization: chemical, cellular, tissue, organ, system, and organismal.
2. Cells are the basic structural and functional living units of an organism and are the smallest living units in the human body.
3. Tissues are groups of cells and the materials surrounding them that work together to perform a particular function.
4. Organs are composed of two or more different types of tissues; they have specific functions and usually have recognizable shapes.
5. Systems consist of related organs that have a common function.
6. An organism is any living individual.
7. **Table 1.2** introduces the 11 systems of the human organism: the integumentary, skeletal, muscular, nervous, endocrine, cardiovascular, lymphatic, respiratory, digestive, urinary, and reproductive systems.

1.3 Characteristics of the Living Human Organism

1. All organisms carry on certain processes that distinguish them from nonliving things.
2. Among the life processes in humans are metabolism, responsiveness, movement, growth, differentiation, and reproduction.

1.4 Homeostasis

1. Homeostasis is the maintenance of relatively stable conditions in the body's internal environment produced by the interplay of all of the body's regulatory processes.
2. Body fluids are dilute, watery solutions. Intracellular fluid (ICF) is inside cells, and extracellular fluid (ECF) is outside cells. Plasma is the ECF within blood vessels. Interstitial fluid is the ECF that fills spaces between tissue cells. Because it surrounds the cells of the body, extracellular fluid is called the body's internal environment.
3. Disruptions of homeostasis come from external and internal stimuli and psychological stresses. When disruption of homeostasis is mild and temporary, responses of body cells quickly restore balance in the internal environment. If disruption is extreme, regulation of homeostasis may fail.
4. Most often, the nervous and endocrine systems acting together or separately regulate homeostasis. The nervous system detects body changes and sends nerve impulses to counteract changes in controlled conditions. The endocrine system regulates homeostasis by secreting hormones.
5. Feedback systems include three components: (1) Receptors monitor changes in a controlled condition and send input to a control center (afferent pathway). (2) The control center sets the value (set point) at which a controlled condition should be maintained, evaluates the input it receives from receptors (efferent pathway), and generates output commands when they are needed. (3) Effectors receive output from the control center and produce a response (effect) that alters the controlled condition.
6. If a response reverses the original stimulus, the system is operating by negative feedback. If a response enhances the original stimulus, the system is operating by positive feedback.
7. One example of negative feedback is the regulation of blood pressure. If a stimulus causes blood pressure (controlled condition) to rise, baroreceptors (pressure-sensitive nerve cells, the receptors) in blood vessels send impulses (input) to the brain (control center). The brain sends impulses (output) to the heart (effector). As a result, heart rate decreases (response) and blood pressure decreases to normal (restoration of homeostasis).
8. One example of positive feedback occurs during the birth of a baby. When labor begins, the cervix of the uterus is stretched (stimulus), and stretch-sensitive nerve cells in the cervix (receptors) send nerve impulses (input) to the brain (control center). The brain responds by releasing oxytocin (output), which stimulates the uterus (effector) to contract more forcefully (response). Movement of the fetus further stretches the cervix, more oxytocin is released, and even more forceful contractions occur. The cycle is broken with the birth of the baby.
9. Disruptions of homeostasis—homeostatic imbalances—can lead to disorders, diseases, and even death. A disorder is a general term for any abnormality of structure or function. A disease is an illness with a definite set of signs and symptoms.

10. Symptoms are subjective changes in body functions that are not apparent to an observer; signs are objective changes that can be observed and measured.

1.5 Basic Anatomical Terminology

1. Descriptions of any region of the body assume the body is in the anatomical position, in which the subject stands erect facing the observer, with the head level and the eyes facing directly forward. The feet are flat on the floor and directed forward, and the upper limbs are at the sides, with the palms turned forward. A body lying facedown is prone; a body lying faceup is supine.

2. Regional names are terms given to specific regions of the body. The principal regions are the head, neck, trunk, upper limbs, and lower limbs. Within the regions, specific body parts have anatomical names and corresponding common names. Examples are thoracic (chest), nasal (nose), and carpal (wrist).

3. Directional terms indicate the relationship of one part of the body to another. **Exhibit 1** summarizes commonly used directional terms.

4. Planes are imaginary flat surfaces that are used to divide the body or organs to visualize interior structures. A midsagittal plane divides the body or an organ into *equal* right and left sides. A parasagittal plane divides the body or an organ into *unequal* right and left sides. A frontal plane divides the body or an organ into anterior and posterior portions. A transverse plane divides the body or an organ into superior and inferior portions. An oblique plane passes through the body or an organ at an oblique angle.

5. Sections are cuts of the body or its organs made along a plane. They are named according to the plane along which the cut is made and include transverse, frontal, and sagittal sections.

6. **Figure 1.10** summarizes body cavities and their membranes. Body cavities are spaces in the body that help protect, separate, and support internal organs. The cranial cavity contains the brain, and the vertebral canal contains the spinal cord. The meninges are protective tissues that line the cranial cavity and vertebral canal. The diaphragm separates the thoracic cavity from the abdominopelvic cavity. Viscera are organs within the thoracic and abdominopelvic cavities. A serous membrane lines the wall of the cavity and adheres to the viscera.

7. The thoracic cavity is subdivided into three smaller cavities: a pericardial cavity, which contains the heart, and two pleural cavities, each of which contains a lung. The central part of the thoracic cavity is an anatomical region called the mediastinum. It is located between the pleural cavities, extending from the sternum to the vertebral column and from the first rib to the diaphragm. It contains all thoracic viscera except the lungs.

8. The abdominopelvic cavity is divided into a superior abdominal and an inferior pelvic cavity. Viscera of the abdominal cavity include the stomach, spleen, liver, gallbladder, small intestine, and most of the large intestine. Viscera of the pelvic cavity include the urinary bladder, portions of the large intestine, and internal organs of the reproductive system.

9. Serous membranes line the walls of the thoracic and abdominal cavities and cover the organs within them. They include the pleura, associated with the lungs; the pericardium, associated with the heart; and the peritoneum, associated with the abdominal cavity.

10. To describe the location of organs more easily, the abdominopelvic cavity is divided into nine regions: right hypochondriac, epigastric, left hypochondriac, right lumbar, umbilical, left lumbar, right inguinal (iliac), hypogastric (pubic), and left inguinal (iliac). To locate the site of an abdominopelvic abnormality in clinical studies, the abdominopelvic cavity is divided into quadrants: right upper quadrant (RUQ), left upper quadrant (LUQ), right lower quadrant (RLQ), and left lower quadrant (LLQ).

1.6 Aging and Homeostasis

1. **Aging** produces observable changes in structure and function and increases vulnerability to stress and disease.

2. Changes associated with aging occur in all body systems.

1.7 Medical Imaging

1. Medical imaging refers to techniques and procedures used to create images of the human body. They allow visualization of internal structures to diagnose abnormal anatomy and deviations from normal physiology.

2. **Table 1.3** summarizes and illustrates several medical imaging techniques.

CLINICAL CONNECTIONS

Noninvasive Diagnostic Techniques (Refer page 3 of Textbook)

Health-care professionals and students of anatomy and physiology commonly use several noninvasive diagnostic techniques to assess certain aspects of body structure and function. A **noninvasive diagnostic technique** is one that does not involve insertion of an instrument or device through the skin or a body opening. In **inspection,** the examiner observes the body for any changes that deviate from normal. For example, a physician may examine the mouth cavity for evidence of disease. Following inspection, one or more additional techniques may be employed. In **palpation** (pal-PĀ-shun; *palp-* = gently touching) the examiner feels body surfaces with the hands. An example is palpating the abdomen to detect enlarged or tender internal organs or abnormal masses. In **auscultation** (aws-kul-TĀ -shun; *auscult-* = listening) the examiner listens to body sounds to evaluate the functioning of certain organs, often using a

stethoscope to amplify the sounds. An example is auscultation of the lungs during breathing to check for crackling sounds associated with abnormal fluid accumulation. In **percussion** (pur-KUSH-un; *percus-* = beat through) the examiner taps on the body surface with the fingertips and listens to the resulting sound. Hollow cavities or spaces produce a different sound than solid organs. For example, percussion may reveal the abnormal presence of fluid in the lungs or air in the intestines. It may also provide information about the size, consistency, and position of an underlying structure. An understanding of anatomy is important for the effective application of most of these diagnostic techniques.

Autopsy (Refer page 7 of Textbook)

An **autopsy** (AW-top-sē = seeing with one's own eyes) or *necropsy* is a postmortem (after death) examination of the body and dissection of its internal organs to confirm or determine the cause of death. An autopsy can uncover the existence of diseases not detected during life, determine the extent of injuries, and explain how those injuries may have contributed to a person's death. It also may provide more information about a disease, assist in the accumulation of statistical data, and educate health-care students. Moreover, an autopsy can reveal conditions that may affect offspring or siblings (such as congenital heart defects). Sometimes an autopsy is legally required, such as during a criminal investigation. It also may be useful in resolving disputes between beneficiaries and insurance companies about the cause of death.

Diagnosis of Disease (Refer page 12 of Textbook)

Diagnosis (dī-ag-NŌ-sis; *dia-* = through; *-gnosis* = knowledge) is the science and skill of distinguishing one disorder or disease from another. The patient's symptoms and signs, his or her medical history, a physical exam, and laboratory tests provide the basis for making a diagnosis. Taking a *medical history* consists of collecting information about events that might be related to a patient's illness. These include the chief complaint (primary reason for seeking medical attention), history of present illness, past medical problems, family medical problems, social history, and review of symptoms. A *physical examination* is an orderly evaluation of the body and its functions. This process includes the noninvasive techniques of inspection, palpation, auscultation, and percussion that you learned about earlier in the chapter, along with measurement of vital signs (temperature, pulse, respiratory rate, and blood pressure), and sometimes laboratory tests.

SELF-QUIZ QUESTIONS

Fill in the blanks in the following statements.

1. A(n) ———— is a group of similar cells and their surrounding materials performing specific functions.

2. The sum of all of the body's chemical processes is ————. It consists of two parts: the phase that builds up new substances is ————, and the phase that breaks down substances is ————.

3. The fluid located within cells is the ————, whereas the fluid located outside of the cells is ————.

Indicate whether the following statements are true or false.

4. In a positive feedback system, the response enhances or intensifies the original stimulus.

5. A person lying face down would be in the supine position.

6. The highest level of structural organization is the system level.

Choose the one best answer to the following questions.

7. A plane that separates the body into unequal right and left sides is a
   (a) transverse plane   (b) frontal plane
   (c) midsagittal plane   (d) coronal plane
   (e) parasagittal plane

8. Midway through a 5-mile workout, a runner begins to sweat profusely. The sweat glands producing the sweat would be considered which part of a feedback loop?
   (a) controlled condition (b) receptors   (c) stimulus
   (d) effectors   (e) control center

9. An unspecialized stem cell becomes a brain cell during fetal development. This is an example of
   (a) differentiation   (b) growth   (c) organization
   (d) responsiveness   (e) homeostasis.

10. A radiography technician needs to x-ray a growth on the urinary bladder. To accomplish this, the camera must be positioned on the ———— region.
    (a) left inguinal   (b) epigastric   (c) hypogastric
    (d) right inguinal   (e) umbilical

11. Which of the following would not be associated with the thoracic cavity? (1) pericardium, (2) mediastinum, (3) peritoneum, (4) pleura,
    (a) 2 and 3   (b) 2   (c) 3
    (d) 1 and 4   (e) 3 and 4

12. Match the following common names and anatomical descriptive adjectives:
    ———— (a) axillary   (1) skull
    ———— (b) inguinal   (2) eye
    ———— (c) cervical   (3) cheek
    ———— (d) cranial   (4) armpit
    ———— (e) oral   (5) arm
    ———— (f) brachial   (6) groin
    ———— (g) orbital   (7) buttock
    ———— (h) gluteal   (8) neck
    ———— (i) buccal   (9) mouth
    ———— (j) coxal   (10) hip

13. Choose the term that best fits the blank in each statement. Some answers may be used more than once.
    ———— (a) Your eyes are ———— to your chin.   (1) superior
    ———— (b) Your skin is ———— to your heart.   (2) inferior
    ———— (c) Your right shoulder is ————   (3) anterior
              and ———— from your umbilicus   (4) posterior
              (belly button).   (5) medial
    ———— (d) In the anatomical position, your   (6) lateral
              thumb is ————.   (7) intermediate
    ———— (e) Your buttocks are ————.   (8) ipsilateral
    ———— (f) Your right foot and right hand   (9) contralateral
              are ————.   (10) proximal
    ———— (g) Your knee is ———— between   (11) distal
              your thigh and toes.   (12) superficial
    ———— (h) Your lungs are ———— to your   (13) deep
              spinal column.
    ———— (i) Your breastbone is ———— to
              your chin.
    ———— (j) Your calf is ———— to your heel.

14. Match the following cavities to their definitions:

—— (a) a fluid-filled space that surrounds the heart

—— (b) the cavity that contains the brain

—— (c) a cavity formed by the ribs, muscles of the chest, sternum, and part of the vertebral column

—— (d) a cavity that contains the stomach, spleen, liver, gallbladder, small intestine, and most of the large intestine

—— (e) fluid-filled space that surrounds a lung

—— (f) the cavity that contains the urinary bladder, part of the large intestine, and the organs of the reproductive system

—— (g) the canal that contains the spinal cord

(1) cranial cavity

(2) vertebral canal

(3) thoracic cavity

(4) pericardial cavity

(5) pleural cavity

(6) abdominal cavity

(7) pelvic cavity

15. Match the following systems with their functions:

—— (a) nervous system

—— (b) endocrine system

—— (c) urinary system

—— (d) cardiovascular system

—— (e) muscular system

—— (f) respiratory system

—— (g) digestive system

—— (h) skeletal system

—— (i) integumentary system

—— (j) lymphatic system and immunity

—— (k) reproductive system

(1) regulates body activities through hormones (chemicals) transported in the blood to various target organs of the body

(2) produces gametes; releases hormones from gonads

(3) protects against disease; returns fluids to blood

(4) protects body by forming a barrier to the outside environment; helps regulate body temperature

(5) transports oxygen and nutrients to cells; protects against disease; carries wastes away from cells

(6) regulates body activities through action potentials (nerve impulses); receives sensory information; interprets and responds to the information

(7) carries out the physical and chemical breakdown of food and absorption of nutrients

(8) transfers oxygen and carbon dioxide between air and blood

(9) supports and protects the body; provides internal framework; provides a place for muscle attachment

(10) powers movements of the body and stabilizes body position

(11) eliminates wastes; regulates the volume and chemical composition of blood

2 THE CHEMICAL LEVEL OF ORGANIZATION

CHAPTER REVIEW

2.1 How Matter Is Organized

1. Chemistry is the science of the structure and interactions of matter.
2. All forms of matter are composed of chemical elements.
3. Oxygen, carbon, hydrogen, and nitrogen make up about 96% of body mass.
4. Each element is made up of small units called atoms. Atoms consist of a nucleus, which contains protons and neutrons, plus electrons that move about the nucleus in regions called electron shells.
5. The number of protons (the atomic number) distinguishes the atoms of one element from those of another element.
6. The mass number of an atom is the sum of its protons and neutrons.
7. Different atoms of an element that have the same number of protons but different numbers of neutrons are called isotopes. Radioactive isotopes are unstable and decay.
8. The atomic mass of an element is the average mass of all naturally occurring isotopes of that element.
9. An atom that *gives up* or *gains* electrons becomes an ion—an atom that has a positive or negative charge because it has unequal numbers of protons and electrons. Positively charged ions are cations; negatively charged ions are anions.
10. If two atoms share electrons, a molecule is formed. Compounds contain atoms of two or more elements.
11. A free radical is an atom or group of atoms with an unpaired electron in its outermost shell. A common example is superoxide, an anion which is formed by the addition of an electron to an oxygen molecule.

2.2 Chemical Bonds

1. Forces of attraction called chemical bonds hold atoms together. These bonds result from gaining, losing, or sharing electrons in the valence shell.
2. Most atoms become stable when they have an octet of eight electrons in their valence (outermost) electron shell.
3. When the force of attraction between ions of opposite charge holds them together, an ionic bond has formed.
4. In a covalent bond, atoms share pairs of valence electrons. Covalent bonds may be single, double, or triple and either nonpolar or polar.
5. An atom of hydrogen that forms a polar covalent bond with an oxygen atom or a nitrogen atom may also form a weaker bond, called a hydrogen bond, with an electronegative atom. The polar covalent bond causes the hydrogen atom to have a partial positive charge (δ^+) that attracts the partial negative charge (δ^-) of neighboring electronegative atoms, often oxygen or nitrogen.

2.3 Chemical Reactions

1. When atoms combine with or break apart from other atoms, a chemical reaction occurs. The starting substances are the reactants, and the ending ones are the products.
2. Energy, the capacity to do work, is of two principal kinds: potential (stored) energy and kinetic energy (energy of motion).
3. Endergonic reactions require energy; exergonic reactions release energy. ATP couples endergonic and exergonic reactions.
4. The initial energy investment needed to start a reaction is the activation energy. Reactions are more likely when the concentrations and the temperatures of the reacting particles are higher.
5. Catalysts accelerate chemical reactions by lowering the activation energy. Most catalysts in living organisms are protein molecules called enzymes.
6. Synthesis reactions involve the combination of reactants to produce larger molecules. The reactions are anabolic and usually endergonic.
7. In decomposition reactions, a substance is broken down into smaller molecules. The reactions are catabolic and usually exergonic.
8. Exchange reactions involve the replacement of one atom or atoms by another atom or atoms.
9. In reversible reactions, end products can revert to the original reactants.

2.4 Inorganic Compounds and Solutions

1. Inorganic compounds usually are small and usually lack carbon. Organic substances always contain carbon, usually contain hydrogen, and always have covalent bonds.
2. Water is the most abundant substance in the body. It is an excellent solvent and suspension medium, participates in hydrolysis and dehydration synthesis reactions, and serves as a lubricant. Because of its many hydrogen bonds, water molecules are cohesive, which causes a high surface tension. Water also has a high capacity for absorbing heat and a high heat of vaporization.
3. Inorganic acids, bases, and salts dissociate into ions in water. An acid ionizes into hydrogen ions (H^+) and anions and is a proton donor; many bases ionize into cations and hydroxide ions (OH^-), and all are proton acceptors. A salt ionizes into neither H^+ nor OH^-.

4. Mixtures are combinations of elements or compounds that are physically blended together but are not bound by chemical bonds. Solutions, colloids, and suspensions are mixtures with different properties.

5. Two ways to express the concentration of a solution are *percentage* (mass per volume), expressed in grams per 100 mL of a solution, and molarity, expressed in moles per liter. A mole (abbreviated mol) is the amount in grams of any substance that has a mass equal to the combined atomic mass of all its atoms.

6. The pH of body fluids must remain fairly constant for the body to maintain homeostasis. On the pH scale, 7 represents neutrality. Values below 7 indicate acidic solutions, and values above 7 indicate alkaline solutions. Normal blood pH is 7.35–7.45.

7. Buffer systems remove or add protons (H^+) to help maintain pH homeostasis.

8. One important buffer system is the carbonic acid–bicarbonate buffer system. The bicarbonate ion (HCO_3^-) acts as a weak base and removes excess H^+, and carbonic acid (H_2CO_3) acts as a weak acid and adds H^+.

2.5 Overview of Organic Compounds

1. Carbon, with its four valence electrons, bonds covalently with other carbon atoms to form large molecules of many different shapes. Attached to the carbon skeletons of organic molecules are functional groups that confer distinctive chemical properties.

2. Small organic molecules are joined together to form larger molecules by dehydration synthesis reactions in which a molecule of water is removed. In the reverse process, called hydrolysis, large molecules are broken down into smaller ones by the addition of water.

2.6 Carbohydrates

1. Carbohydrates include monosaccharides, disaccharides, and polysaccharides.

2. Carbohydrates provide most of the chemical energy needed to generate ATP.

2.7 Lipids

1. Lipids are a diverse group of compounds that include fatty acids, triglycerides (fats and oils), phospholipids, steroids, and eicosanoids.

2. Fatty acids are the simplest lipids; they are used to synthesize triglycerides and phospholipids.

3. Triglycerides protect, insulate, and provide energy.

4. Phospholipids are important cell membrane components.

5. Steroids are important in cell membrane structure, regulating sexual functions, maintaining normal blood sugar level, aiding lipid digestion and absorption, and helping bone growth.

6. Eicosanoids (prostaglandins and leukotrienes) modify hormone responses, contribute to inflammation, dilate airways, and regulate body temperature.

2.8 Proteins

1. Proteins are constructed from amino acids.

2. Proteins give structure to the body, regulate processes, provide protection, help muscles contract, transport substances, and serve as enzymes.

3. Levels of structural organization among proteins include primary, secondary, tertiary, and (sometimes) quaternary. Variations in protein structure and shape are related to their diverse functions.

2.9 Nucleic Acids

1. Deoxyribonucleic acid (DNA) and ribonucleic acid (RNA) are nucleic acids consisting of nitrogenous bases, five-carbon (pentose) sugars, and phosphate groups.

2. DNA is a double helix and is the primary chemical in genes. RNA takes part in protein synthesis.

2.10 Adenosine Triphosphate

1. Adenosine triphosphate (ATP) is the principal energy-transferring molecule in living systems.

2. When ATP transfers energy to an endergonic reaction, it is decomposed to adenosine diphosphate (ADP) and a phosphate group.

3. ATP is synthesized from ADP and a phosphate group using the energy supplied by various decomposition reactions, particularly those of glucose.

CLINICAL CONNECTIONS

Harmful and Beneficial Effects of Radiation (Refer page 28 of Textbook)

Radioactive isotopes may have either harmful or helpful effects. Their radiations can break apart molecules, posing a serious threat to the human body by producing tissue damage or causing various types of cancer. Although the decay of naturally occurring radioactive isotopes typically releases just a small amount of radiation into the environment, localized accumulations can occur. Radon-222, a colorless and odorless gas that is a naturally occurring radioactive breakdown product of uranium, may seep out of the soil and accumulate in buildings. It is not only associated with many cases of lung cancer in smokers but also has been implicated in many cases of lung cancer in nonsmokers. Beneficial effects of certain radioisotopes include their use in medical imaging procedures to diagnose and treat certain disorders. Some radioisotopes can be used as **tracers** to follow the movement of certain substances through the body. Thallium-201 is used to monitor blood flow through the heart during an exercise stress test. Iodine-131 is used to detect cancer of the thyroid gland and to assess its size and activity, and may also be used to destroy part of an overactive thyroid gland. Cesium-137 is used to treat advanced cervical cancer, and iridium-192 is used to treat prostate cancer.

Free Radicals and Antioxidants (Refer page 29 of Textbook)

There are several sources of **free radicals**, including exposure to ultraviolet radiation in sunlight, exposure to x-rays, and some reactions that occur during normal metabolic processes. Certain harmful substances, such as carbon tetrachloride (a solvent used in dry cleaning), also give rise to free radicals when they participate in metabolic reactions in the body. Among the many disorders, diseases, and conditions linked to oxygen-derived free radicals are cancer, atherosclerosis, Alzheimer's disease, emphysema, diabetes mellitus, cataracts, macular degeneration, rheumatoid arthritis, and deterioration associated with aging. Consuming more **antioxidants**—substances that inactivate oxygen-derived free radicals—is thought to slow the pace of damage caused by free radicals. Important dietary antioxidants include selenium, zinc, beta-carotene, and vitamins C and E. Red, blue, or purple fruits and vegetables contain high levels of antioxidants.

Artificial Sweeteners (Refer page 42 of Textbook)

Some individuals use **artificial sweeteners** to limit their sugar consumption for medical reasons, while others do so to avoid calories that might result in weight gain. Examples of artificial sweeteners include aspartame (trade names NutraSweet® and Equal®), saccharin (Sweet 'N Low®), and sucralose (Splenda®). Aspartame is 200 times sweeter than sucrose and it adds essentially no calories to the diet because only small amounts of it are used to produce a sweet taste. Saccharin is about 400 times sweeter than sucrose, and sucralose is 600 times sweeter than sucrose. Both saccharin and sucralose have zero calories because they pass through the body without being metabolized. Artificial sweeteners are also used as sugar substitutes because they do not cause tooth decay. In fact, studies have shown that using artificial sweeteners in the diet helps reduce the incidence of dental cavities.

Fatty Acids in Health and Disease (Refer page 44 of Textbook)

As its name implies, a group of fatty acids called **essential fatty acids (EFAs)** is essential to human health. However, they cannot be made by the human body and must be obtained from foods or supplements. Among the more important EFAs are *omega-3 fatty acids, omega-6 fatty acids,* and cis-*fatty acids.*

Omega-3 and omega-6 fatty acids are polyunsaturated fatty acids that are believed to work together to promote health. They may have a protective effect against heart disease and stroke by lowering total cholesterol, raising HDL (high-density lipoproteins or "good cholesterol") and lowering LDL (low-density lipoproteins or "bad cholesterol"). In addition, omega-3 and omega-6

fatty acids decrease bone loss by increasing calcium utilization by the body; reduce symptoms of arthritis due to inflammation; promote wound healing; improve certain skin disorders (psoriasis, eczema, and acne); and improve mental functions. Primary sources of omega-3 fatty acids include flaxseed, fatty fish, oils that have large amounts of polyunsaturated fatty acids, fish oils, and walnuts. Primary sources of omega-6 fatty acids include most processed foods (cereals, breads, white rice), eggs, baked goods, oils with large amounts of polyunsaturated fatty acids, and meats (especially organ meats, such as liver).

Note in **Figure 2.17a** that the hydrogen atoms on either side of the double bond in oleic acid are on the same side of the unsaturated fatty acid. Such cis-fatty acids are nutritionally beneficial unsaturated fatty acids that are used by the body to produce hormonelike regulators and cell membranes. However, when cis-fatty acids are heated, pressurized, and combined with a catalyst in a process called *hydrogenation,* they are changed to unhealthy *trans*-fatty acids. In *trans*-fatty acids the hydrogen atoms are on opposite sides of the double bond of an unsaturated fatty acid. Hydrogenation is used by manufacturers to make vegetable oils solid at room temperature and less likely to turn rancid. If oil used for frying is reused (like in fast food french fry machines), cis-fatty acids are converted to *trans*-fatty acids. Among the adverse effects of *trans*-fatty acids are an increase in total cholesterol, a decrease in HDL, an increase in LDL, and an increase in triglycerides. These effects, which can increase the risk of heart disease and other cardiovascular diseases, are similar to those caused by saturated fats.

DNA Fingerprinting (Refer page 51 of Textbook)

A technique called **DNA fingerprinting** is used in research and in courts of law to ascertain whether a person's DNA matches the DNA obtained from samples or pieces of legal evidence such as blood stains or hairs. In each person, certain DNA segments contain base sequences that are repeated several times. Both the number of repeat copies in one region and the number of regions subject to repeat are different from one person to another. DNA fingerprinting can be done with minute quantities of DNA—for example, from a single strand of hair, a drop of semen, or a spot of blood. It also can be used to identify a crime victim or a child's biological parents and even to determine whether two people have a common ancestor.

SELF-QUIZ QUESTIONS

Fill in the blanks in the following statements.

1. An atom with a mass number of 18 that contains 10 neutrons would have an atomic number of _____ .

2. Matter exists in three forms: _____ , _____ , and _____ .

3. The building blocks of carbohydrates are the monomers _____ while the building blocks of proteins are the monomers _____ .

Indicate whether the following statements are true or false.

4. The elements that compose most of the body's mass are carbon, hydrogen, oxygen, and nitrogen.

5. Ionic bonds are created when atoms share electrons in the valence shell.

6. Human blood has a normal pH between 7.35 and 7.45 and is considered slightly alkaline.

Choose the one best answer to the following questions.

7. Which of the following would be considered a compound?
   (1) $C_6H_{12}O_6$, (2) O_2, (3) Fe, (4) H_2, (5) CH_4.
   (a) all are compounds
   (b) 1, 2, 4, and 5
   (c) 1 and 5
   (d) 2 and 4
   (e) 3

8. The monosaccharides glucose and fructose combine to form the disaccharide sucrose by a process known as
   (a) dehydration synthesis
   (b) hydrolysis
   (c) decomposition
   (d) hydrogen bonding
   (e) ionization.

9. Which of the following is *not* a function of proteins?
   (a) provide structural framework
   (b) bring about contraction
   (c) transport materials throughout the body
   (d) store energy
   (e) regulate many physiological processes

10. Which of the following organic compounds are classified as lipids?
    (1) polysaccharides, (2) triglycerides, (3) steroids, (4) enzymes, (5) eicosanoids.
    (a) 1, 2, and 4
    (b) 2, 3, and 5
    (c) 2 and 5
    (d) 2, 3, 4, and 5
    (e) 2 and 3

11. A compound dissociates in water and forms a cation other than H^+ and an anion other than OH^-. This substance most likely is a(n)
    (a) acid
    (b) base
    (c) enzyme
    (d) buffer
    (e) salt.

12. Which of the following statements regarding ATP are *true*? (1) ATP is the energy currency for the cell. (2) The energy supplied by the hydrolysis of ATP is constantly being used by cells. (3) Energy is required to produce ATP. (4) The production of ATP involves both aerobic and anaerobic phases. (5) The process of producing energy in the form of ATP is termed the law of conservation of energy.

    (a) 1, 2, 3, and 4 (b) 1, 2, 3, and 5 (c) 2, 4, and 5

    (d) 1, 2, and 4 (e) 3, 4, and 5

13. During the course of analyzing an unknown chemical, a chemist determines that the chemical is composed of carbon, hydrogen, and oxygen in the proportion of 1 carbon to 2 hydrogens to 1 oxygen. The chemical is probably

    (a) an amino acid (b) DNA

    (c) a triglyceride (d) a protein

    (e) a monosaccharide.

14. Match the following reactions with the term that describes them:

    _____ (a) $H_2 + Cl_2 \longrightarrow 2HCl$

    _____ (b) $3\,NaOH + H_3PO_4 \longrightarrow$ $Na_3PO_4 + 3\,H_2O$

    _____ (c) $CaCO_3 + CO_2 + H_2O \longrightarrow$ $Ca(HCO_3)_2$

    _____ (d) $NH_3 + H_2O \rightleftharpoons NH_4^+ + OH^-$

    _____ (e) $C_{12}H_{22}O_{11} + H_2O \longrightarrow$ $C_6H_{12}O_6 + C_6H_{12}O_6$

    (1) synthesis reaction
    (2) exchange reaction
    (3) decomposition reaction
    (4) reversible reaction

15. Match the following:

    _____ (a) an abundant polar covalent molecule that serves as a solvent, has a high heat capacity, creates a high surface tension, and serves as a lubricant

    _____ (b) a substance that dissociates into one or more hydrogen ions and one or more anions

    (1) acid
    (2) free radical
    (3) base
    (4) buffer
    (5) enzyme
    (6) ion
    (7) pH

    _____ (c) a substance that dissociates into cations and anions, neither of which is a hydrogen ion or a hydroxide ion

    _____ (d) a proton acceptor

    _____ (e) a measure of hydrogen ion concentration

    _____ (f) a chemical compound that can convert strong acids and bases into weak ones

    _____ (g) a catalyst for chemical reactions that is specific, efficient, and under cellular control

    _____ (h) a single-stranded compound that contains a five-carbon sugar, and the bases adenine, cytosine, guanine, and uracil

    _____ (i) a compound that functions to temporarily store and then transfer energy liberated in exergonic reactions to cellular activities that require energy

    _____ (j) a double-stranded compound that contains a five-carbon sugar, the bases adenine, thymine, cytosine, and guanine, and the body's genetic material

    _____ (k) a charged atom

    _____ (l) a charged atom with an unpaired electron in its outermost shell

    (8) salt
    (9) RNA
    (10) ATP
    (11) water
    (12) DNA

3 THE CELLULAR LEVEL OF ORGANIZATION

CHAPTER REVIEW

Introduction

1. A cell is the basic, living, structural and functional unit of the body.
2. Cell biology is the scientific study of cellular structure and function.

3.1 Parts of a Cell

1. **Figure 3.1** provides an overview of the typical structures in body cells.
2. The principal parts of a cell are the plasma membrane; the cytoplasm, the cellular contents between the plasma membrane and nucleus; and the nucleus.

3.2 The Plasma Membrane

1. The plasma membrane, which surrounds and contains the cytoplasm of a cell, is composed of proteins and lipids.
2. According to the fluid mosaic model, the membrane is a mosaic of proteins floating like icebergs in a lipid bilayer sea.
3. The lipid bilayer consists of two back-to-back layers of phospholipids, cholesterol, and glycolipids. The bilayer arrangement occurs because the lipids are amphipathic, having both polar and nonpolar parts.
4. Integral proteins extend into or through the lipid bilayer; peripheral proteins associate with membrane lipids or integral proteins at the inner or outer surface of the membrane.
5. Many integral proteins are glycoproteins, with sugar groups attached to the ends that face the extracellular fluid. Together with glycolipids, the glycoproteins form a glycocalyx on the extracellular surface of cells.
6. Membrane proteins have a variety of functions. Integral proteins are channels and carriers that help specific solutes cross the membrane; receptors that serve as cellular recognition sites; enzymes that catalyze specific chemical reactions; and linkers that anchor proteins in the plasma membranes to protein filaments inside and outside the cell. Peripheral proteins serve as enzymes and linkers; support the plasma membrane; anchor integral proteins; and participate in mechanical activities. Membrane glycoproteins function as cell-identity markers.
7. Membrane fluidity is greater when there are more double bonds in the fatty acid tails of the lipids that make up the bilayer.

Cholesterol makes the lipid bilayer stronger but less fluid at normal body temperature. Its fluidity allows interactions to occur within the plasma membrane, enables the movement of membrane components, and permits the lipid bilayer to self-seal when torn or punctured.

8. The membrane's selective permeability permits some substances to pass more readily than others. The lipid bilayer is permeable to most nonpolar, uncharged molecules. It is impermeable to ions and charged or polar molecules other than water and urea. Channels and carriers increase the plasma membrane's permeability to small and medium-sized polar and charged substances, including ions, that cannot cross the lipid bilayer.
9. The selective permeability of the plasma membrane supports the existence of concentration gradients, differences in the concentrations of chemicals between one side of the membrane and the other.

3.3 Transport across the Plasma Membrane

1. In passive processes, a substance moves down its concentration gradient across the membrane using its own kinetic energy of motion. In active processes, cellular energy is used to drive the substance "uphill" against its concentration gradient.
2. In diffusion, molecules or ions move from an area of higher concentration to an area of lower concentration until an equilibrium is reached. The rate of diffusion across a plasma membrane is affected by the steepness of the concentration gradient, temperature, mass of the diffusing substance, surface area available for diffusion, and the distance over which diffusion must occur.
3. Nonpolar, hydrophobic molecules such as oxygen, carbon dioxide, nitrogen, steroids, and fat-soluble vitamins (A, E, D, and K) plus small, polar, uncharged molecules such as water, urea, and small alcohols diffuse through the lipid bilayer of the plasma membrane via simple diffusion.
4. In channel-mediated facilitated diffusion, a solute moves down its concentration gradient across the lipid bilayer through a membrane channel. Examples include ion channels that allow specific ions such as K^+, Cl^-, Na^+, or Ca^{2+} (which are too hydrophilic to penetrate the membrane's nonpolar interior) to move across the plasma membrane. In carrier-mediated facilitated diffusion, a solute such as glucose binds to a specific carrier protein

on one side of the membrane and is released on the other side after the carrier undergoes a change in shape.

5. Osmosis is a type of diffusion in which there is net movement of water through a selectively permeable membrane from an area of higher water concentration to an area of lower water concentration. In an isotonic solution, red blood cells maintain their normal shape; in a hypotonic solution, they swell and undergo hemolysis; in a hypertonic solution, they shrink and undergo crenation.

6. Substances can cross the membrane against their concentration gradient by active transport. Actively transported substances include ions such as Na^+, K^+, H^+, Ca^{2+}, I^-, and Cl^-; amino acids; and monosaccharides. Two sources of energy drive active transport: Energy obtained from hydrolysis of ATP is the source in primary active transport, and energy stored in a Na^+ or H^+ concentration gradient is the source in secondary active transport. The most prevalent primary active transport pump is the sodium–potassium pump, also known as Na^+–K^+ ATPase. Secondary active transport mechanisms include both symporters and antiporters that are powered by either a Na^+ or H^+ concentration gradient. Symporters move two substances in the same direction across the membrane; antiporters move two substances in opposite directions.

7. In endocytosis, tiny vesicles detach from the plasma membrane to move materials across the membrane into a cell; in exocytosis, vesicles merge with the plasma membrane to move materials out of a cell. Receptor-mediated endocytosis is the selective uptake of large molecules and particles (ligands) that bind to specific receptors in membrane areas called clathrin-coated pits. In bulk-phase endocytosis (pinocytosis), the ingestion of extracellular fluid, a vesicle surrounds the fluid to take it into the cell.

8. Phagocytosis is the ingestion of solid particles. Some white blood cells destroy microbes that enter the body in this way.

9. In transcytosis, vesicles undergo endocytosis on one side of a cell, move across the cell, and undergo exocytosis on the opposite side.

3.4 Cytoplasm

1. Cytoplasm—all the cellular contents within the plasma membrane except for the nucleus—consists of cytosol and organelles. Cytosol is the fluid portion of cytoplasm, containing water, ions, glucose, amino acids, fatty acids, proteins, lipids, ATP, and waste products. It is the site of many chemical reactions required for a cell's existence. Organelles are specialized structures with characteristic shapes that have specific functions.

2. Components of the cytoskeleton, a network of several kinds of protein filaments that extend throughout the cytoplasm, include microfilaments, intermediate filaments, and microtubules. The cytoskeleton provides a structural framework for the cell and is responsible for cell movements.

3. The centrosome consists of the pericentriolar matrix and a pair of centrioles. The pericentriolar matrix organizes microtubules in nondividing cells and the mitotic spindle in dividing cells.

4. Cilia and flagella, motile projections of the cell surface, are formed by basal bodies. Cilia move fluid along the cell surface; flagella move an entire cell.

5. Ribosomes consist of two subunits made in the nucleus that are composed of ribosomal RNA and ribosomal proteins. They serve as sites of protein synthesis.

6. Endoplasmic reticulum (ER) is a network of membranes that form flattened sacs or tubules; it extends from the nuclear envelope throughout the cytoplasm. Rough ER is studded with ribosomes that synthesize proteins; the proteins then enter the space within the ER for processing and sorting. Rough ER produces secretory proteins, membrane proteins, and organelle proteins; forms glycoproteins; synthesizes phospholipids; and attaches proteins to phospholipids. Smooth ER lacks ribosomes. It synthesizes fatty acids and steroids; inactivates or detoxifies drugs and other potentially harmful substances; removes phosphate from glucose-6-phosphate; and releases calcium ions that trigger contraction in muscle cells.

7. The Golgi complex consists of flattened sacs called cisterns. The entry, medial, and exit regions of the Golgi complex contain different enzymes that permit each to modify, sort, and package proteins for transport in secretory vesicles, membrane vesicles, or transport vesicles to different cellular destinations.

8. Lysosomes are membrane-enclosed vesicles that contain digestive enzymes. Endosomes, phagosomes, and pinocytic vesicles deliver materials to lysosomes for degradation. Lysosomes function in digestion of worn-out organelles (autophagy), digestion of a host cell (autolysis), and extracellular digestion.

9. Peroxisomes contain oxidases that oxidize amino acids, fatty acids, and toxic substances; the hydrogen peroxide produced in the process is destroyed by catalase. The proteases contained in proteasomes, another kind of organelle, continually degrade unneeded, damaged, or faulty proteins by cutting them into small peptides.

10. Mitochondria consist of a smooth external mitochondrial membrane, an internal mitochondrial membrane containing mitochondrial cristae, and a fluid-filled cavity called the mitochondrial matrix. These so-called powerhouses of the cell produce most of a cell's ATP and can play an important early role in apoptosis.

3.5 Nucleus

1. The nucleus consists of a double nuclear envelope; nuclear pores, which control the movement of substances between the nucleus and cytoplasm; nucleoli, which produce ribosomes; and genes arranged on chromosomes, which control cellular structure and direct cellular activities.

2. Human somatic cells have 46 chromosomes, 23 inherited from each parent. The total genetic information carried in a cell or an organism is its genome.

3.6 Protein Synthesis

1. Cells make proteins by transcribing and translating the genetic information contained in DNA.

2. The genetic code is the set of rules that relates the base triplet sequences of DNA to the corresponding codons of RNA and the amino acids they specify.

3. In transcription, the genetic information in the sequence of base triplets in DNA serves as a template for copying the information into a complementary sequence of codons in messenger RNA. Transcription begins on DNA in a region called a promoter. Regions of DNA that code for protein synthesis are called exons; those that do not are called introns.

4. Newly synthesized pre-mRNA is modified before leaving the nucleus.

5. In the process of translation, the nucleotide sequence of mRNA specifies the amino acid sequence of a protein. The mRNA binds to a ribosome, specific amino acids attach to tRNA, and anticodons of tRNA bind to codons of mRNA, bringing specific amino acids into position on a growing polypeptide. Translation begins at the start codon and ends at the stop codon.

3.7 Cell Division

1. Cell division, the process by which cells reproduce themselves, consists of nuclear division (mitosis or meiosis) and cytoplasmic division (cytokinesis). Cell division that replaces cells or adds new ones is called somatic cell division and involves mitosis and cytokinesis. Cell division that results in the production of gametes (sperm and ova) is called reproductive cell division and consists of meiosis and cytokinesis.

2. The cell cycle, an orderly sequence of events in which a somatic cell duplicates its contents and divides in two, consists of interphase and a mitotic phase. Human somatic cells contain 23 pairs of homologous chromosomes and are thus diploid ($2n$). Before the mitotic phase, the DNA molecules, or chromosomes, replicate themselves so that identical sets of chromosomes can be passed on to the next generation of cells.

3. A cell between divisions that is carrying on every life process except division is said to be in interphase, which consists of three phases: G_1, S, and G_2. During the G_1 phase, the cell replicates its organelles and cytosolic components, and centrosome replication begins; during the S phase, DNA replication occurs; during the G_2 phase, enzymes and other proteins are synthesized and centrosome replication is completed.

4. Mitosis is the splitting of the chromosomes and the distribution of two identical sets of chromosomes into separate and equal nuclei; it consists of prophase, metaphase, anaphase, and telophase.

5. In cytokinesis, which usually begins in late anaphase and ends once mitosis is complete, a cleavage furrow forms at the cell's metaphase plate and progresses inward, pinching in through the cell to form two separate portions of cytoplasm.

6. A cell can either remain alive and functioning without dividing, grow and divide, or die. The control of cell division depends on specific cyclin-dependent protein kinases and cyclins.

7. Apoptosis is normal, programmed cell death. It first occurs during embryological development and continues throughout the lifetime of an organism.

8. Certain genes regulate both cell division and apoptosis. Abnormalities in these genes are associated with a wide variety of diseases and disorders.

9. In sexual reproduction each new organism is the result of the union of two different gametes, one from each parent. Gametes contain a single set of chromosomes (23) and thus are haploid (n).

10. Meiosis is the process that produces haploid gametes; it consists of two successive nuclear divisions, called meiosis I and meiosis II. During meiosis I, homologous chromosomes undergo synapsis (pairing) and crossing-over; the net result is two haploid cells that are genetically unlike each other and unlike the starting diploid parent cell that produced them. During meiosis II, two haploid cells divide to form four haploid cells.

3.8 Cellular Diversity

1. The sizes of cells are measured in micrometers. One micrometer (μm) equals 10^{-6} m (1/25,000 of an inch). Cells in the human body range in size from 8 μm to 140 μm.

2. A cell's shape is related to its function.

3.9 Aging and Cells

1. Aging is a normal process accompanied by progressive alteration of the body's homeostatic adaptive responses.

2. Many theories of aging have been proposed, including genetically programmed cessation of cell division, buildup of free radicals, and an intensified autoimmune response.

CLINICAL CONNECTIONS

Medical Uses of Isotonic, Hypertonic, and Hypotonic Solutions (Refer page 63 of Textbook)

RBCs and other body cells may be damaged or destroyed if exposed to hypertonic or hypotonic solutions. For this reason, most **intravenous (IV) solutions**, liquids infused into the blood of a vein, are isotonic. Examples are isotonic saline (0.9% NaCl) and D5W, which stands for dextrose 5% in water. Sometimes infusion of a hypertonic solution such as mannitol (sugar alcohol) is useful to treat patients who have *cerebral edema*, excess interstitial fluid in the brain. Infusion of such a solution relieves fluid overload by causing osmosis of water from interstitial fluid into the blood. The kidneys then excrete the excess water from the blood into the urine. Hypotonic solutions, given either orally or through an IV, can be used to treat people who are dehydrated. The water in the hypotonic solution moves from the blood into interstitial fluid and then into body cells to rehydrate them. Water and most sports drinks that you consume to "rehydrate" after a workout are hypotonic relative to your body cells.

Digitalis Increases Ca^{2+} in Heart Muscle Cells (Refer page 66 of Textbook)

Digitalis often is given to patients with *heart failure*, a condition of weakened pumping action by the heart. Digitalis exerts its effect by slowing the action of the sodium–potassium pumps, which lets more Na^+ accumulate inside heart muscle cells. The result is a decreased Na^+ concentration gradient across the plasma membrane, which causes the Na^+–Ca^{2+} antiporters to slow down. As a result, more Ca^{2+} remains inside heart muscle cells. The slight increase in the level of Ca^{2+} in the cytosol of heart muscle cells increases the force of their contractions and thus strengthens the force of the heartbeat.

Viruses and Receptor-Mediated Endocytosis (Refer page 67 of Textbook)

Although receptor-mediated endocytosis normally imports needed materials, some viruses are able to use this mechanism to enter and infect body cells. For example, the human immunodeficiency virus (HIV), which causes acquired immunodeficiency syndrome (AIDS), can attach to a receptor called CD4. This receptor is present in the plasma membrane of white blood cells called helper T cells. After binding to CD4, HIV enters the helper T cell via receptor-mediated endocytosis.

Phagocytosis and Microbes (Refer page 67 of Textbook)

Phagocytosis is a vital defense mechanism that helps protect the body from disease. Macrophages dispose of invading microbes and billions of aged, worn-out red blood cells every day; neutrophils also help rid the body of invading microbes. **Pus** is a mixture of dead neutrophils, macrophages, and tissue cells and fluid in an infected wound.

Cilia and Smoking (Refer page 72 of Textbook)

The movement of cilia is paralyzed by nicotine in cigarette smoke. For this reason, smokers cough often to remove foreign particles from their airways. Cells that line the uterine (fallopian) tubes also have cilia that sweep oocytes (egg cells) toward the uterus, and females who smoke have an increased risk of ectopic (outside the uterus) pregnancy.

Smooth ER and Drug Tolerance (Refer page 73 of Textbook)

One of the functions of smooth ER, as noted earlier, is to detoxify certain drugs. Individuals who repeatedly take such drugs, such as the sedative phenobarbital, develop changes in the smooth ER in their liver cells. Prolonged administration of phenobarbital results in increased tolerance to the drug; the same dose no longer produces the same degree of sedation. With repeated exposure to the drug, the amount of smooth ER and its enzymes increases to protect the cell from its toxic effects. As the amount of smooth ER increases, higher and higher dosages of the drug are needed to achieve the original effect. This could result in an increased possibility of overdose and increased drug dependence.

Tay-Sachs Disease (Refer page 77 of Textbook)

Some disorders are caused by faulty or absent lysosomal enzymes. For instance, **Tay-Sachs disease** (TĀ-SAKS), which most often affects children of Ashkenazi (Eastern European Jewish) descent, is an inherited condition characterized by the absence of a single lysosomal enzyme called Hex A. This enzyme normally breaks down a membrane glycolipid called ganglioside G_{M2} that is especially prevalent in nerve cells. As the excess ganglioside G_{M2} accumulates, the nerve cells function less efficiently. Children with Tay-Sachs disease typically experience seizures and muscle rigidity. They gradually become blind, demented, and uncoordinated and usually die before the age of 5. Tests can now reveal whether an adult is a carrier of the defective gene.

Proteasomes and Disease (Refer page 77 of Textbook)

Some diseases could result from failure of proteasomes to degrade abnormal proteins. For example, clumps of misfolded proteins accumulate in brain cells of people with Parkinson's disease and Alzheimer's disease. Discovering why the proteasomes fail to clear these abnormal proteins is a goal of ongoing research.

Genomics (Refer page 79 of Textbook)

In the last decade of the twentieth century, the genomes of humans, mice, fruit flies, and more than 50 microbes were sequenced. As a result, research in the field of **genomics**, the study of the relationships between the genome and the biological functions of an organism, has flourished. The Human Genome Project began in 1990 as an effort to sequence all of the nearly 3.2 billion nucleotides of our genome and was completed in April 2003. Scientists now know that the total number of genes in the human genome is about 30,000. Information regarding the human genome and how it is affected by the environment seeks to identify and discover the functions of the specific genes that play a role in genetic diseases. Genomic medicine also aims to design new drugs and to provide screening tests to enable physicians to provide more effective counseling and treatment for disorders with significant genetic components such as hypertension (high blood pressure), obesity, diabetes, and cancer.

Recombinant DNA (Refer page 85 of Textbook)

Scientists have developed techniques for inserting genes from other organisms into a variety of host cells. Manipulating the cell in this way can cause the host organism to produce proteins it normally does not synthesize. Organisms so altered are called **recombinants** (rē-KOM-bi-nants), and their DNA—a combination of DNA from different sources—is called **recombinant DNA**. When recombinant DNA functions properly, the host will synthesize the protein specified by the new gene it has acquired. The technology that has arisen from the manipulation of genetic material is referred to as **genetic engineering**.

The practical applications of recombinant DNA technology are enormous. Strains of recombinant bacteria produce large quantities of many important therapeutic substances, including *human growth hormone (hGH)*, required for normal growth and metabolism; *insulin*, a hormone that helps regulate blood glucose level and is used by diabetics; *interferon (IFN)*, an antiviral (and possibly anticancer) substance; and *erythropoietin (EPO)*, a hormone that stimulates production of red blood cells.

Mitotic Spindle and Cancer (Refer page 87 of Textbook)

One of the distinguishing features of cancer cells is uncontrolled division, which results in the formation of a mass of cells called a *neoplasm* or *tumor*. One of the ways to treat cancer is by *chemotherapy*, the use of anticancer drugs. Some of these drugs stop cell division by inhibiting the formation of the mitotic spindle. Unfortunately, these types of anticancer drugs also kill all types of rapidly dividing cells in the body, causing side effects such as nausea, diarrhea, hair loss, fatigue, and decreased resistance to disease.

Free Radicals and Antioxidants (Refer page 93 of Textbook)

Free radicals produce oxidative damage in lipids, proteins, or nucleic acids by "stealing" an electron to accompany their unpaired electrons. Some effects are wrinkled skin, stiff joints, and hardened arteries. Normal metabolism—for example, aerobic cellular respiration in mitochondria—produces some free radicals. Others are present in air pollution, radiation, and certain foods we eat. Naturally occurring enzymes in peroxisomes and in the cytosol normally dispose of free radicals. Certain dietary substances, such as vitamin E, vitamin C, beta-carotene, zinc, and selenium, are referred to as **antioxidants** because they inhibit the formation of free radicals.

DISORDERS: HOMEOSTATIC IMBALANCES

Most chapters in the text are followed by concise discussions of major diseases and disorders that illustrate departures from normal homeostasis. These discussions provide answers to many questions that you might ask about medical problems.

Cancer

Cancer is a group of diseases characterized by uncontrolled or abnormal cell division. When cells in a part of the body divide without control, the excess tissue that develops is called a **tumor** or *neoplasm* (NĒ-ō-plazm; *neo-* = new). The study of tumors is called **oncology** (on-KOL-ō-jē; *onco-* = swelling or mass). Tumors may be cancerous and often fatal, or they may be harmless. A cancerous neoplasm is called a **malignant tumor** (ma-LIG-nant) or *malignancy*. One property of most malignant tumors is their ability to undergo **metastasis** (me-TAS-ta-sis), the spread of cancerous cells to other parts of the body. A **benign tumor** (be-NĪN) is a neoplasm that does not metastasize. An example is a wart. Most benign tumors may be removed surgically if they interfere with normal body function or become disfiguring. Some benign tumors can be inoperable and perhaps fatal.

Types of Cancer

The name of a cancer is derived from the type of tissue in which it develops. Most human cancers are **carcinomas** (kar-si-NŌ-maz; *carcin* = cancer; *-omas* = tumors), malignant tumors that arise from epithelial cells. **Melanomas** (mel-a-NŌ-maz; *melan-* = black), for example, are cancerous growths of melanocytes, skin epithelial cells that produce the pigment melanin. **Sarcoma** (sar-KŌ-ma; *sarc-* = flesh) is a general term for any cancer arising from muscle cells or connective tissues. For example, **osteogenic sarcoma** (*osteo-* = bone; *-genic* = origin), the most frequent type of childhood cancer, destroys normal bone tissue. **Leukemia** (loo-KĒ-mē-a; *leuk-* = white; *-emia* = blood) is a cancer of blood-forming organs characterized by rapid growth of abnormal leukocytes (white blood cells). **Lymphoma** (lim-FŌ-ma) is a malignant disease of lymphatic tissue—for example, of lymph nodes.

Growth and Spread of Cancer

Cells of malignant tumors duplicate rapidly and continuously. As malignant cells invade surrounding tissues, they often trigger **angiogenesis** (an′-jē-o-JEN-e-sis), the growth of new networks of blood vessels. Proteins that stimulate angiogenesis in tumors are called *tumor angiogenesis factors (TAFs)*. The formation of new blood vessels can occur either by overproduction of TAFs or by the lack of naturally occurring angiogenesis inhibitors. As the cancer grows, it begins to compete with normal tissues for space and nutrients. Eventually, the normal tissue decreases in size and dies. Some malignant cells may detach from the initial (primary) tumor and invade a body cavity or enter the blood or lymph, then circulate to and invade other body tissues, establishing secondary tumors. Malignant cells resist the antitumor defenses of the body. The pain associated with cancer develops when the tumor presses on nerves or blocks a passageway in an organ so that secretions build up pressure, or as a result of dying tissue or organs.

Causes of Cancer

Several factors may trigger a normal cell to lose control and become cancerous.

CARCINOGENS One cause is environmental agents: substances in the air we breathe, the water we drink, and the food we eat. A chemical agent or radiation that produces cancer is called a **carcinogen** (car-SIN-o-- jen). Carcinogens induce **mutations** (mū-TA--shuns), permanent changes in the DNA base sequence of a gene. The World Health Organization estimates that carcinogens are associated with 60–90% of all human cancers. Examples of carcinogens are hydrocarbons found in cigarette tar, radon gas from the earth, and ultraviolet (UV) radiation in sunlight.

ONCOGENES Intensive research efforts are directed toward studying cancer-causing genes, or **oncogenes** (ON-ko-jēnz). When inappropriately activated, these genes have the ability to transform a normal cell into a cancerous cell. Most oncogenes derive from normal genes called **proto-oncogenes** that regulate growth and development. The proto-oncogene undergoes some change that causes it (1) to be expressed inappropriately, (2) to make its products in excessive amounts, or (3) to make its products at the wrong time. Some oncogenes cause excessive production of growth factors, chemicals that stimulate cell growth. Others may trigger changes in a cell-surface receptor, causing it to send signals as though it were being activated by a growth factor. As a result, the growth pattern of the cell becomes abnormal.

Proto-oncogenes in every cell carry out normal cellular functions until a malignant change occurs. It appears that some proto-oncogenes are activated to oncogenes by mutations in which the DNA of the proto-oncogene is altered. Other proto-oncogenes are activated by a rearrangement of the chromosomes so that segments of DNA are exchanged. Rearrangement activates proto-oncogenes by placing them near genes that enhance their activity.

ONCOGENIC VIRUSES Some cancers have a viral origin. Viruses are tiny packages of nucleic acids, either RNA or DNA, that can reproduce only while inside the cells they infect. Some viruses, termed oncogenic viruses, cause cancer by stimulating abnormal proliferation of cells. For instance, the human papillomavirus (HPV) causes virtually all cervical cancers in women. The virus produces a protein that causes proteasomes to destroy p53, a protein that normally suppresses unregulated cell division. In the absence of this suppressor protein, cells proliferate without control.

Some studies suggest that certain cancers may be linked to a cell having abnormal numbers of chromosomes. As a result, the cell could potentially have extra copies of oncogenes or too few copies of tumor-suppressor genes, which in either case could lead to uncontrolled cell proliferation. There is also evidence suggesting that cancer may be caused by normal stem cells that develop into cancerous stem cells capable of forming malignant tumors.

Later in the book, we will discuss the process of inflammation, which is a defensive response to tissue damage. It appears that inflammation contributes to various steps in the development of cancer. Some evidence suggests that chronic inflammation stimulates the proliferation of mutated cells and enhances their survival, promotes angiogenesis, and contributes to invasion and metastasis of cancer cells. There is a clear relationship between certain chronic inflammatory conditions and the transformation of inflamed tissue into a malignant tissue. For example, chronic gastritis (inflammation of the stomach lining) and peptic ulcers may be a causative factor in 60–90% of stomach cancers. Chronic hepatitis (inflammation of the liver) and cirrhosis of the liver are believed to be responsible for about 80% of liver cancers. Colorectal cancer is 10 times more likely to occur in patients with chronic inflammatory diseases of the colon, such as ulcerative colitis and Crohn's disease. And the relationship between asbestosis and silicosis, two chronic lung inflammatory conditions, and lung cancer has long been recognized. Chronic inflammation is also an underlying contributor to rheumatoid arthritis, Alzheimer's disease, depression, schizophrenia, cardiovascular disease, and diabetes.

Carcinogenesis: A Multistep Process

Carcinogenesis (kar′-si-nō-JEN-e-sis) is a multistep process of cancer development in which as many as 10 distinct mutations may have to accumulate in a cell before it becomes cancerous. The progression of genetic changes leading to cancer is best understood for colon (colorectal) cancer. Such cancers, as well as lung and breast cancer, take years or decades to develop. In colon cancer, the tumor begins as an area of increased cell proliferation that results from one mutation. This growth then progresses to abnormal, but noncancerous, growths called adenomas. After two or three additional mutations, a mutation of the tumor-suppressor gene *p53* occurs and a carcinoma develops. The fact that so many mutations are needed for a cancer to develop indicates that cell growth is normally controlled with many sets of checks and balances. Thus, it is not surprising that a compromised immune system contributes significantly to carcinogenesis.

Treatment of Cancer

Many cancers are removed surgically. However, cancer that is widely distributed throughout the body or exists in organs with essential functions, such as the brain, which might be greatly harmed by surgery, may be treated with chemotherapy and radiation therapy instead. Sometimes surgery, chemotherapy, and radiation therapy are used in combination. Chemotherapy involves administering drugs that cause death of cancerous cells. Radiation therapy breaks chromosomes, thus blocking cell division. Because cancerous cells divide rapidly, they are more vulnerable to the destructive effects of chemotherapy and radiation therapy than are normal cells. Unfortunately for the patients, hair follicle cells, red bone marrow cells, and cells lining the gastrointestinal tract also are rapidly dividing. Hence, the side effects of chemotherapy and radiation therapy include hair loss due to death of hair follicle cells, vomiting and nausea due to death of cells lining the stomach and intestines, and susceptibility to infection due to slowed production of white blood cells in red bone marrow.

Treating cancer is difficult because it is not a single disease and because the cells in a single tumor population rarely behave all in the same way. Although most cancers are thought to derive from a single abnormal cell, by the time a tumor reaches a clinically detectable size, it may contain a diverse population of abnormal cells. For example, some cancerous cells metastasize readily, and others do not. Some are sensitive to chemotherapy drugs and some are drug-resistant. Because of differences in drug resistance, a single chemotherapeutic agent may destroy susceptible cells but permit resistant cells to proliferate.

Another potential treatment for cancer that is currently under development is *virotherapy*, the use of viruses to kill cancer cells. The viruses employed in this strategy are designed so that they specifically target cancer cells without affecting the healthy cells of the body. For example, proteins (such as antibodies) that specifically bind to receptors found only in cancer cells are attached to viruses. Once inside the body, the viruses bind to cancer cells and then infect them. The cancer cells are eventually killed once the viruses cause cellular lysis.

Researchers are also investigating the role of *metastasis regulatory genes* that control the ability of cancer cells to undergo metastasis. Scientists hope to develop therapeutic drugs that can manipulate these genes and, therefore, block metastasis of cancer cells.

MEDICAL TERMINOLOGY

Most chapters in this text are followed by a glossary of key medical terms that include both normal and pathological conditions. You should familiarize yourself with these terms because they will play an essential role in your medical vocabulary.

Some of these conditions, as well as ones discussed in the text, are referred to as local or systemic. A *local disease* is one that affects one part or a limited area of the body. A *systemic disease* affects the entire body or several parts.

Anaplasia (an′-a-PLĀ-zē-a; *an-* = not; *-plasia* = to shape) The loss of tissue differentiation and function that is characteristic of most malignancies.

Atrophy (AT-rō-fē; *a-* = without; *-trophy* = nourishment) A decrease in the size of cells, with a subsequent decrease in the size of the affected tissue or organ; wasting away.

Dysplasia (dis-PLĀ-zē-a; *dys-* = abnormal) Alteration in the size, shape, and organization of cells due to chronic irritation or inflammation; may progress to neoplasia (tumor formation, usually malignant) or revert to normal if the irritation is removed.

Hyperplasia (hī-per-PLĀ-zē-a; *hyper-* = over) Increase in the number of cells of a tissue due to an increase in the frequency of cell division.

Hypertrophy (hī-PER-trō-fē) Increase in the size of cells without cell division.

Metaplasia (met′-a-PLĀ-zē-a; *meta-* = change) The transformation of one type of cell into another.

Progeny (PROJ-e-nē; *pro-* = forward; *-geny* = production) Offspring or descendants.

Proteomics (prō′-tē-Ō-miks; *proteo-* = protein) The study of the proteome (all of an organism's proteins) in order to identify all of the proteins produced; it involves determining the three-dimensional structure of proteins so that drugs can be designed to alter protein activity to help in the treatment and diagnosis of disease.

Tumor marker A substance introduced into circulation by tumor cells that indicates the presence of a tumor, as well as the specific type. Tumor markers may be used to screen, diagnose, make a prognosis, evaluate a response to treatment, and monitor for recurrence of cancer.

SELF-QUIZ QUESTIONS

Fill in the blanks in the following statements.

1. The three major parts of the cell are the _____, _____, and _____.

2. Cell death that is genetically programmed is known as _____, while cell death which is due to tissue injury is known as _____.

3. _____ are special DNA sequences located at the ends of chromosomes and whose erosion contributes to cellular aging and death.

4. The mRNA base sequence that is complementary to the DNA base sequence ATC would be _____.

Indicate whether the following statements are true or false.

5. A small membrane surface area will increase the rate of diffusion across the cell membrane.

6. The cells created during meiosis are genetically different from the original cell.

7. An important and abundant active mechanism that helps maintain cellular tonicity is the Na^+/K^+ ATPase pump.

Choose the one best answer to the following questions.

8. If the concentration of solutes in the ECF and ICFs are equal, the cell is in a(n) _____ solution.
   - (a) hypertonic
   - (b) hydrophobic
   - (c) saturated
   - (d) hypotonic
   - (e) isotonic

9. Which membrane protein is *incorrectly* matched with its function?
   - (a) receptor: allows recognition of specific molecules
   - (b) ion channel: allows passage of specific ions through the membrane
   - (c) carrier: allows cells to recognize each other and foreign cells
   - (d) linker: allows binding of one cell to another and provides stability and shape to a cell
   - (e) enzyme: catalyzes cellular reactions

10. Place the following steps in protein synthesis in the correct order.
    - (a) binding of anticodons of tRNA to codons of mRNA
    - (b) modification of newly synthesized pre-mRNA by snRNPs before leaving the nucleus and entering the cytoplasm
    - (c) attachment of RNA polymerase at promoter
    - (d) binding of mRNA to a ribosome's small subunit
    - (e) amino acids joined by peptide bonds
    - (f) joining of large and small ribosomal subunits to create a functional ribosome
    - (g) transcription of a segment of DNA onto mRNA
    - (h) detachment of protein from ribosome when ribosome reaches stop codon on mRNA
    - (i) detachment of RNA polymerase after reaching terminator
    - (j) attachment of specific amino acids to tRNA
    - (k) binding of initiator tRNA to start codon on mRNA

11. Which of the following organelles function primarily in decomposition reactions? (1) ribosomes, (2) proteasomes, (3) lysosomes, (4) centrosomes, (5) peroxisomes
    - (a) 2, 3, and 5
    - (b) 3 and 5
    - (c) 2, 4, and 5
    - (d) 1 and 4
    - (e) 2 and 5

12. Which of the following statements regarding the nucleus are true? (1) Nucleoli within the nucleus are the sites of ribosome synthesis. (2) The nucleus contains the cell's hereditary units. (3) The nuclear membrane is a solid, impermeable membrane. (4) Protein synthesis occurs within the nucleus. (5) In nondividing cells, DNA is found in the nucleus in the form of chromatin.
    - (a) 1, 2, and 3
    - (b) 1, 2, and 4
    - (c) 1, 2, and 5
    - (d) 2, 4, and 5
    - (e) 2, 3, and 4

13. Match the following:
    - _____ (a) mitosis
    - _____ (b) meiosis
    - _____ (c) prophase
    - _____ (d) metaphase
    - _____ (e) anaphase
    - _____ (f) telophase
    - _____ (g) cytokinesis
    - _____ (h) interphase

    - (1) cytoplasmic division
    - (2) somatic cell division resulting in the formation of two identical cells
    - (3) reproductive cell division that reduces the number of chromosomes by half
    - (4) stage of cell division when replication of DNA occurs
    - (5) stage when chromatin fibers condense and shorten to form chromosomes
    - (6) stage when centromeres split and chromatids move to opposite poles of the cell
    - (7) stage when centromeres of chromatid pairs line up at the center of the mitotic spindle
    - (8) stage when chromosomes uncoil and revert to chromatin

14. Match the following:

_____ (a) cytoskeleton
_____ (b) centrosome
_____ (c) ribosomes
_____ (d) rough ER
_____ (e) smooth ER
_____ (f) Golgi complex
_____ (g) lysosomes
_____ (h) peroxisomes
_____ (i) mitochondria
_____ (j) cilia
_____ (k) flagellum
_____ (l) proteasomes
_____ (m) vesicles

(1) membrane-enclosed vesicles formed in the Golgi complex that contain strong hydrolytic and digestive enzymes

(2) network of protein filaments that extend throughout the cytoplasm, providing cellular shape, organization, and movement

(3) sites of protein synthesis

(4) contain enzymes that break apart unneeded, damaged, or faulty proteins into small peptides

(5) site where secretory proteins and membrane molecules are synthesized

(6) membrane-enclosed vesicles that contain enzymes that oxidize various organic substances

(7) short microtubular structures extending from the plasma membrane and involved in movement of materials along the cell's surface

(8) modifies, sorts, packages, and transports molecules synthesized in the rough ER

(9) an organizing center for growth of the mitotic spindle

(10) function in ATP generation

(11) functions in synthesizing fatty acids and steroids, helping liver cells release glucose into the bloodstream, and detoxification

(12) membrane-bound sacs that transport, transfer, or secrete proteins

(13) long microtubular structure extending from the plasma membrane and involved in movement of a cell

15. Match the following:

_____ (a) diffusion
_____ (b) osmosis
_____ (c) facilitated diffusion
_____ (d) primary active transport
_____ (e) secondary active transport
_____ (f) vesicular transport
_____ (g) phagocytosis
_____ (h) pinocytosis
_____ (i) exocytosis
_____ (j) receptor-mediated endocytosis
_____ (k) transcytosis

(1) passive transport in which a solute binds to a specific carrier on one side of the membrane and is released on the other side

(2) movement of materials out of the cell by fusing of secretory vesicles with the plasma membrane

(3) the random mixing of particles in a solution due to the kinetic energy of the particles; substances move from high to low concentrations until equilibrium is reached

(4) transport of substances either into or out of the cell by means of small, spherical membranous sacs formed by budding off from existing membranes

(5) uses energy derived from hydrolysis of ATP to change the shape of a carrier protein, which "pumps" a substance across a cellular membrane against its concentration gradient

(6) vesicular movement involving endocytosis on one side of a cell and subsequent exocytosis on the opposite side of the cell

(7) type of endocytosis that involves the nonselective uptake of tiny droplets of extracellular fluid

(8) type of endocytosis in which large solid particles are taken in

(9) movement of water from an area of higher to an area of lower water concentration through a selectively permeable membrane

(10) process that allows a cell to take specific ligands from the ECF by forming vesicles

(11) indirectly uses energy obtained from the breakdown of ATP; involves symporters and antiporters

4 THE TISSUE LEVEL OF ORGANIZATION

4.1 Types of Tissues

1. A tissue is a group of cells, usually with similar embryological origin, specialized for a particular function.
2. The tissues of the body are classified into four basic types: epithelial, connective, muscular, and nervous.

4.2 Cell Junctions

1. Cell junctions are points of contact between adjacent plasma membranes.
2. Tight junctions form fluid-tight seals between cells; adherens junctions, desmosomes, and hemidesmosomes anchor cells to one another or to the basement membrane; and gap junctions permit electrical and chemical signals to pass between cells.

4.3 Comparison between Epithelial and Connective Tissues

1. Epithelial tissue has many cells tightly packed together and is avascular.
2. Connective tissue has relatively few cells with lots of extracellular material.

4.4 Epithelial Tissue

1. The subtypes of epithelial tissue include covering and lining epithelium (surface epithelium) and glandular epithelium.
2. Epithelial tissue consists mostly of cells with little extracellular material between adjacent plasma membranes. The apical, lateral, and basal surfaces of epithelial cells are modified in various ways to carry out specific functions. Although epithelial tissue is avascular, it has a nerve supply. The high rate of cell division gives epithelial tissue a high capacity for renewal.
3. Covering and lining epithelium can be simple, pseudostratified, or stratified. The cell shapes may be squamous (flat), cuboidal (cubelike), columnar (rectangular), or transitional (variable). The subtypes of epithelial tissue include covering and lining epithelium and glandular epithelium.
4. Simple squamous epithelium, a single layer of flat cells (**Table 4.1A**), is found in parts of the body where filtration or diffusion is a priority process. Endothelium lines the heart and blood vessels. Mesothelium forms the serous membranes that line the thoracic and abdominopelvic cavities and covers the organs within them.
5. Simple cuboidal epithelium, a single layer of cube-shaped cells that function in secretion and absorption (**Table 4.1B**), is found covering the ovaries, in the kidneys and eyes, and lining some glandular ducts.
6. Nonciliated simple columnar epithelium, a single layer of nonciliated rectangular cells (**Table 4.1C**), lines most of the gastrointestinal tract and contains specialized cells that perform absorption and secrete mucus. Ciliated simple columnar epithelium, a single layer of ciliated rectangular cells (**Table 4.1D**), is found in a few portions of the upper respiratory tract, where it moves foreign particles trapped in mucus out of the respiratory tract. A nonciliated variety has no goblet cells and lines ducts of many glands, the epididymis, and part of the male urethra (**Table 4.1E**) and a ciliated variety of pseudostratified columnar epithelium (**Table 4.1F**) contains goblet cells and lines most of the upper respiratory tract. The ciliated variety moves mucus in the respiratory tract. The nonciliated variety functions in absorption and protection.
7. Stratified epithelium consists of several layers of cells: Cells of the apical layer of stratified squamous epithelium and several layers deep to it are flat (**Table 4.1G**); a nonkeratinized variety lines the mouth, and a keratinized variety forms the epidermis. Cells at the apical layer of stratified cuboidal epithelium are cube-shaped (**Table 4.1H**); found in adult sweat glands and in a portion of the male urethra, stratified cuboidal epithelium protects and provides limited secretion and absorption. Cells of the apical layer of stratified columnar epithelium have a columnar shape (**Table 4.1I**); this type is found in a portion of the male urethra and in large excretory ducts of some glands, and functions in protection and secretion.
8. Transitional epithelium (urothelium) consists of several layers of cells whose appearance varies with the degree of stretching (**Table 4.1J**). It lines the urinary bladder.
9. A gland is a single cell or a group of epithelial cells adapted for secretion. There are two types of glandular epithelium: endocrine and exocrine. Endocrine glands secrete hormones into interstitial fluid and then into the blood (**Table 4.2A**). Exocrine glands secrete into ducts or directly onto a free surface (**Table 4.2B**).
10. The structural classification of exocrine glands includes unicellular and multicellular glands. The functional classification of exocrine glands includes merocrine, apocrine, and holocrine glands.

4.5 Connective Tissue

1. Connective tissue, one of the most abundant body tissues, consists of relatively few cells and an abundant extracellular matrix of ground substance and protein fibers. It usually has a nerve supply, and it is usually highly vascular.

2. Cells in connective tissue proper are derived primarily from mesenchymal cells. Cell types include fibroblasts (secrete extracellular matrix), macrophages (perform phagocytosis), plasma cells (secrete antibodies), mast cells (produce histamine), adipocytes (store fat), and white blood cells (respond to infections).

3. The ground substance and fibers make up the extracellular matrix. The ground substance supports and binds cells together, provides a medium for the exchange of materials, stores water, and actively influences cell functions. Substances found in the ground substance include water and polysaccharides. Also present are proteoglycans and adhesion proteins.

4. The fibers in the extracellular matrix provide strength and support and are of three types: (a) Collagen fibers are found in large amounts in bone, tendons, and ligaments. (b) Elastic fibers are found in skin, blood vessel walls, and lungs. (c) Reticular fibers are found around fat cells, nerve fibers, and skeletal and smooth muscle cells.

5. Two major subclasses of connective tissue are embryonic (found in embryo and fetus) and mature (present in the newborn). Embryonic connective tissues (see **Table 4.3**) are mesenchyme, which forms almost all other connective tissues, and mucous connective tissue, found in the umbilical cord of the fetus, where it gives support. Mature connective tissue differentiates from mesenchyme and is subdivided into several types: connective tissue proper (loose and dense), supporting connective tissue (cartilage and bone), and liquid connective tissue (blood and lymph).

6. Loose connective tissue includes areolar connective tissue, adipose tissue, and reticular connective tissue. Areolar connective tissue consists of the three types of fibers (collagen, elastic, and reticular), several types of cells, and a semifluid ground substance (**Table 4.4A**); it is found in the subcutaneous layer, in mucous membranes, and around blood vessels, nerves, and body organs. Adipose tissue consists of adipocytes, which store triglycerides (**Table 4.4B**); it is found in the subcutaneous layer, around organs, and in yellow bone marrow. Brown adipose tissue (BAT) generates heat. Reticular connective tissue consists of reticular fibers and reticular cells and is found in the liver, spleen, and lymph nodes (**Table 4.4C**).

7. Dense connective tissue includes dense regular, dense irregular, and elastic. Dense regular connective tissue consists of parallel bundles of collagen fibers and fibroblasts (**Table 4.5A**); it forms tendons, most ligaments, and aponeuroses. Dense irregular connective tissue usually consists of collagen fibers and a few fibroblasts (**Table 4.5B**); it is found in fasciae, the dermis of skin, and membrane capsules around organs. Elastic connective tissue consists of branching elastic fibers and fibroblasts (**Table 4.5C**) and is found in the walls of large arteries, lungs, trachea, and bronchial tubes.

8. Cartilage is a supporting connective tissue that contains chondrocytes and has a rubbery extracellular matrix (chondroitin sulfate) containing collagen and elastic fibers. Hyaline cartilage, which consists of a gel-like ground substance and appears bluish white in the body, is found in the embryonic skeleton, at the ends of bones, in the nose, and in respiratory structures (**Table 4.6A**); it is flexible, allows movement, provides support, and is usually surrounded by a perichondrium. Fibrocartilage is found in the pubic symphysis, intervertebral discs, and menisci (cartilage pads) of the knee joint (**Table 4.6B**); it contains chondrocytes scattered among clearly visible bundles of collagen fibers. Elastic cartilage maintains the shape of organs such as the epiglottis of the larynx, auditory (eustachian) tubes, and external ear (**Table 4.6C**); its chondrocytes are located within a threadlike network of elastic fibers, and it has a perichondrium.

9. Bone or osseous tissue is a supporting connective tissue that consists of an extracellular matrix of mineral salts and collagen fibers that contribute to the hardness of bone, and osteocytes that are located in lacunae (**Table 4.7**). It supports and protects the body, provides a surface area for muscle attachment, helps the body move, stores minerals, and houses blood-forming tissue.

10. There are two types of liquid connective tissue: blood and lymph. Blood consists of blood plasma and formed elements—red blood cells, white blood cells, and platelets (**Table 4.8**); its cells transport oxygen and carbon dioxide, carry on phagocytosis, participate in allergic reactions, provide immunity, and bring about blood clotting. Lymph, the extracellular fluid that flows in lymphatic vessels, is a clear fluid similar to blood plasma but with less protein.

4.6 Membranes

1. An epithelial membrane consists of an epithelial layer overlying a connective tissue layer. Types include mucous, serous, and cutaneous membranes.

2. Mucous membranes line cavities that open to the exterior, such as the gastrointestinal tract.

3. Serous membranes line closed cavities (pleura, pericardium, peritoneum) and cover the organs in the cavities. These membranes consist of parietal and visceral layers.

4. The cutaneous membrane is the skin. It covers the entire body and consists of a superficial epidermis (epithelium) and a deep dermis (connective tissue).

5. Synovial membranes line joint cavities and consist of areolar connective tissue; they do not have an epithelial layer.

4.7 Muscular Tissue

1. Muscular tissue consists of cells called muscle fibers or myocytes that are specialized for contraction. It provides motion, maintenance of posture, heat production, and protection.

2. Skeletal muscle tissue is attached to bones and is striated and voluntary (**Table 4.9A**).

3. The action of cardiac muscle tissue, which forms most of the heart wall and is striated, is involuntary (**Table 4.9B**).

4. Smooth muscle tissue is found in the walls of hollow internal structures (blood vessels and viscera) and is nonstriated and involuntary (**Table 4.9C**).

4.8 Nervous Tissue

1. The nervous system is composed of neurons (nerve cells) and neuroglia (protective and supporting cells) (**Table 4.10**).

2. Neurons respond to stimuli by converting the stimuli into electrical signals called nerve action potentials (nerve impulses), and conducting nerve impulses to other cells.

3. Most neurons consist of a cell body and two types of processes: dendrites and axons.

4.9 Excitable Cells

1. Electrical excitability is the ability to respond to certain stimuli by producing electrical signals such as action potentials.
2. Because neurons and muscle fibers exhibit electrical excitability, they are considered excitable cells.

4.10 Tissue Repair: Restoring Homeostasis

1. Tissue repair is the replacement of worn-out, damaged, or dead cells by healthy ones.
2. Stem cells may divide to replace lost or damaged cells.

3. If the injury is superficial, tissue repair involves parenchymal regeneration; if damage is extensive, granulation tissue is involved.
4. Good nutrition and blood circulation are vital to tissue repair.

4.11 Aging and Tissues

1. Tissues heal faster and leave less obvious scars in the young than in the aged; surgery performed on fetuses leaves no scars.
2. The extracellular components of tissues, such as collagen and elastic fibers, also change with age.

CLINICAL CONNECTIONS

Biopsy (Refer page 94 of Textbook)

A **biopsy** (BĪ-op-sē; *bio-* = life; *-opsy* = to view) is the removal of a sample of living tissue for microscopic examination. This procedure is used to help diagnose many disorders, especially cancer, and to discover the cause of unexplained infections and inflammations. Both normal and potentially diseased tissues are removed for purposes of comparison. Once the tissue samples are removed, either surgically or through a needle and syringe, they may be preserved, stained to highlight special properties, or cut into thin sections for microscopic observation. Sometimes a biopsy is conducted while a patient is anesthetized during surgery to help a physician determine the most appropriate treatment. For example, if a biopsy of thyroid tissue reveals malignant cells, the surgeon can proceed immediately with the most appropriate procedure.

Basement Membranes and Disease (Refer page 97 of Textbook)

Under certain conditions, basement membranes become markedly thickened, due to increased production of collagen and laminin. In untreated cases of diabetes mellitus, the basement membrane of small blood vessels (capillaries) thickens, especially in the eyes and kidneys. Because of this the blood vessels cannot function properly, and blindness and kidney failure may result.

Papanicolaou Test (Refer page 106 of Textbook)

A **Papanicolaou test** (pa-pa-NI-kō-lō), also called a *Pap test* or *Pap smear*, involves collection and microscopic examination of epithelial cells that have been scraped off the apical layer of a tissue. A very common type of Pap test involves examining the cells from the nonkeratinized stratified squamous epithelium of the vagina and cervix (inferior portion) of the uterus. This type of Pap test is performed mainly to detect early changes in the cells of the female reproductive system that may indicate a precancerous condition or cancer. In performing a Pap smear, the cells are scraped from the tissue and then smeared on a microscope slide. The slides are then sent to a laboratory for analysis. It is recommended that Pap tests should be performed every three years beginning at age 21. It is further recommended that females aged 30 to 65 should have Pap testing and HPV (human papillomavirus) testing (cotesting) every five years or a Pap test alone every three years. Females with certain high risk factors may need more frequent screening or even continue screening beyond age 65.

Chondroitin Sulfate, Glucosamine, and Joint Disease (Refer page 110 of Textbook)

Chondroitin sulfate and **glucosamine** (a proteoglycan) have been used as nutritional supplements either alone or in combination to promote and maintain the structure and function of joint cartilage, to provide pain relief from osteoarthritis, and to reduce joint inflammation. Although these supplements have benefited some individuals with moderate to severe osteoarthritis, the benefit is minimal in lesser cases. More research is needed to determine how they act and why they help some people and not others.

Sprain (Refer page 111 of Textbook)

Despite their strength, ligaments may be stressed beyond their normal capacity. This results in **sprain**, a stretched or torn ligament. The ankle joint is most frequently sprained. Because of their poor blood supply, the healing of even partially torn ligaments is a very slow process; completely torn ligaments require surgical repair.

Liposuction and Cryolipolysis (Refer page 111 of Textbook)

A surgical procedure called **liposuction** (LIP-ō-suk′-shun; *lip-* = fat) or *suction lipectomy* (*-ectomy* = to cut out) involves suctioning out small amounts of adipose tissue from various areas of the body. In one type of liposuction, an incision is made in the skin, the fat is removed through a stainless steel tube, called a *cannula*, with the assistance of a powerful vacuum-pressure unit that suctions out the fat. Ultrasound and laser can also be used to liquify fat for removal. The technique can be used as a body-contouring procedure in regions such as the things, buttocks, arms, breasts, and abdomen, and to transfer fat to another area of the body. Postsurgical complications that may develop include fat that may enter blood vessels broken during the procedure and obstruct blood flow, infection, loss of feeling in the area, fluid depletion, injury to internal structures, and severe postoperative pain.

There are several types of liposuction available. One is called *tumescent liposuction*. In this variation, large amounts of fluid are injected during the procedure and the area to be treated becomes engorged with fluid or swollen (tumescent). This creates more space between the skin and subcutaneous layer and helps separate the adipose cells, allowing the cannula to move more easily through fat. Another alternative is *ultrasound-assisted liposuction (UAL)*. In this procedure, a special cannula delivers high-frequency sound waves that liquefy the fat cells and the liquid is removed by suction. In still another type of liposuction, called **laser-assisted liposuction**, a special cannula delivers laser energy that liquefies that fat cells and the liquid is removed by suction.

Cryolipolysis (*cryo*=cold) or **CoolSculpting** refers to the destruction of fat cells by the external application of controlled cooling. Since fat crystallizes faster than cells surrounding adipose tissue, the cold temperature kills the fat cells while sparing damage to nerve cells, blood vessels, and other structures. Within a few days of the procedure apoptosis (genetically programmed death) begins, and within several months the fat cells are removed.

Adhesions (Refer page 125 of Textbook)

Scar tissue can form **adhesions** (ad-HĒ-zhuns; *adhaero* = to stick to), abnormal joining of tissues. Adhesions commonly form in the abdomen around a site of previous inflammation such as an inflamed appendix, and they can develop after surgery. Although adhesions do not always cause problems, they can decrease tissue flexibility, cause obstruction (such as in the intestine), and make a subsequent operation, such as a cesarean section (C-section), more difficult. In rare cases adhesions can result in infertility. An *adhesiotomy*, the surgical release of adhesions, may be required.

DISORDERS: HOMEOSTATIC IMBALANCES

Disorders of epithelial tissue are mainly specific to individual organs, such as peptic ulcer disease (PUD), which erodes the epithelial lining of the stomach or small intestine. For this reason, epithelial disorders are described along with their relevant body systems throughout the text. The most prevalent disorders of connective tissues are **autoimmune diseases**—diseases in which antibodies produced by the immune system fail to distinguish what is foreign from what is self and attack the body's own tissues. One of the most common autoimmune disorders is rheumatoid arthritis, which attacks the synovial membranes of joints. Because connective tissue is one of the most abundant and widely distributed of the four main types of tissues, disorders related to them often affect multiple body systems. Common disorders of muscular tissues and nervous tissue are described at the ends of Chapters 10 and 12, respectively.

Systemic Lupus Erythematosus

Systemic lupus erythematosus (SLE), (er-i-thē-ma-TŌ-sus), or simply *lupus*, is a chronic inflammatory disease of connective tissue occurring mostly in nonwhite women during their childbearing years. It is an autoimmune disease that can cause tissue damage in every body system. The disease, which can range from a mild condition in most patients to a rapidly fatal disease, is marked by periods of exacerbation and remission. The prevalence of SLE is about 1 in 2000, with females more likely to be afflicted than males by a ratio of 8 or 9 to 1.

Although the cause of SLE is unknown, genetic, environmental, and hormonal factors all have been implicated. The genetic component is suggested by studies of twins and family history. Environmental factors include viruses, bacteria, chemicals, drugs, exposure to excessive sunlight, and emotional stress. Sex hormones, such as estrogens, may also trigger SLE.

Signs and symptoms of SLE include painful joints, low-grade fever, fatigue, mouth ulcers, weight loss, enlarged lymph nodes and spleen, sensitivity to sunlight, rapid loss of large amounts of scalp hair, and anorexia. A distinguishing feature of lupus is an eruption across the bridge of the nose and cheeks called a "butterfly rash." Other skin lesions may occur, including blistering and ulceration. The erosive nature of some SLE skin lesions was thought to resemble the damage inflicted by the bite of a wolf—thus, the name *lupus* (= wolf). The most serious complications of the disease involve inflammation of the kidneys, liver, spleen, lungs, heart, brain, and gastrointestinal tract. Because there is no cure for SLE, treatment is supportive, including anti-inflammatory drugs such as aspirin, and immunosuppressive drugs.

MEDICAL TERMINOLOGY

Atrophy (AT-rō-fē; *a-* = without; *-trophy* = nourishment) A decrease in the size of cells, with a subsequent decrease in the size of the affected tissue or organ.

Hypertrophy (hī-PER-trō-fē; *hyper-* = above) Increase in the size of a tissue because its cells enlarge without undergoing cell division.

Tissue rejection An immune response of the body directed at foreign proteins in a transplanted tissue or organ; immunosuppressive drugs, such as cyclosporine, have largely overcome tissue rejection in heart-, kidney-, and liver-transplant patients.

Tissue transplantation The replacement of a diseased or injured tissue or organ. The most successful transplants involve use of a person's own tissues or those from an identical twin.

Xenotransplantation (zen'-ō-trans-plan-TĀ-shun; *xeno-* = strange, foreign) The replacement of a diseased or injured tissue or organ with cells or tissues from an animal. Porcine (from pigs) and bovine (from cows) heart valves are used for some heart-valve replacement surgeries.

SELF-QUIZ QUESTIONS

Fill in the blanks in the following statements.

1. The four types of tissues are ——— , ——— , ——— , and ——— .

2. Epithelial tissue tends to be classified according to two criteria: ——— and ——— .

Indicate whether the following statements are true or false.

3. Epithelial tissue cells have an apical surface at the top and are attached to a basement membrane at the bottom.

4. Connective tissue fibers that are arranged in bundles and lend strength and flexibility to a tissue are collagen fibers.

Choose the one best answer to the following questions.

5. Which of the following muscular tissues can be voluntarily controlled?
   (1) cardiac, (2), smooth, (3) skeletal.
   (a) 1, 2, and 3   (b) 2   (c) 1
   (d) 1 and 3   (e) 3

6. Which of the following tissues is avascular?
   (a) cardiac muscle   (b) stratified squamous epithelial
   (c) compact bone   (d) skeletal muscle
   (e) adipose

7. If the lining of an organ produces and releases mucus, which of the following cells would likely be found in the tissue lining the organ?
   (a) goblet cells   (b) mast cells   (c) macrophages
   (d) osteoblasts   (e) fibroblasts

8. Why does damaged cartilage heal slowly?
   (a) Damaged cartilage undergoes fibrosis, which interferes with less the movement of materials needed for repair.
   (b) Cartilage does not contain fibroblasts, which are needed to produce the fibers in cartilage tissue.
   (c) Cartilage is avascular, so materials needed for repair must diffuse from surrounding tissue.

(d) Chondrocytes cannot be replaced once they are damaged.

(e) Chondrocytes undergo mitosis slowly, which delays healing.

9. Which of the following is *true* concerning serous membranes?

(a) A serous membrane lines a body part that opens directly to the body's exterior.

(b) The parietal portion of a serous membrane attaches to the organ.

(c) The visceral portion of a serous membrane attaches to a body cavity wall.

(d) The serous membrane covering the heart is known as the peritoneum.

(e) The serous membrane covering the lungs is known as the pleura.

10. The type of exocrine gland that forms its secretory product and simply releases it from the cell by exocytosis is the

(a) apocrine gland   (b) merocrine gland   (c) holocrine gland

(d) endocrine gland   (e) tubular gland.

11. Tissue changes that occur with aging can be due to (1) cross-links forming between glucose and proteins, (2) a decrease in the amount of collagen fibers, (3) a decreased blood supply, (4) improper nutrition, (5) a higher cellular metabolic rate.

(a) 1, 2, 3, 4, and 5   (b) 1, 2, 3, and 4   (c) 1 and 4

(d) 1, 3, and 4   (e) 1, 2, and 3

12. What type of cell junction would be required for cells to communicate with one another?

(a) adherens junction   (b) desmosome   (c) gap junction

(d) tight junction   (e) hemidesmosome

13. Match the following epithelial tissues to their descriptions:

_____ (a) contains a single layer of flat cells; found in the body where filtration (kidney) or diffusion (lungs) are priority processes

_____ (b) found in the superficial part of skin; provides protection from heat, microbes, and chemicals

_____ (c) contains cube-shaped cells functioning in secretion and absorption

_____ (d) lines the lower respiratory tract and uterine tubes; wavelike motion of cilia propels materials through the lumen

_____ (e) contains cells with microvilli and goblet cells; found in linings of the digestive, reproductive, and urinary tracts

_____ (f) found in the urinary bladder; contains cells that can change shape (stretch or relax)

_____ (g) contains cells that are all attached to the basement membrane, although some do not reach the surface; those cells that do extend to the surface secrete mucus or contain cilia

_____ (h) a fairly rare type of epithelium that has a mainly protective function

(1) pseudostratified ciliated columnar epithelium

(2) ciliated simple columnar epithelium

(3) transitional epithelium

(4) simple squamous epithelium

(5) simple cuboidal epithelium

(6) nonciliated simple columnar epithelium

(7) stratified cuboidal epithelium

(8) keratinized stratified squamous epithelium

14. For each of the following items, indicate the tissue type with which they are associated. Use **E** for epithelial tissues, **C** for connective tissues, **M** for muscle tissues, and **N** for nervous tissue.

_____ (a) bind, support

_____ (b) contains elongated cells that generate force

_____ (c) neuroglia

_____ (d) avascular

_____ (e) may contain fibroblasts

_____ (f) tightly packed cells

_____ (g) intercalated discs

_____ (h) goblet cells

_____ (i) contain extracellular matrix

_____ (j) striated

_____ (k) generate action potentials

_____ (l) cilia

_____ (m) ground substance

_____ (n) apical surface

_____ (o) excitable

15. Match the following connective tissues to their descriptions:

_____ (a) the tissue from which all other connective tissues eventually arise

_____ (b) connective tissue with a clear, liquid matrix that flows in lymphatic vessels

_____ (c) connective tissue consisting of several kinds of cells, containing all three fiber types randomly arranged, and found in the subcutaneous layer deep to the skin

_____ (d) a loose connective tissue specialized for triglyceride storage

_____ (e) tissue that contains reticular fibers and reticular cells and forms the stroma of certain organs such as the spleen

_____ (f) tissue with irregularly arranged collagen fibers found in the dermis of the skin

_____ (g) tissue found in the lungs that is strong and can recoil back to its original shape after being stretched

_____ (h) tissue that affords flexibility at joints and reduces joint friction

_____ (i) tissue that provides strength and rigidity and is the strongest of the three types of cartilage

_____ (j) bundles of collagen arranged in parallel patterns; compose tendons and ligaments

_____ (k) tissue that forms the internal framework of the body and works with skeletal muscle to generate movement

_____ (l) tissue that contains a network of elastic fibers, providing strength, elasticity, and maintenance of shape; located in the external ear

_____ (m) connective tissue with formed elements suspended in a liquid matrix called plasma

(1) blood

(2) fibrocartilage

(3) mesenchyme

(4) dense regular connective tissue

(5) lymph

(6) hyaline cartilage

(7) dense irregular connective tissue

(8) areolar connective tissue

(9) reticular connective tissue

(10) bone (osseous tissue)

(11) elastic connective tissue

(12) elastic cartilage

(13) adipose tissue)

5 THE INTEGUMENTARY SYSTEM

5.1 Structure of the Skin

1. The integumentary system consists of the skin, hair, oil and sweat glands, nails, and sensory receptors.
2. The skin is the largest organ of the body in weight. The principal parts of the skin are the epidermis (superficial) and dermis (deep).
3. The subcutaneous layer (hypodermis) is deep to the dermis and not part of the skin. It anchors the dermis to underlying tissues and organs, and it contains lamellated corpuscles.
4. The types of cells in the epidermis are keratinocytes, melanocytes, intraepidermal macrophages, and tactile epithelial cells.
5. The epidermal layers, from deep to superficial, are the stratum basale, stratum spinosum, stratum granulosum, stratum lucidum (in thick skin only), and stratum corneum (see **Table 5.1**). Stem cells in the stratum basale undergo continuous cell division, producing keratinocytes for the other layers.
6. The dermis is composed of dense irregular connective tissue containing collagen and elastic fibers. It is divided into papillary and reticular regions. The papillary region contains thin collagen and fine elastic fibers, dermal papillae, and corpuscles of touch. The reticular region contains bundles of thick collagen and some coarse elastic fibers, fibroblasts and macrophages, adipose tissue, hair follicles, nerves, sebaceous (oil) glands, and sudoriferous (sweat) glands. (See **Table 5.2**.)
7. Epidermal ridges provide the basis for fingerprints and footprints.
8. The color of skin is due to melanin, carotene, and hemoglobin.
9. In tattooing, a pigment is deposited with a needle in the dermis. Body piercing is the insertion of jewelry through an artificial opening.

5.2 Accessory Structures of the Skin

1. Accessory structures of the skin—hair, skin glands, and nails—develop from the embryonic epidermis.
2. A hair consists of a hair shaft, most of which is superficial to the surface, a hair root that penetrates the dermis and sometimes the subcutaneous layer, and a hair follicle.
3. Associated with each hair follicle is a sebaceous (oil) gland, an arrector pili muscle, and a hair root plexus.
4. New hairs develop from division of hair matrix cells in the hair bulb; hair replacement and growth occur in a cyclical pattern consisting of growth, regression, and resting stages.

5. Hairs offer a limited amount of protection—from the sun, heat loss, and entry of foreign particles into the eyes, nose, and ears. They also function in sensing light touch.
6. Lanugo of the fetus is shed before birth. Most body hair on males is terminal (coarse, pigmented); most body hair on females is vellus (fine).
7. Sebaceous (oil) glands are usually connected to hair follicles; they are absent from the palms and soles. Sebaceous glands produce sebum, which moistens hairs and waterproofs the skin. Clogged sebaceous glands may produce acne.
8. There are two types of sudoriferous (sweat) glands: eccrine and apocrine. Eccrine sweat glands have an extensive distribution; their ducts terminate at pores at the surface of the epidermis. Eccrine sweat glands are involved in thermoregulation and waste removal and are stimulated during emotional stress. Apocrine sweat glands are limited to the skin of the axillae, groin, and areolae; their ducts open into hair follicles. Apocrine sweat glands are stimulated during emotional stress and sexual excitement. (See **Table 5.3**.)
9. Ceruminous glands are modified sudoriferous glands that secrete cerumen. They are found in the external auditory canal (ear canal).
10. Nails are hard, dead keratinized epidermal cells over the dorsal surfaces of the distal portions of the digits. The principal parts of a nail are the nail body, free edge, nail root, lunula, hyponychium, nail bed, eponychium, and nail matrix. Cell division of the nail matrix cells produces new nails.

5.3 Types of Skin

1. Thin skin covers all parts of the body except for the palms, palmar surfaces of the digits, and the soles.
2. Thick skin covers the palms, palmar surfaces of the digits, and soles. (See **Table 5.4**.)

5.4 Functions of the Skin

1. Skin functions include body temperature regulation, blood storage, protection, sensation, excretion and absorption, and synthesis of vitamin D.
2. The skin participates in thermoregulation by liberating sweat at its surface and by adjusting the flow of blood in the dermis.
3. The skin provides physical, chemical, and biological barriers that help protect the body.
4. Cutaneous sensations include tactile sensations, thermal sensations, and pain.

5.5 Maintaining Homeostasis: Skin Wound Healing

1. In an epidermal wound, the central portion of the wound usually extends down to the dermis; the wound edges involve only superficial damage to the epidermal cells.
2. Epidermal wounds are repaired by enlargement and migration of basal cells, contact inhibition, and division of migrating and stationary basal cells.
3. During the inflammatory phase of deep wound healing, a blood clot unites the wound edges, epithelial cells migrate across the wound, vasodilation and increased permeability of blood vessels enhance delivery of phagocytes, and mesenchymal cells develop into fibroblasts.
4. During the migratory phase, fibroblasts migrate along fibrin threads and begin synthesizing collagen fibers and glycoproteins.
5. During the proliferative phase, epithelial cells grow extensively.
6. During the maturation phase, the scab sloughs off, the epidermis is restored to normal thickness, collagen fibers become more

organized, fibroblasts begin to disappear, and blood vessels are restored to normal.

5.6 Development of the Integumentary System

1. The epidermis develops from the embryonic ectoderm, and the accessory structures of the skin (hair, nails, and skin glands) are epidermal derivatives.
2. The dermis is derived from mesodermal cells.

5.7 Aging and the Integumentary System

1. Most effects of aging begin to occur when people reach their late 40s.
2. Among the effects of aging are wrinkling, loss of subcutaneous adipose tissue, atrophy of sebaceous glands, and decrease in the number of melanocytes and intraepidermal macrophages.

CLINICAL CONNECTIONS

Skin Grafts (Refer page 130 of Textbook)

New skin cannot regenerate if an injury destroys a large area of the stratum basale and its stem cells. Skin wounds of this magnitude require skin grafts in order to heal. A **skin graft** is the transfer of a patch of healthy skin taken from a donor site to cover a wound. A skin graft is performed to protect against fluid loss and infection, to promote tissue healing, to reduce scar formation, to prevent loss of function, and for cosmetic reasons. To avoid tissue rejection, the transplanted skin is usually taken from the same individual (*autograft*) or an identical twin (*isograft*). If skin damage is so extensive that an autograft would cause harm, a self-donation procedure called *autologous skin transplantation* (aw-TOL-ō-gus) may be used. In this procedure, performed most often for severely burned patients, small amounts of an individual's epidermis are removed, and the keratinocytes are cultured in the laboratory to produce thin sheets of skin. The new skin is transplanted back to the patient so that it covers the burn wound and generates a permanent skin. Also available as skin grafts for wound coverage are products (Apligraft and Transite) grown in the laboratory from the foreskins of circumcised infants.

Psoriasis (Refer page 131 of Textbook)

Psoriasis (sō-RĪ-a-sis) is a common and chronic skin disorder in which keratinocytes divide and move more quickly than normal from the stratum basale to the stratum corneum. They are shed prematurely in as little as 7 to 10 days. The immature keratinocytes make an abnormal keratin, which forms flaky, silvery scales at the skin surface, most often on the knees, elbows, and scalp (dandruff). Effective treatments—various topical ointments and ultraviolet phototherapy—suppress cell division, decrease the rate of cell growth, or inhibit keratinization.

Stretch Marks (Refer page 132 of Textbook)

Because of the collagenous, vascular structure of the dermis, **striae** (STRĪ-ē = streaks) or *stretch marks*, a form of internal scarring, can result from the internal damage to this layer that occurs when the skin is stretched too much. When the skin is overstretched, the lateral bonding between adjacent collagen fibers is disrupted and small dermal blood vessels rupture. This is why stretch marks initially appear as reddish streaks at these sites. Later, after scar tissue

(which is poorly vascularized) forms at these sites of dermal breakdown, the stretch marks appear as silvery white streaks. Stretch marks often occur in the abdominal skin during pregnancy, on the skin of weight-lifters where the skin is stretched by a rapid increase in muscle mass, and in the stretched skin accompanying gross obesity.

Tension Lines and Surgery (Refer page 132 of Textbook)

In certain regions of the body, collagen fibers within the reticular region tend to orient more in one direction than another because of natural tension experienced by these regions of the skin resulting from bony projections, orientation of muscles, and movements of joints. **Tension lines** (*lines of cleavage*) in the skin indicate the predominant direction of underlying collagen fibers. Knowledge of tension lines is especially important to plastic surgeons. For example, a surgical incision running parallel to the collagen fibers will heal with only a fine scar. A surgical incision made across the rows of fibers disrupts the collagen, and the wound tends to gape open and heal in a broad, thick scar.

Albinism and Vitiligo (Refer page 133 of Textbook)

Albinism (AL-bin-izm; *albin-* = white) is the inherited inability of an individual to produce melanin. Most **albinos** (al-BĪ-nōs), people affected by albinism, have melanocytes that are unable to synthesize tyrosinase. Melanin is missing from their hair, eyes, and skin. This results in problems with vision and a tendency of the skin to burn easily on overexposure to sunlight.

In another condition, called **vitiligo** (vit-i-LĪ-gō), the partial or complete loss of melanocytes from patches of skin produces irregular white spots. The loss of melanocytes is related to an immune system malfunction in which antibodies attack the melanocytes.

Skin Color as a Diagnostic Clue (Refer page 133 of Textbook)

The color of skin and mucous membranes can provide clues for diagnosing certain conditions. When blood is not picking up an adequate amount of oxygen from the lungs, as in someone who has stopped breathing, the mucous membranes, nail beds, and skin appear bluish or **cyanotic** (sī-a-NOT-ik; *cyan-* = blue). **Jaundice** (JON-dis; *jaund-* = yellow) is due to a buildup of the yellow pigment bilirubin in the skin. This condition gives a yellowish appearance to the skin and the whites of the eyes, and usually indicates liver disease. **Ery-thema** (er-e-THĒ-ma; *eryth-* = red), redness of the skin, is caused by engorgement of capillaries in the dermis with blood due to skin injury, exposure to heat, infection, inflammation, or allergic reactions. **Pallor** (PAL-or), or paleness of the skin, may occur in conditions such as shock and anemia. All skin color changes are observed most readily in people with light-colored skin and may be more difficult to discern in people with darker skin. However, examination of the nail beds and gums can provide some information about circulation in individuals with darker skin.

Hair Removal (Refer page 134 of Textbook)

A substance that removes hair is called a **depilatory** (de-PIL-a-tō-rē). It dissolves the protein in the hair shaft, turning it into a gelatinous mass that can be wiped away. Because the hair root is not affected, regrowth of the hair occurs. In **electrolysis**, an electric current is used to destroy the hair matrix so the hair cannot regrow. **Laser treatments** may also be used to remove hair.

Chemotherapy and Hair Loss (Refer page 134 of Textbook)

Chemotherapy is the treatment of disease, usually cancer, by means of chemical substances or drugs. Chemotherapeutic agents interrupt the life cycle of rapidly dividing cancer cells. Unfortunately, the drugs also affect other rapidly dividing cells in the body, such as the hair matrix cells of a hair. It is for this reason that individuals undergoing chemotherapy experience hair loss. Since about 15% of the hair matrix cells of scalp hairs are in the resting stage, these cells are not affected by chemotherapy. Once chemotherapy is stopped, the hair matrix cells replace lost hair follicles and hair growth resumes.

Hair and Hormones (Refer page 134 of Textbook)

At puberty, when the testes begin secreting significant quantities of androgens (masculinizing sex hormones), males develop the typical male pattern of hair growth throughout the body, including a beard and a hairy chest. In females at puberty, the ovaries and the adrenal glands produce small quantities of androgens, which promote hair growth throughout the body including the axillae and pubic region. Occasionally, a tumor of the adrenal glands, testes, or ovaries produces an excessive amount of androgens. The result in females or prepubertal males is **hirsutism** (HER-soo-tizm; *hirsut-* = shaggy), excessive body hair or body hair in areas that usually are not hairy.

Surprisingly, androgens also must be present for occurrence of the most common form of baldness, **androgenic alopecia** (an'-drō-JEN-ik al'-ō-PĒ-shē-a) or *male-pattern baldness*. In genetically predisposed adults, androgens inhibit hair growth. In men, hair loss usually begins with a receding hairline followed by hair loss in the temples and crown. Women are more likely to have thinning of hair on top of the head. The first drug approved for enhancing scalp hair growth was minoxidil (Rogaine). It causes vasodilation (widening of blood vessels), thus increasing circulation; direct stimulation of hair follicle cells to pass into growth stage follicles; and inhibition of androgens. In about a third of the people who try it, minoxidil improves hair growth, causing scalp follicles to enlarge and lengthening the growth cycle. For many, however, the hair growth is meager. Minoxidil does not help people who already are bald.

Acne (Refer page 136 of Textbook)

During childhood, sebaceous glands are relatively small and inactive. At puberty, androgens from the testes, ovaries, and adrenal glands stimulate sebaceous glands to grow in size and increase their production of sebum. **Acne** is an inflammation of sebaceous glands that usually begins at puberty, when the sebaceous glands are stimulated by androgens. Acne occurs predominantly in sebaceous follicles that have been colonized by bacteria, some of which thrive in the lipid-rich sebum. The infection may cause a cyst or sac

of connective tissue cells to form, which can destroy and displace epidermal cells. This condition, called **cystic acne**, can permanently scar the epidermis. Treatment consists of gently washing the affected areas once or twice daily with a mild soap, topical antibiotics (such as clindamycin and erythromycin), topical drugs such as benzoyl peroxide or tretinoin, and oral antibiotics (such as tetracycline, minocycline, erythromycin, and isotretinoin). Contrary to popular belief, foods such as chocolate or fried foods do not cause or worsen acne.

Impacted Cerumen (Refer page 137 of Textbook)

Some people produce an abnormally large amount of cerumen in the external auditory canal. If it accumulates until it becomes impacted (firmly wedged), sound waves may be prevented from reaching the eardrum. Treatments for **impacted cerumen** include periodic ear irrigation with enzymes to dissolve the wax and removal of wax with a blunt instrument by trained medical personnel. The use of cotton-tipped swabs or sharp objects is not recommended for this purpose because they may push the cerumen further into the external auditory canal and damage the eardrum.

Transdermal Drug Administration (Refer page 140 of Textbook)

Most drugs are either absorbed into the body through the digestive system or injected into subcutaneous tissue or muscle. An alternative route, **transdermal** (*transcutaneous*) **drug administration**, enables a drug contained within an adhesive skin patch to pass across the epidermis and into the blood vessels of the dermis. The drug is released continuously at a controlled rate over a period of one to several days. This method is especially useful for drugs that are quickly eliminated from the body because such drugs, if taken in other forms, would have to be taken quite frequently. Because the major barrier to penetration is the stratum corneum, transdermal absorption is most rapid in regions where this layer is thin, such as the scrotum, face, and scalp. A growing number of drugs are available for transdermal administration, including nitroglycerin, for prevention of angina pectoris (chest pain associated with heart disease); scopolamine, for motion sickness; estradiol, used for estrogen-replacement therapy during menopause; ethinyl estradiol and norelgestromin in contraceptive patches; nicotine, used to help people stop smoking; and fentanyl, used to relieve severe pain in cancer patients.

Sun Damage, Sunscreens, and Sunblocks (Refer page 144 of Textbook)

Although basking in the warmth of the sun may feel good, it is not a healthy practice. There are two forms of ultraviolet radiation that affect the health of the skin. Longer wavelength ultraviolet A (UVA) rays make up nearly 95% of the ultraviolet radiation that reaches the earth. UVA rays are not absorbed by the ozone layer. They penetrate the furthest into the skin, where they are absorbed by melanocytes and thus are involved in sun tanning. UVA rays also depress the immune system. Shorter wavelength ultraviolet B (UVB) rays are partially absorbed by the ozone layer and do not penetrate the skin as deeply as UVA rays. UVB rays cause sunburn and are responsible for most of the tissue damage (production of oxygen free radicals which disrupt collagen and elastic fibers) that results in wrinkling and aging of the skin and cataract formation. Both UVA and UVB rays are thought to cause skin cancer. Long-term overexposure to sunlight results in dilated blood vessels, age spots, freckles, and changes in skin texture.

Exposure to ultraviolet radiation (either natural sunlight or the artificial light of a tanning booth) may also produce **photosensitivity**, a heightened reaction of the skin after consumption of certain medications or contact with certain substances. Photosensitivity is characterized by redness, itching, blistering, peeling, hives, and even shock. Among the medications or substances that may cause a photosensitivity reaction are certain antibiotics (tetracycline), nonsteroidal anti-inflammatory drugs (ibuprofen or naproxen), certain herbal supplements (St. John's wort), some birth control pills, some high blood pressure medications, some antihistamines, and certain artificial sweeteners, perfumes, aftershaves, lotions, detergents, and medicated cosmetics.

Self-tanning lotions (*sunless tanners*), topically applied substances, contain a color additive (dihydroxyacetone) that produces a tanned appearance by interacting with proteins in the skin.

Sunscreens are topically applied preparations that contain various chemical agents (such as benzophenone or one of its derivatives) that absorb UVB rays but let most of the UVA rays pass through.

Sunblocks are topically applied preparations that contain substances such as zinc oxide that reflect and scatter both UVB and UVA rays.

Both sunscreens and sunblocks are graded according to a *sun protection factor (SPF)* rating, which measures the level of protection they supposedly provide against UV rays. The higher the rating, presumably the greater the degree of protection. As a precautionary measure, individuals who plan to spend a significant amount of time in the sun should use a sunscreen or a sunblock with an SPF of 15 or higher. Although sunscreens protect against sunburn, there is considerable debate as to whether they actually protect against skin cancer. In fact, some studies suggest that sunscreens increase the incidence of skin cancer because of the false sense of security they provide.

DISORDERS: HOMEOSTATIC IMBALANCES

Skin Cancer

Excessive exposure to ultraviolet radiation from the sun or tanning beds causes virtually all of the one million cases of **skin cancer** diagnosed annually in the United States. One-half of all cancers in the United States are skin cancers. There are three common forms of skin cancer. **Basal cell carcinomas** account for about 78% of all skin cancers. The tumors arise from cells in the stratum basale of the epidermis and rarely metastasize. **Squamous cell carcinomas**, which account for about 20% of all skin cancers, arise from the stratum spinosum of the epidermis, and they have a variable tendency to metastasize. Basal and squamous cell carcinomas are together known as *nonmelanoma skin cancer.*

Malignant melanomas arise from melanocytes and account for about 2% of all skin cancers. The estimated lifetime risk of developing melanoma is now 1 in 75, double the risk only 20 years ago. In part, this increase is due to depletion of the ozone layer, which absorbs some UV light high in the atmosphere. But the main reason for the increase is that more people are spending more time in the sun and in tanning beds. Malignant melanomas metastasize rapidly and can kill a person within months of diagnosis.

The key to successful treatment of malignant melanoma is early detection. The early warning signs of malignant melanoma are identified by the acronym ABCDE (**Figure SG5.1**). *A* is for *asymmetry*; malignant melanomas tend to lack symmetry. This means that they have irregular shapes, such as two very different looking halves. *B* is for *border*; malignant melanomas have irregular—notched, indented, scalloped, or indistinct—borders. *C* is for *color*; malignant melanomas have uneven coloration and may contain several colors. *D* is for *diameter*; ordinary moles typically are smaller than 6 mm (0.25 in.), about the size of a pencil eraser. *E* is for *evolving*; malignant melanomas change in size, shape, and color. Once a malignant melanoma has the characteristics of A, B, and C, it is usually larger than 6 mm.

Among the risk factors for skin cancer are the following:

1. **Skin type.** Individuals with light-colored skin who never tan but always burn are at high risk.
2. **Sun exposure.** People who live in areas with many days of sunlight per year and at high altitudes (where ultraviolet light is more intense) have a higher risk of developing skin cancer. Likewise, people who engage in outdoor occupations and those who have suffered three or more severe sunburns have a higher risk.
3. **Family history.** Skin cancer rates are higher in some families than in others.
4. **Age.** Older people are more prone to skin cancer owing to longer total exposure to sunlight.
5. **Immunological status.** Immunosuppressed individuals have a higher incidence of skin cancer.

Figure SG5.1 Comparison of a normal nevus (mole) and a malignant melanoma.

Excessive exposure to ultraviolet radiation from the sun or tanning beds accounts for almost all cases of skin cancer.

Publiphoto/Science Source

(a) Normal nevus (mole)

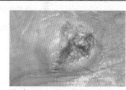

Biophoto Associates/Science Source

(c) Squamous cell carcinoma

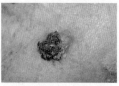

Biophoto Associates/Science Source

(b) Basal cell carcinoma

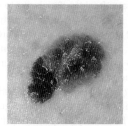

Biophoto Associates/Science Source

(d) Malignant melanoma

Q Which is the most common type of skin cancer?

Burns

A burn is tissue damage caused by excessive heat, electricity, radioactivity, or corrosive chemicals that denature (break down) proteins in the skin. Burns destroy some of the skin's important contributions to homeostasis—protection against microbial invasion and dehydration, and thermoregulation.

Burns are graded according to their severity. A *first-degree burn* involves only the epidermis (**Figure SG5.2a**). It is characterized by mild pain and erythema (redness) but no blisters. Skin functions remain intact. Immediate flushing with cold water may lessen the pain and damage caused by a first-degree burn. Generally, healing of a first-degree burn will occur in 3 to 6 days and may be accompanied by flaking or peeling. One example of a first-degree burn is mild sunburn.

A *second-degree burn* destroys the epidermis and part of the dermis (**Figure SG5.2b**). Some skin functions are lost. In a second-degree burn,

Figure SG5.2 Burns.

A burn is tissue damage caused by agents that destroy the proteins in the skin.

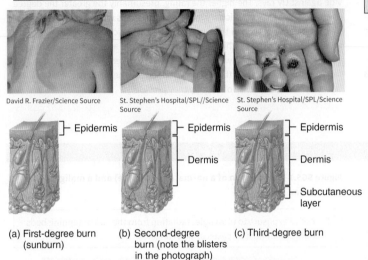

David R. Frazier/Science Source   St. Stephen's Hospital/SPL//Science Source   St. Stephen's Hospital/SPL/Science Source

(a) First-degree burn (sunburn)   (b) Second-degree burn (note the blisters in the photograph)   (c) Third-degree burn

Q What factors determine the seriousness of a burn?

Figure SG5.3 Rule-of-nines method for determining the extent of a burn.
The percentages are the approximate proportions of the body surface area.

The rule of nines is a quick way for estimating the surface area affected by a burn in an adult.

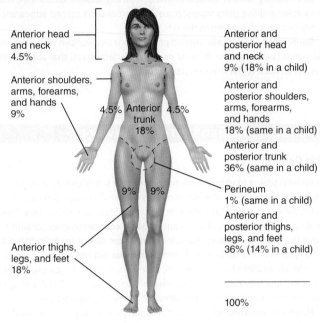

Anterior head and neck 4.5%

Anterior shoulders, arms, forearms, and hands 9%

4.5% Anterior trunk 18% 4.5%

Anterior thighs, legs, and feet 18%

9%   9%

Anterior and posterior head and neck 9% (18% in a child)

Anterior and posterior shoulders, arms, forearms, and hands 18% (same in a child)

Anterior and posterior trunk 36% (same in a child)

Perineum 1% (same in a child)

Anterior and posterior thighs, legs, and feet 36% (14% in a child)

100%

Anterior view

Q What percentage of the body would be burned if only the anterior trunk and anterior left upper limb were involved?

redness, blister formation, edema, and pain result. In a blister the epidermis separates from the dermis due to the accumulation of tissue fluid between them. Associated structures, such as hair follicles, sebaceous glands, and sweat glands, usually are not injured. If there is no infection, second-degree burns heal without skin grafting in about 3 to 4 weeks, but scarring may result. First- and second-degree burns are collectively referred to as *partial-thickness burns*.

A *third-degree burn* or *full-thickness burn* destroys the epidermis, dermis, and subcutaneous layer (**Figure SG5.2c**). Most skin functions are lost. Such burns vary in appearance from marble-white to mahogany colored to charred, dry wounds. There is marked edema, and the burned region is numb because sensory nerve endings have been destroyed. Regeneration occurs slowly, and much granulation tissue forms before being covered by epithelium. Skin grafting may be required to promote healing and to minimize scarring.

The injury to the skin tissues directly in contact with the damaging agent is the *local effect* of a burn. Generally, however, the *systemic effects* of a major burn are a greater threat to life. The systemic effects of a burn may include (1) a large loss of water, plasma, and plasma proteins, which causes shock; (2) bacterial infection; (3) reduced circulation of blood; (4) decreased production of urine; and (5) diminished immune responses.

The seriousness of a burn is determined by its depth and extent of area involved, as well as the person's age and general health. According to the American Burn Association's classification of burn injury, a major burn includes third-degree burns over 10% of body surface area; or second-degree burns over 25% of body surface area; or any third-degree burns on the face, hands, feet, or *perineum* (per´-i-NĒ-um, which includes the anal and urogenital regions). When the burn area exceeds 70%, more than half the victims die. A quick means for estimating the surface area affected by a burn in an adult is the **rule of nines** (**Figure SG5.3**):

1. Count 9% if both the anterior and posterior surfaces of the head and neck are affected.

2. Count 9% for both the anterior and posterior surfaces of each upper limb (total of 18% for both upper limbs).

3. Count four times nine, or 36%, for both the anterior and posterior surfaces of the trunk, including the buttocks.

4. Count 9% for the anterior and 9% for the posterior surfaces of each lower limb as far up as the buttocks (total of 36% for both lower limbs).

5. Count 1% for the perineum.

In severely burned patients with full-thickness or deep partial-thickness burns where there is not sufficient autograft, a tissue engineered product called Integra® Dermal Regeneration Template (DRT) is available. It is designed to promote organized regeneration of the dermis while providing a protective barrier against fluid loss and microbes. Integra® DRT consists of two layers, just like human skin. The bottom layer, called the matrix layer, is composed of bovine (cow) collagen and the carbohydrate glycosaminoglycan (GAG). It mimics the dermis, functions as an extracellular layer, and induces the body's own dermal cells to migrate into the area and regenerate a new dermis. The outer layer, called the silicone layer, consists of a thin layer of silicone that mimics the epidermis. Its role is to close the wound, control fluid loss, and serve as a protective barrier. Once the dermis has regenerated sufficiently (about three weeks), the silicone layer is removed and a thin sheet of the patient's own epidermal cells is applied.

Many people who have been burned in fires also inhale smoke. If the smoke is unusually hot or dense or if inhalation is prolonged, serious problems can develop. The hot smoke can damage the trachea (windpipe), causing its lining to swell. As the swelling narrows the trachea, airflow into the lungs is obstructed. Further, small airways inside the lungs can also narrow, producing wheezing or shortness of breath. A person who has inhaled smoke is given oxygen through a face mask, and a tube may be inserted into the trachea to assist breathing.

Pressure Ulcers

Pressure ulcers, also known as *decubitus ulcers* (dē-KŪ-bi-tus) or *bedsores*, are caused by a constant deficiency of blood flow to tissues (**Figure SG5.4**). Typically the affected tissue overlies a bony projection that has been subjected to prolonged pressure against an object such as a bed, cast, or splint. If the pressure is relieved in a few hours, redness occurs but no lasting tissue damage results. Blistering of the affected area may indicate superficial damage; a reddish-blue discoloration may indicate deep tissue damage. Prolonged pressure causes tissue ulceration. Small breaks in the epidermis become infected, and the sensitive subcutaneous layer and deeper tissues are damaged. Eventually, the tissue dies. Pressure ulcers occur most often in bedridden patients. With proper care, pressure ulcers are preventable, but they can develop very quickly in patients who are very old or very ill.

Figure SG5.4 Pressure ulcer.

> A pressure ulcer is a shedding of epithelium caused by a constant deficiency of blood flow to tissues.

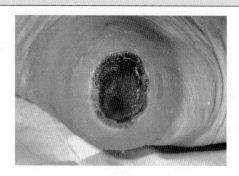

Pressure ulcer on heel

Q What parts of the body are usually affected by pressure ulcers?

MEDICAL TERMINOLOGY

Abrasion (a-BRĀ-shun; *ab-* = away; *-rasion* = scraped) An area where skin has been scraped away.

Blister A collection of serous fluid within the epidermis or between the epidermis and dermis, due to short-term but severe friction. The term **bulla** (BUL-a) refers to a large blister.

Callus (KAL-lus = hard skin) An area of hardened and thickened skin that is usually seen in palms and soles and is due to persistent pressure and friction.

Cold sore A lesion, usually in an oral mucous membrane, caused by type 1 herpes simplex virus (HSV) transmitted by oral or respiratory routes. The virus remains dormant until triggered by factors such as ultraviolet light, hormonal changes, and emotional stress. Also called a fever blister.

Comedo (KOM-ē-dō = to eat up) A collection of sebaceous material and dead cells in the hair follicle and excretory duct of the sebaceous (oil) gland. Usually found over the face, chest, and back, and more commonly during adolescence. Also called a **blackhead**.

Contact dermatitis (der-ma-TĪ-tis; *dermat-* = skin; *-itis* = inflammation of) Inflammation of the skin characterized by redness, itching, and swelling and caused by exposure of the skin to chemicals that bring about an allergic reaction, such as poison ivy toxin.

Contusion (kon-TOO-shun; *contundere* = to bruise) Condition in which tissue deep to the skin is damaged, but the epidermis is not broken.

Corn A painful conical thickening of the stratum corneum of the epidermis found principally over toe joints and between the toes, often caused by friction or pressure. Corns may be hard or soft, depending on their location. Hard corns are usually found over toe joints, and soft corns are usually found between the fourth and fifth toes.

Cyst (SIST = sac containing fluid) A sac with a distinct connective tissue wall, containing a fluid or other material.

Eczema (EK-ze-ma; *ekzeo-* = to boil over) An inflammation of the skin characterized by patches of red, blistering, dry, extremely itchy skin. It occurs mostly in skin creases in the wrists, backs of the knees, and fronts of the elbows. It typically begins in infancy and many children outgrow the condition. The cause is unknown but is linked to genetics and allergies.

Frostbite Local destruction of skin and subcutaneous tissue on exposed surfaces as a result of extreme cold. In mild cases, the skin is blue and swollen and there is slight pain. In severe cases there is considerable swelling, some bleeding, no pain, and blistering. If untreated, gangrene may develop. Frostbite is treated by rapid rewarming.

Hemangioma (hē-man′-jē-Ō-ma; *hem-* = blood; *-angi-* = blood vessel; *-oma* = tumor) Localized benign tumor of the skin and subcutaneous layer that results from an abnormal increase in the number of blood vessels. One type is a **portwine stain**, a flat, pink, red, or purple lesion present at birth, usually at the nape of the neck.

Hives Reddened elevated patches of skin that are often itchy. Most commonly caused by infections, physical trauma, medications, emotional stress, food additives, and certain food allergies. Also called **urticaria** (ūr-ti-KAR-ē-a).

Keloid (KĒ-loid; *kelis* = tumor) An elevated, irregular darkened area of excess scar tissue caused by collagen formation during healing. It extends beyond the original injury and is tender and frequently painful. It occurs in the dermis and underlying subcutaneous tissue, usually after trauma, surgery, a burn, or severe acne; more common in people of African descent.

Keratosis (ker′-a-TŌ-sis; *kera-* = horn) Formation of a hardened growth of epidermal tissue, such as *solar keratosis*, a premalignant lesion of the sun-exposed skin of the face and hands.

Laceration (las-er-Ā-shun; *lacer-* = torn) An irregular tear of the skin.

Lice Contagious arthropods that include two basic forms. **Head lice** are tiny, jumping arthropods that suck blood from the scalp. They lay eggs, called nits, and their saliva causes itching that may lead to complications. **Pubic lice** are tiny arthropods that do not jump; they look like miniature crabs.

Papule (PAP-ūl; *papula* = pimple) A small, round skin elevation less than 1 cm in diameter. One example is a pimple.

Pruritus (proo-RĪ-tus; *pruri-* = to itch) Itching, one of the most common dermatological disorders. It may be caused by skin disorders (infections), systemic disorders (cancer, kidney failure), psychogenic factors (emotional stress), or allergic reactions.

Tinea corporis (TIN-ē-a KOR-po-ris) A fungal infection characterized by scaling, itching, and sometimes painful lesions that may appear on any part of the body; also known as **ringworm**. Fungi thrive in warm, moist places such as skin folds of the groin, where it is known as **tinea cruris** (KROO-ris) (*jock itch*) or between the toes, where it is called **tinea pedis** (PE-dis) (*athlete's foot*).

Topical In reference to a medication, applied to the skin surface rather than ingested or injected.

Wart Mass produced by uncontrolled growth of epithelial skin cells; caused by a papillomavirus. Most warts are noncancerous.

SELF-QUIZ QUESTIONS

Fill in the blanks in the following statements.

1. The epidermal layer that is found in thick skin but not in thin skin is the _____.

2. The most common sweat glands that release a watery secretion are _____ sweat glands; modified sweat glands in the ear are _____ glands; sweat glands located in the axillae, groin, areolae, and beards of males and that release a slightly viscous, lipid-rich secretion are _____ sweat glands.

Indicate whether the following statements are true or false.

3. An individual with a dark skin color has more melanocytes than a fair-skinned person.

4. In order to permanently prevent growth of an unwanted hair, you must destroy the hair matrix.

Choose the one best answer to the following questions.

5. The layer of the epidermis that contains stem cells undergoing mitosis is the
   (a) stratum corneum.
   (b) stratum lucidum.
   (c) stratum basale.
   (d) stratum spinosum.
   (e) stratum granulosum.

6. The substance that helps promote mitosis in epidermal skin cells is
   (a) keratohyalin. (b) melanin. (c) carotene.
   (d) collagen. (e) epidermal growth factor.

7. Which of the following is *not* a function of skin?
   (a) calcium production (b) vitamin D synthesis
   (c) protection (d) excretion of wastes
   (e) temperature regulation

8. To expose underlying tissues in the bottom of the foot, a foot surgeon must first cut through the skin. Place the following layers in the order that the scalpel would cut. (1) stratum lucidum, (2) stratum corneum, (3) stratum basale, (4) stratum granulosum, (5) stratum spinosum.
   (a) 3, 5, 4, 1, 2 (b) 2, 1, 5, 4, 3 (c) 2, 1, 4, 5, 3
   (d) 1, 3, 5, 4, 2 (e) 3, 4, 5, 1, 2

9. Aging of the skin can result in
   (a) an increase in collagen and elastic fibers.
   (b) a decrease in the activity of sebaceous glands.
   (c) a thickening of the skin.
   (d) an increased blood flow to the skin.
   (e) an increase in toenail growth.

10. Which of the following is *not* true?
    (a) Albinism is an inherited inability of melanocytes to produce melanin.
    (b) Striae occurs when the dermis is overstretched to the point of tearing.
    (c) In order to prevent excessive scarring, surgeons should cut parallel to the lines of cleavage.
    (d) The papillary layer of the dermis is responsible for fingerprints.
    (e) Much of the body's fat is located in the dermis of the skin.

11. A patient is brought into the emergency room suffering from a burn. The patient does not feel any pain at the burn site. Using a gentle pull on a hair, the examining physician can remove entire hair follicles from the patient's arm. This patient is suffering from what type of burn?
    (a) third degree
    (b) second degree
    (c) first degree
    (d) partial thickness
    (e) localized

12. Which of the following statements are *true*? (1) Nails are composed of tightly packed, hard, keratinized cells of the epidermis that form a clear, solid covering over the dorsal surface of the terminal end of digits. (2) The free edge of the nail is white due to the absence of capillaries. (3) Nails help us grasp and manipulate small objects. (4) Nails protect the ends of digits from trauma. (5) Nail color is due to a combination of melanin and carotene.
    (a) 1, 2, and 3 (b) 1, 3, and 4 (c) 1, 2, 3, and 4
    (d) 2, 3, and 4 (e) 1, 3, and 5

13. Match the following:
    _____ (a) produce the protein that help protect the skin and underlying tissues from light, heat, microbes, and many chemicals
    _____ (b) produce a pigment that contributes to skin color and absorbs ultraviolet light
    _____ (c) cells that arise from red bone marrow, migrate to the epidermis, and participate in immune responses
    _____ (d) cells thought to function in the sensation of touch
    _____ (e) located in the dermis, they function in the sensations of warmth, coolness, pain, itching, and tickling
    _____ (f) smooth muscles associated with the hair follicles; when contracted, they pull the hair shafts perpendicular to the skin's surface
    _____ (g) an abnormal thickening of the epidermis
    _____ (h) release a lipid-rich secretion that functions as a water-repellent sealant in the stratum granulosum
    _____ (i) pressure-sensitive cells found mostly in the subcutaneous layer
    _____ (j) a fatty substance that covers and protects the skin of the fetus from the constant exposure to amniotic fluid
    _____ (k) associated with hair follicles, these secrete an oily substance that helps prevent hair from becoming brittle, prevents evaporation of water from the skin's surface, and inhibits the growth of certain bacteria

    (1) Merkel cells
    (2) callus
    (3) keratinocytes
    (4) Langerhans cells
    (5) melanocytes
    (6) free nerve endings
    (7) sebaceous glands
    (8) lamellar granules
    (9) pacinian (lamellated) corpuscles
    (10) vernix caseosa
    (11) arrector pili

14. Match the following:

_____ (a) deep region of the dermis composed primarily of dense irregular connective tissue

_____ (b) composed of keratinized stratified squamous epithelial tissue

_____ (c) not considered part of the skin, it contains areolar and adipose tissues and blood vessels; attaches skin to underlying tissues and organs

_____ (d) superficial region of the dermis; composed of areolar connective tissue

(1) subcutaneous layer (hypodermis)
(2) papillary region
(3) reticular region
(4) epidermis

15. Match the following and place the phases of deep wound healing in the correct order:

_____ (a) epithelial cells migrate under scab to bridge the wound; formation of granulation tissue

_____ (b) sloughing of scab; reorganization of collagen fibers; blood vessels return to normal

_____ (c) vasodilation and increased permeability of blood vessels to deliver cells involved in phagocytosis; clot formation

_____ (d) extensive growth of epithelial cells beneath scab; random deposition of collagen fibers; continued growth of blood vessels

(1) proliferative phase
(2) inflammatory phase
(3) maturation phase
(4) migratory phase

6 THE SKELETAL SYSTEM: BONE TISSUE

CHAPTER REVIEW

Introduction

1. A bone is made up of several different tissues: bone or osseous tissue, cartilage, dense connective tissue, epithelium, adipose tissue, and nervous tissue.
2. The entire framework of bones and their cartilages constitutes the skeletal system.

6.1 Functions of Bone and the Skeletal System

1. The skeletal system functions in support, protection, movement, mineral homeostasis, blood cell production, and triglyceride storage.

6.2 Structure of Bone

1. Parts of a typical long bone are the diaphysis (shaft), proximal and distal epiphyses (ends), metaphyses, articular cartilage, periosteum, medullary (marrow) cavity, and endosteum.

6.3 Histology of Bone Tissue

1. Bone tissue consists of widely separated cells surrounded by large amounts of extracellular matrix.
2. The four principal types of cells in bone tissue are osteoprogenitor cells, osteoblasts (bone-building cells), osteocytes (maintain daily activity of bone), and osteoclasts (bone-destroying cells).
3. The extracellular matrix of bone contains abundant mineral salts (mostly hydroxyapatite) and collagen fibers.
4. Compact bone tissue consists of osteons (haversian systems) with little space between them.
5. Compact bone tissue lies over spongy bone tissue in the epiphyses and makes up most of the bone tissue of the diaphysis. Functionally, compact bone tissue is the strongest form of bone and protects, supports, and resists stress.
6. Spongy bone tissue does not contain osteons. It consists of trabeculae surrounding many red bone marrow–filled spaces.
7. Spongy bone tissue forms most of the structure of short, flat, and irregular bones, and the interior of the epiphyses in long bones. Functionally, spongy bone tissue trabeculae offer resistance along lines of stress, support and protect red bone marrow, and make bones lighter for easier movement.

6.4 Blood and Nerve Supply of Bone

1. Long bones are supplied by periosteal, nutrient, metaphyseal, and epiphyseal arteries; veins accompany the arteries.
2. Nerves accompany blood vessels in bone; the periosteum is rich in sensory neurons.

6.5 Bone Formation

1. The process by which bone forms, called ossification, occurs in four principal situations: (1) the initial formation of bones in an embryo and fetus; (2) the growth of bones during infancy, childhood, and adolescence until their adult sizes are reached; (3) the remodeling of bone (replacement of old bone by new bone tissue throughout life); and (4) the repair of fractures (breaks in bones) throughout life.
2. Bone development begins during the sixth or seventh week of embryonic development. The two types of ossification, intramembranous and endochondral, involve the replacement of a preexisting connective tissue with bone. Intramembranous ossification refers to bone formation directly within mesenchyme arranged in sheetlike layers that resemble membranes. Endochondral ossification refers to bone formation within hyaline cartilage that develops from mesenchyme. The primary ossification center of a long bone is in the diaphysis. Cartilage degenerates, leaving cavities that merge to form the medullary cavity. Osteoblasts lay down bone. Next, ossification occurs in the epiphyses, where bone replaces cartilage, except for the epiphyseal (growth) plate.
3. The epiphyseal plate consists of four zones: zone of resting cartilage, zone of proliferating cartilage, zone of hypertrophic cartilage, and zone of calcified cartilage. Because of the cell division in the epiphyseal (growth) plate, the diaphysis of a bone increases in length.
4. Bone grows in thickness or diameter due to the addition of new bone tissue by periosteal osteoblasts around the outer surface of the bone (appositional growth).
5. Bone remodeling is an ongoing process in which osteoclasts carve out small tunnels in old bone tissue and then osteoblasts rebuild it.
6. In bone resorption, osteoclasts release enzymes and acids that degrade collagen fibers and dissolve mineral salts.
7. Dietary minerals (especially calcium and phosphorus) and vitamins (A, C, D, K, and B_{12}) are needed for bone growth and maintenance. Insulin-like growth factors (IGFs), growth hormone, thyroid hormones, and insulin stimulate bone growth.
8. Sex hormones slow resorption of old bone and promote new bone deposition.

6.6 Fracture and Repair of Bone

1. A fracture is any break in a bone. Types of fractures include closed (simple), open (compound), comminuted, greenstick, impacted, stress, Pott, and Colles.
2. Fracture repair involves formation of a fracture hematoma during the reactive phase, fibrocartilaginous callus and bony callus formation during the reparative phase, and a bone remodeling phase.

6.7 Bone's Role in Calcium Homeostasis

1. Bone is the major reservoir for calcium in the body.
2. Parathyroid hormone (PTH) secreted by the parathyroid glands increases blood Ca^{2+} level. Calcitonin (CT) from the thyroid gland has the potential to decrease blood Ca^{2+} level. Vitamin D enhances absorption of calcium and phosphate and thus raises the blood levels of these substances.

6.8 Exercise and Bone Tissue

1. Mechanical stress increases bone strength by increasing deposition of mineral salts and production of collagen fibers.
2. Removal of mechanical stress weakens bone through demineralization and collagen fiber reduction.

6.9 Aging and Bone Tissue

1. The principal effect of aging is demineralization, a loss of calcium from bones, which is due to reduced osteoblast activity.
2. Another effect is decreased production of extracellular matrix proteins (mostly collagen fibers), which makes bones more brittle and thus more susceptible to fracture.

CLINICAL CONNECTIONS

Remodeling and Orthodontics (Refer page 158 of Textbook)

Orthodontics (or-thō-DON-tiks) is the branch of dentistry concerned with the prevention and correction of poorly aligned teeth. The movement of teeth by braces places a stress on the bone that forms the sockets that anchor the teeth. In response to this artificial stress, osteoclasts and osteoblasts remodel the sockets so that the teeth align properly.

Paget's Disease (Refer page 160 of Textbook)

A delicate balance exists between the actions of osteoclasts and osteoblasts. Should too much new tissue be formed, the bones become abnormally thick and heavy. If too much mineral material is deposited in the bone, the surplus may form thick bumps, called *spurs*, on the bone that interfere with movement at joints. Excessive loss of calcium or tissue weakens the bones, and they may break, as occurs in osteoporosis, or they may become too flexible, as in rickets and osteomalacia. In **Paget's disease**, there is an excessive proliferation of osteoclasts so that bone resorption occurs faster than bone deposition. In response, osteoblasts attempt to compensate, but the new bone is weaker because it has a higher proportion of spongy to compact bone, mineralization is decreased, and the newly synthesized extracellular matrix contains abnormal proteins. The newly formed bone, especially that of the pelvis, limbs, lower vertebrae, and skull, becomes enlarged, hard, and brittle and fractures easily.

Hormonal Abnormalities That Affect Height (Refer page 160 of Textbook)

Excessive or deficient secretion of hormones that normally control bone growth can cause a person to be abnormally tall or short. Oversecretion of growth hormone (GH) during childhood produces **giantism**, in which a person becomes much taller and heavier than normal. **Dwarfism** is a condition of small stature in which the height of an individual is typically under 4 feet 10 inches, usually averaging 4 feet. Generally, there are two types of dwarfism: proportionate and disproportionate. In **proportionate dwarfism**, all parts of the body are small but they are proportionate to each other. One cause of proportionate dwarfism is a hyposecretion of GH during childhood and the condition is appropriately called **pituitary dwarfism**. The condition can be treated medically with administration of GH until epiphyseal plate closure. In **disproportionate dwarfism**, some parts of the body are normal size or larger than normal while others are smaller than normal. For example, the trunk can be average size while the limbs are short and the head may be large in relation to the rest of the body, with a prominent forehead and flattened nose at the bridge. The most common cause of this type of dwarfism is a condition called **achondroplasia** (a-kon-drō-PLĀ-zē-a; *a* = without; *chondro* = cartilage; -*plasai* = to mold), an inherited condition in which the conversion of hyaline cartilage to bone is abnormal and the long bones of the limbs stop growing in childhood. Other bones are unaffected, and thus the person has short stature but a normal size head and trunk. This type of dwarfism is called **achondroplastic dwarfism**. The condition is essentially untreatable, although some individuals opt for limb-lengthening surgery.

Treatments for Fractures (Refer page 161 of Textbook)

Treatments for fractures vary according to age, type of fracture, and the bone involved. The ultimate goals of fracture treatment are realignment of the bone fragments, immobilization to maintain realignment, and restoration of function. For bones to unite properly, the fractured ends must be brought into alignment. This process, called **reduction**, is commonly referred to as setting a fracture. In **closed reduction**, the fractured ends of a bone are brought into alignment by manual manipulation, and the skin remains intact. In **open reduction**, the fractured ends of a bone are brought into alignment by a surgical procedure using internal fixation devices such as screws, plates, pins, rods, and wires. Following reduction, a fractured bone may be kept immobilized by a cast, sling, splint, elastic bandage, external fixation device, or a combination of these devices.

DISORDERS: HOMEOSTATIC IMBALANCES

Bone Scan

A **bone scan** is a diagnostic procedure that takes advantage of the fact that bone is living tissue. A small amount of a radioactive tracer compound that is readily absorbed by bone is injected intravenously. The degree of uptake of the tracer is related to the amount of blood flow to the bone. A scanning device (gamma camera) measures the radiation emitted from the bones, and the information is translated into a photograph that can be read like an x-ray on a monitor. Normal bone tissue is identified by a consistent gray color throughout because of its uniform uptake of the radioactive tracer. Darker or lighter areas may indicate bone abnormalities. Darker areas or "hot spots" are areas of increased metabolism that absorb more of the radioactive tracer due to increased blood flow. Hot spots may indicate bone cancer, abnormal healing of fractures, or abnormal bone growth. Lighter areas or "cold spots" are areas of decreased metabolism that absorb less of the radioactive tracer due to decreased blood flow. Cold spots may indicate problems such as degenerative bone disease, decalcified bone, fractures, bone infections, Paget's disease, or rheumatoid arthritis. A bone scan detects abnormalities 3 to 6 months sooner than standard x-ray procedures and exposes the patient to less radiation. A bone scan is the standard test for bone density screening, particularly important in screening for osteoporosis in females.

Osteoporosis

Osteoporosis (os'-tē-Ō-pō-RŌ-sis; *-por-* = passageway; *-osis* = condition), literally a condition of porous bones, affects 10 million people a year in the United States (**Figure SG6.1**). In addition, 18 million people have low bone mass (*osteopenia*), which puts them at risk for osteoporosis. The basic problem is that bone resorption (breakdown) outpaces bone deposition (formation). In large part this is due to depletion of calcium from the body—more calcium is lost in urine, feces, and sweat than is absorbed from the diet. Bone mass becomes so depleted that bones fracture, often spontaneously, under the mechanical stresses of everyday living. For example, a hip fracture might result from simply sitting down too quickly. In the United States, osteoporosis causes more than 1.5 million fractures a year, mainly in the hips, wrists, and vertebrae. Osteoporosis afflicts the entire skeletal system. In addition to fractures, osteoporosis causes shrinkage of vertebrae, height loss, hunched backs, and bone pain.

Osteoporosis primarily affects middle-aged and elderly people, 80% of them women. Older women suffer from osteoporosis more often than men for two reasons: (1) Women's bones are less massive than men's bones, and

Figure SG6.1 Comparison of spongy bone tissue from (a) a normal young adult and (b) a person with osteoporosis. Notice the weakened trabeculae in (b). Compact bone tissue is similarly affected by osteoporosis.

In osteoporosis, bone resorption outpaces bone formation, so bone mass decreases.

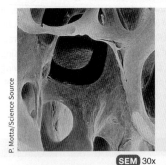 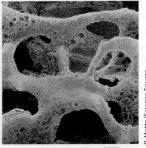

| **SEM** 30x | **SEM** 30x |
| (a) Normal bone | (b) Osteoporotic bone |

Q If you wanted to develop a drug to lessen the effects of osteoporosis, would you look for a chemical that inhibits the activity of osteoblasts or that of osteoclasts?

(2) production of estrogens in women declines dramatically at menopause, whereas production of the main androgen, testosterone, in older men wanes gradually and only slightly. Estrogens and testosterone stimulate osteoblast activity and synthesis of bone matrix. Besides gender, risk factors for developing osteoporosis include a family history of the disease, European or Asian ancestry, thin or small body build, an inactive lifestyle, cigarette smoking, a diet low in calcium and vitamin D, more than two alcoholic drinks a day, and the use of certain medications.

Osteoporosis is diagnosed by taking a family history and undergoing a *bone mineral density* (BMD) test. Performed like x-rays, BMD tests measure bone density. They can also be used to confirm a diagnosis of osteoporosis, determine the rate of bone loss, and monitor the effects of treatment. There is also a relatively new tool called *FRAX®* that incorporates risk factors besides bone mineral density to accurately estimate fracture risk. Patients fill out an online survey of risk factors such as age, gender, height, weight, ethnicity, prior fracture history, parental history of hip fracture, use of glucocorticoids (for example, cortisone), smoking, alcohol intake, and rheumatoid arthritis. Using the data, FRAX® provides an estimate of the probability that a person will suffer a fracture of the hip or other major bone in the spine, shoulder, or forearm due to osteoporosis within 10 years.

Treatment options for osteoporosis are varied. With regard to nutrition, a diet high in calcium is important to reduce the risk of fractures. Vitamin D is necessary for the body to utilize calcium. In terms of exercise, regular performance of weight-bearing exercises has been shown to maintain and build bone mass. These exercises include walking, jogging, hiking, climbing stairs, playing tennis, and dancing. Resistance exercises, such as weight lifting, also build bone strength and muscle mass.

Medications used to treat osteoporosis are generally of two types: (1) **antireabsorptive drugs** slow down the progression of bone loss and (2) **bone-building drugs** promote increasing bone mass. Among the antireabsorptive drugs are (1) *bisphosphonates,* which inhibit osteoclasts (Fosamax®, Actonel®, Boniva®, and calcitonin); (2) *selective estrogen receptor modulators*, which mimic the effects of estrogens without unwanted side effects (Raloxifene®, Evista®); and (3) estrogen replacement therapy (ERT), which replaces estrogens lost during and after menopause (Premarin®), and hormone replacement therapy (HRT), which replaces estrogens and progesterone lost during and after menopause (Prempro®). ERT helps maintain and increase bone mass after menopause. Women on ERT have a slightly increased risk of stroke and blood clots. HRT also helps maintain and increase bone mass. Women on HRT have increased risks of heart disease, breast cancer, stroke, blood clots, and dementia.

Among the bone-building drugs is parathyroid hormone (PTH), which stimulates osteoblasts to produce new bone (Forteo®). Others are under development.

Rickets and Osteomalacia

Rickets and **osteomalacia** (os'-tē-ō-ma-LĀ-shē-a; *malacia* = softness) are two forms of the same disease that result from inadequate calcification of the extracellular bone matrix, usually caused by a vitamin D deficiency. Rickets is a disease of children in which the growing bones become "soft" or rubbery and are easily deformed. Because new bone formed at the epiphyseal (growth) plates fails to ossify, bowed legs and deformities of the skull, rib cage, and pelvis are common. Osteomalacia is the adult counterpart of rickets, sometimes called *adult rickets*. New bone formed during remodeling fails to calcify, and the person experiences varying degrees of pain and tenderness in bones, especially the hip and legs. Bone fractures also result from minor trauma. Prevention and treatment for rickets and osteomalacia consists of the administration of adequate vitamin D and exposure to moderate amounts of sunlight.

MEDICAL TERMINOLOGY

Osteoarthritis (os′-tē-ō-ar-THRĪ-tis; *-arthr-* = joint) The degeneration of articular cartilage such that the bony ends touch; the resulting friction of bone against bone worsens the condition. Usually associated with the elderly.

Osteomyelitis (os′-tē-ō-mī-e-LĪ-tis) An infection of bone characterized by high fever, sweating, chills, pain, nausea, pus formation, edema, and warmth over the affected bone and rigid overlying muscles. It is often caused by bacteria, usually *Staphylococcus aureus.* The bacteria may reach the bone from outside the body (through open fractures, penetrating wounds, or orthopedic surgical procedures); from other sites of infection in the body (abscessed teeth, burn infections, urinary tract infections, or upper respiratory infections) via the blood; and from adjacent soft tissue infections (as occurs in diabetes mellitus).

Osteopenia (os′-tē-ō-PĒ-nē-a; *penia* = poverty) Reduced bone mass due to a decrease in the rate of bone synthesis to a level too low to compensate for normal bone resorption; any decrease in bone mass below normal. An example is osteoporosis.

Osteosarcoma (os′-tē-ō-sar-KŌ-ma; *sarcoma* = connective tissue tumor) Bone cancer that primarily affects osteoblasts and occurs most often in teenagers during their growth spurt; the most common sites are the metaphyses of the thigh bone (femur), shin bone (tibia), and arm bone (humerus). Metastases occur most often in lungs; treatment consists of multidrug chemotherapy and removal of the malignant growth, or amputation of the limb.

SELF-QUIZ QUESTIONS

Fill in the blanks in the following statements.

1. Bone growth in length is called ———— growth, and bone growth in diameter (thickness) is called ———— growth.

2. The crystallized inorganic mineral salts in bone contribute to bone's ————, while the collagen fibers and other organic molecules provide bone with ————.

Indicate whether the following statements are true or false.

3. Bone resorption involves increased activity of osteoclasts.

4. The formation of bone from cartilage is known as endochondral ossification.

5. The growth of bone is controlled primarily by hormones.

Choose the one best answer to the following questions.

6. Place in order the steps involved in intramembranous ossification. (1) Bony matrices fuse to form trabeculae. (2) Clusters of osteoblasts form a center of ossification that secretes the organic extracellular matrix. (3) Spongy bone is replaced with compact bone on the bone's surface. (4) Periosteum develops on the bone's periphery. (5) The extracellular matrix hardens by deposition of calcium and mineral salts.
   (a) 2, 4, 5, 1, 3   (b) 4, 3, 5, 1, 2   (c) 1, 2, 5, 4, 3
   (d) 2, 5, 1, 4, 3   (e) 5, 1, 3, 4, 2

7. Place in order the steps involved in endochondral ossification. (1) Nutrient artery invades the perichondrium. (2) Osteoclasts create a marrow cavity. (3) Chondrocytes enlarge and calcify. (4) Secondary ossification centers appear at epiphyses. (5) Osteoblasts become active in the primary ossification center.
   (a) 3, 1, 5, 2, 4   (b) 3, 1, 5, 4, 2   (c) 1, 3, 5, 2, 4
   (d) 1, 2, 3, 5, 4   (e) 2, 5, 4, 3, 1

8. Spongy bone differs from compact bone because spongy bone
   (a) is composed of numerous osteons (haversian systems).
   (b) is found primarily in the diaphyses of long bones, and compact bone is found primarily in the epiphyses of long bones.
   (c) contains osteons all aligned in the same direction along lines of stress.
   (d) does not contain osteocytes contained in lacunae.
   (e) is composed of trabeculae that are oriented along lines of stress.

9. A primary effect of weight-bearing exercise on bones is to
   (a) provide oxygen for bone development.
   (b) increase the demineralization of bone.

   (c) maintain and increase bone mass.
   (d) stimulate the release of sex hormones for bone growth.
   (e) utilize the stored triglycerides from the yellow bone marrow.

10. Place in order the steps involved in the repair of a bone fracture. (1) Osteoblast production of trabeculae and bony callus formation; (2) formation of a hematoma at the site of fracture; (3) resorption of remaining bone fragments and remodeling of bone; (4) migration of fibroblasts to the fracture site; (5) bridging of broken ends of bones by a fibrocartilaginous callus.
    (a) 2, 4, 5, 1, 3   (b) 2, 5, 4, 1, 3   (c) 1, 2, 5, 4, 3
    (d) 2, 5, 1, 3, 4   (e) 5, 2, 4, 1, 3

11. Match the following:
    ———— (a) column-like layer of maturing chondrocytes
    ———— (b) layer of small, scattered chondrocytes anchoring the epiphyseal (growth) plate to the bone
    ———— (c) layer of actively dividing chondrocytes
    ———— (d) region of dead chondrocytes

    (1) zone of hypertrophic cartilage
    (2) zone of calcified cartilage
    (3) zone of proliferating cartilage
    (4) zone of resting cartilage

12. Match the following:
    ———— (a) decreases blood calcium levels by accelerating calcium deposition in bones and inhibiting osteoclasts
    ———— (b) required for collagen synthesis
    ———— (c) during childhood, it promotes growth at epiphyseal plate; production stimulated by human growth hormone
    ———— (d) involved in bone growth by increasing osteoblast activity; causes long bones to stop growing in length
    ———— (e) required for protein synthesis
    ———— (f) active form of vitamin D; raises blood calcium levels by increasing absorption of calcium from digestive tract
    ———— (g) raises blood calcium levels by increasing bone resorption

    (1) PTH
    (2) CT
    (3) calcitriol
    (4) insulinlike growth factors
    (5) sex hormones
    (6) vitamin C
    (7) vitamin K

13. Match the following:

—— (a) space within the shaft of the bone that contains yellow bone marrow

—— (b) triglyceride storage tissue

—— (c) hemopoietic tissue

—— (d) thin layer of hyaline cartilage covering the ends of bones where they form a joint

—— (e) proximal and distal ends of bones

—— (f) the long, cylindrical main portion of the bone; the shaft

—— (g) in a growing bone, the region that contains the epiphyseal (growth) plate

—— (h) the tough covering that surrounds the bone surface wherever cartilage is not present

—— (i) a layer of hyaline cartilage in the area between the shaft and end of a growing bone

—— (j) membrane lining the medullary cavity

—— (k) a remnant of the active epiphyseal (growth) plate; a sign that the bone has stopped growing in length

—— (l) bundles of collagen fibers that attach periosteum to bone

(1) articular cartilage
(2) endosteum
(3) medullary cavity
(4) diaphysis
(5) epiphyses
(6) metaphysis
(7) periosteum
(8) red bone marrow
(9) yellow bone marrow
(10) perforating (Sharpey's) fibers
(11) epiphyseal line
(12) epiphyseal (growth) plate

14. Match the following:

—— (a) a broken bone in which one end of the fractured bone is driven into the other end

—— (b) a condition of porous bones characterized by decreased bone mass and increased susceptibility to fractures

—— (c) splintered bone, with smaller fragments lying between main fragments

—— (d) a broken bone that does not break through the skin

—— (e) a partial break in a bone in which one side of the bone is broken and the other side bends

—— (f) a broken bone that protrudes through the skin

—— (g) microscopic bone breaks resulting from inability to withstand repeated stressful impact

—— (h) a degeneration of articular cartilage allowing the bony ends to touch; worsens due to friction between the bones

—— (i) condition characterized by failure of new bone formed by remodeling to calcify in adults

—— (j) an infection of bone

(1) closed (simple) fracture
(2) open (compound) fracture
(3) impacted fracture
(4) greenstick fracture
(5) stress fracture
(6) comminuted fracture
(7) osteoporosis
(8) osteomalacia
(9) osteoarthritis
(10) osteomyelitis

15. Match the following:

—— (a) small spaces between lamellae that contain osteocytes

—— (b) tiny canals that penetrate compact bone; carry blood vessels, lymphatic vessels, and nerves from the periosteum

—— (c) areas between osteons; fragments of old osteons

—— (d) cells that secrete the components required to build bone

—— (e) microscopic unit of compact bone tissue

—— (f) interconnected, tiny canals filled with extracellular fluid; connect lacunae to each other and to the central canal

—— (g) canals that extend longitudinally through the bone and carry blood vessels and nerves to the central canal

—— (h) large cells derived from numerous monocytes; involved in resorption by release of lysosomal enzymes and acids

—— (i) irregular lattice of thin columns of bone found in spongy bone tissue

—— (j) rings of hard calcified matrix found just beneath the periosteum and lining the medullary cavity

—— (k) mature cells that maintain the daily metabolism of bone

—— (l) an opening in the shaft of the bone allowing an artery to pass into the bone

—— (m) unspecialized stem cells derived from mesenchyme

(1) osteogenic cells
(2) osteocytes
(3) osteon (haversian system)
(4) perforating (Volkmann's) canals
(5) circumferential lamellae
(6) osteoblasts
(7) trabeculae
(8) interstitial lamellae
(9) canaliculi
(10) osteoclasts
(11) nutrient foramen
(12) lacunae
(13) haversian (central) canals

7 THE SKELETAL SYSTEM: THE AXIAL SKELETON

Introduction

1. Bones protect soft body parts and make movement possible; they also serve as landmarks for locating parts of other body systems.
2. The musculoskeletal system is composed of the bones, joints, and muscles working together.

7.1 Divisions of the Skeletal System (see **Table 7.1**)

1. The axial skeleton consists of bones arranged along the longitudinal axis. The parts of the axial skeleton are the skull, auditory ossicles (ear bones), hyoid bone, vertebral column, sternum, and ribs.
2. The appendicular skeleton consists of the bones of the girdles and the upper and lower limbs (extremities). The parts of the appendicular skeleton are the pectoral (shoulder) girdles, bones of the upper limbs, pelvic (hip) girdles, and bones of the lower limbs.

7.2 Types of Bones

1. On the basis of shape, bones are classified as long, short, flat, irregular, or sesamoid. Sesamoid bones develop in tendons or ligaments.
2. Sutural bones are found within the sutures of some cranial bones.

7.3 Bone Surface Markings

1. Surface markings are structural features visible on the surfaces of bones.
2. Each marking—whether a depression, an opening, or a process—is structured for a specific function, such as joint formation, muscle attachment, or passage of nerves and blood vessels (see **Table 7.2**).

7.4 Skull: An Overview

1. The 22 bones of the skull include cranial bones and facial bones.
2. The eight cranial bones are the frontal, parietal (2), temporal (2), occipital, sphenoid, and ethmoid.
3. The 14 facial bones are the nasal (2), maxillae (2), zygomatic (2), lacrimal (2), palatine (2), inferior nasal conchae (2), vomer, and mandible.

7.5 Cranial Bones

1. The frontal bone forms the forehead (the anterior part of the cranium).
2. The frontal bone also forms the roofs of the orbits and most of the anterior part of the cranial floor.
3. The parietal bones form the greater portion of the sides of the cranial cavity.
4. The parietal bones also form most of the roof of the cranial cavity.
5. The temporal bones form the inferior lateral aspects of the cranium.
6. The temporal bones also form part of the cranial floor.
7. The occipital bone forms the posterior part of the cranium.
8. The occipital bone also forms of the base of the cranium.
9. The sphenoid bone lies at the middle part of the base of the skull.
10. The sphenoid bone is known as the keystone of the cranial floor because it articulates with all the other cranial bones, holding them together.
11. The ethmoid bone is located in the anterior part of the cranial floor medial to the orbits.
12. The ethmoid bone is anterior to the sphenoid and posterior to the nasal bones.

7.6 Facial Bones

1. The nasal bones from the bridge of the nose.
2. The lacrimal bones are posterior and lateral to the nasal bones and form a part of the medial wall of each orbit.
3. The palatine bones form the posterior portion of hard palate, part of the floor and lateral wall of the nasal cavity, and a small portion of the floors of the orbits.
4. The inferior nasal conchae form a part of the inferior lateral wall of the nasal cavity and project into the nasal cavity.
5. The vomer forms the inferior portion of the nasal septum.
6. The maxillae form the upper jawbone.
7. The zygomatic bones (cheekbones) form the prominences of the cheeks and part of the lateral wall and floor of each orbit.
8. The mandible is the lower jawbone, the largest and strongest facial bone.

7.7 Special Features of the Skull

1. The nasal septum consists of the vomer, perpendicular plate of the ethmoid, and septal cartilage. The nasal septum divides the nasal cavity into left and right sides.
2. Seven skull bones form each of the orbits (eye sockets).
3. The foramina of the skull bones provide passages for nerves and blood vessels.

4. Sutures are immovable joints that connect most bones of the skull. Examples are the coronal, sagittal, lambdoid, and squamous sutures.
5. Paranasal sinuses are cavities in bones of the skull that are connected to the nasal cavity. The frontal, sphenoid, and ethmoid bones and the maxillae contain paranasal sinuses.
6. Fontanels are mesenchyme-filled spaces between the cranial bones of fetuses and infants. The major fontanels are the anterior, posterior, anterolaterals (2), and posterolaterals (2). After birth, the fontanels fill in with bone and become sutures.

7.8 Hyoid Bone

1. The hyoid bone is a U-shaped bone that does not articulate with any other bone.
2. It supports the tongue and provides attachment for some tongue muscles and for some muscles of the pharynx and neck.

7.9 Vertebral Column

1. The vertebral column, sternum, and ribs constitute the skeleton of the body's trunk.
2. The 26 bones of the adult vertebral column are the cervical vertebrae (7), the thoracic vertebrae (12), the lumbar vertebrae (5), the sacrum (5 fused vertebrae), and the coccyx (usually 4 fused vertebrae).
3. The adult vertebral column contains four normal curves (cervical, thoracic, lumbar, and sacral) that provide strength, support, and balance.
4. Each vertebra usually consists of a body, vertebral arch, and seven processes. Vertebrae in the different regions of the column vary in size, shape, and detail.

7.10 Vertebral Regions

1. The cervical vertebrae (C1–C7) are smaller than all other vertebrae except those that form the coccyx.

2. The first two cervical vertebrae are the atlas (C1) and the axis (C2).
3. The thoracic vertebrae (T1–T12) are considerably larger and stronger than cervical vertebrae.
4. The thoracic vertebrae articulate with the ribs.
5. The lumbar vertebrae (L1–L5) are the largest and strongest of the unfused bones in the vertebral column.
6. The various projections of the lumbar vertebrae are short and thick.
7. The sacrum is a triangular bone formed by the union of the five sacral vertebrae (S1–S5).
8. The coccyx is formed by the fusion of usually four coccygeal vertebrae (Co1–Co4).

7.11 Thorax

1. The thoracic skeleton consists of the sternum, ribs, costal cartilages, and thoracic vertebrae.
2. The thoracic cage protects vital organs in the chest area and upper abdomen.

Sternum

3. The sternum (breastbone) is located in the center of the anterior thoracic wall.
4. The sternum consists of the manubrium, body, and xiphoid process.

Ribs

5. The twelve pairs of ribs give structural support to the sides of the thoracic cavity.
6. The three types of ribs are the true (vertebrosternal) ribs, vertebrochondral, ribs, and floating (vertebral) ribs.

CLINICAL CONNECTIONS

Black Eye (Refer page 172 of Textbook)

A **black eye** is a bruising around the eye, commonly due to an injury to the face, rather than an eye injury. In response to trauma, blood and other fluids accumulate in the space around the eye, causing the swelling and dark discoloration. One cause might be a blow to the sharp ridge just superior to the supraorbital margin that fractures the frontal bone, resulting in bleeding. Another is a blow to the nose. Certain surgical procedures (face lift, eyelid surgery, jaw surgery, or nasal surgery) can also result in black eyes.

Cleft Palate and Cleft Lip (Refer page 180 of Textbook)

Usually the palatine processes of the maxillary bones unite during weeks 10 to 12 of embryonic development. Failure to do so can result in one type of **cleft palate**. The condition may also involve incomplete fusion of the horizontal plates of the palatine bones (see **Figure 7.7**). Another form of this condition, called cleft lip, involves a split in the upper lip. **Cleft lip** and cleft palate often occur together. Depending on the extent and position of the cleft, speech and swallowing may be affected. In addition, children with cleft palate tend to have many ear infections, which can lead to hearing loss. Facial and oral surgeons recommend closure of cleft lip during the first few weeks following

birth, and surgical results are excellent. Repair of cleft palate typically is completed between 12 and 18 months of age, ideally before the child begins to talk. Because the palate is important for pronouncing consonants, speech therapy may be required, and orthodontic therapy may be needed to align the teeth. Recent research strongly suggests that supplementation with folic acid (one of the B vitamins) during early pregnancy decreases the incidence of cleft palate and cleft lip. The mechanism behind this is not yet understood.

Temporomandibular Joint Syndrome (Refer page 181 of Textbook)

One problem associated with the temporomandibular joint is **temporomandibular joint (TMJ) syndrome**. It is characterized by dull pain around the ear, tenderness of the jaw muscles, a clicking or popping noise when opening or closing the mouth, limited or abnormal opening of the mouth, headache, tooth sensitivity, and abnormal wearing of the teeth. TMJ syndrome can be caused by improperly aligned teeth, grinding or clenching the teeth, trauma to the head and neck, or arthritis. Treatments include application of moist heat or ice, limiting the diet to soft foods, administration of pain relievers such as aspirin, muscle retraining, use of a splint or bite plate to reduce clenching and teeth grinding (especially when worn at night), adjustment or reshaping of the teeth (orthodontic treatment), and surgery.

Deviated Nasal Septum (Refer page 181 of Textbook)

A **deviated nasal septum** is one that does not run along the midline of the nasal cavity. It deviates (bends) to one side. A blow to the nose can easily damage, or break, this delicate septum of bone and displace and damage the cartilage. Often, when a broken nasal septum heals, the bones and cartilage deviate to one side or the other. This deviated septum can block airflow into the constricted side of the nose, making it difficult to breathe through that half of the nasal cavity. The deviation usually occurs at the junction of the vomer bone with the septal cartilage. Septal deviations may also occur due to developmental abnormality. If the deviation is severe, it may block the nasal passageway entirely. Even a partial blockage may lead to infection. If inflammation occurs, it may cause nasal congestion, blockage of the paranasal sinus openings, chronic sinusitis, headache, and nosebleeds. The condition usually can be corrected or improved surgically.

Sinusitis (Refer page 185 of Textbook)

Sinusitis (sīn-ū-SĪ-tis) is an inflammation of the mucous membrane of one or more paranasal sinuses. It may be caused by a microbial infection (virus, bacterium, or fungus), allergic reactions, nasal polyps, or a severely deviated nasal septum. If the inflammation or an obstruction blocks the drainage of mucus into the nasal cavity, fluid pressure builds up in the paranasal sinuses, and a sinus headache may develop. Other symptoms may include nasal congestion, inability to smell, fever, and cough. Treatment options include decongestant sprays or drops, oral decongestants, nasal corticosteroids, antibiotics, analgesics to relieve pain, warm compresses, and surgery.

Caudal Anesthesia (Refer page 193 of Textbook)

Anesthetic agents that act on the sacral and coccygeal nerves are sometimes injected through the sacral hiatus, a procedure called **caudal anesthesia**. While this approach is not as common as lumbar epidural block, it is preferred when sacral nerve spread of the anesthetics is preferred over lumbar nerve spread. Because the sacral hiatus is between the sacral cornua, the cornua are important bony landmarks for locating the hiatus. Anesthetic agents also may be injected through the posterior sacral foramina. Since the hiatal and foraminal injection sites are inferior to the lowest portion of the spinal cord, there is little danger of damaging the cord. The lumbar approach is preferred because there is considerable variability in the anatomy of the sacral hiatus, and with advancing age the dorsal ligaments and cornua thicken, making it difficult to identify the hiatal margins.

Rib Fractures, Dislocations, and Separations (Refer page 199 of Textbook)

Rib fractures are the most common chest injuries. They usually result from direct blows, most often from impact with a steering wheel, falls, or crushing injuries to the chest. Ribs tend to break at the point where the greatest force is applied, but they may also break at their weakest point—the site of greatest curvature, just anterior to the costal angle. The middle ribs are the most commonly fractured. In some cases, fractured ribs may puncture the heart, great vessels of the heart, lungs, trachea, bronchi, esophagus, spleen, liver, and kidneys. Rib fractures are usually quite painful. Rib fractures are no longer bound with bandages because of the pneumonia that would result from lack of proper lung ventilation.

Dislocated ribs, which are common in body contact sports, involve displacement of a costal cartilage from the sternum, with resulting pain, especially during deep inhalations.

Separated ribs involve displacement of a rib and its costal cartilage; as a result, a rib may move superiorly, overriding the rib above and causing severe pain.

DISORDERS: HOMEOSTATIC IMBALANCES

Herniated (Slipped) Disc

In their function as shock absorbers, intervertebral discs are constantly being compressed. If the anterior and posterior ligaments of the discs become injured or weakened, the pressure developed in the nucleus pulposus may be great enough to rupture the surrounding fibrocartilage (annulus fibrosus). If this occurs, the nucleus pulposus may herniate (protrude) posteriorly or into one of the adjacent vertebral bodies (**Figure SG7.1**). This condition is called a **herniated** (*slipped*) **disc**. Because the lumbar region bears much of the weight of the body, and is the region of the most flexing and bending, herniated discs most often occur in the lumbar area.

Frequently, the nucleus pulposus slips posteriorly toward the spinal cord and spinal nerves. This movement exerts pressure on the spinal nerves, causing local weakness and acute pain. If the roots of the sciatic nerve, which passes from the spinal cord to the foot, are compressed, the pain radiates down the posterior thigh, through the calf, and occasionally into the foot. If pressure is exerted on the spinal cord itself, some of its neurons may be destroyed. Treatment options include bed rest, medications for pain, physical therapy and exercises, and *percutaneous endoscopic discectomy* (removal of disc material using a laser). A person with a herniated disc may also undergo a laminectomy, a procedure in which parts of the laminae of the vertebra and intervertebral disc are removed to relieve pressure on the nerves.

Abnormal Curves of the Vertebral Column

Various conditions may exaggerate the normal curves of the vertebral column, or the column may acquire a lateral bend, resulting in **abnormal curves of the vertebral column**.

Figure SG7.1 Herniated (slipped) disc.

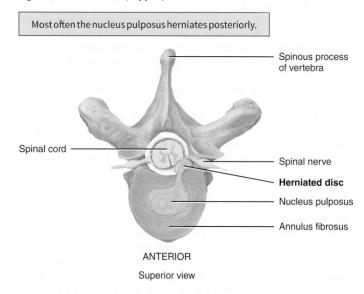

Most often the nucleus pulposus herniates posteriorly.

- Spinous process of vertebra
- Spinal cord
- Spinal nerve
- **Herniated disc**
- Nucleus pulposus
- Annulus fibrosus

ANTERIOR

Superior view

Q Why do most herniated discs occur in the lumbar region?

Figure SG7.2 Abnormal curves of the vertebral column.

An abnormal curve is the result of the exaggeration of a normal curve.

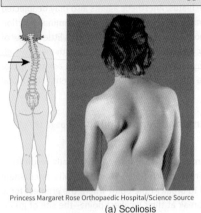

Princess Margaret Rose Orthopaedic Hospital/Science Source

(a) Scoliosis

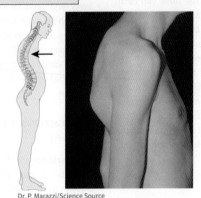

Dr. P. Marazzi/Science Source

(b) Kyphosis

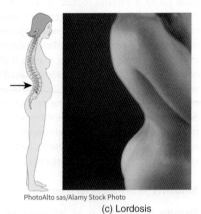

PhotoAlto sas/Alamy Stock Photo

(c) Lordosis

Q Which abnormal curve is common in women with advanced osteoporosis?

Scoliosis (skō-lē-Ō-sis; *scolio-* = crooked), the most common of the abnormal curves, is a lateral bending of the vertebral column, usually in the thoracic region (**Figure SG7.2a**). It may result from congenitally (present at birth) malformed vertebrae, chronic sciatica (pain in the lower back and lower limb), paralysis of muscles on one side of the vertebral column, poor posture, or one leg being shorter than the other.

Signs of scoliosis include uneven shoulders and waist, one shoulder blade more prominent than the other, one hip higher than the other, and leaning to one side. In severe scoliosis (a curve greater than 70 degrees), breathing is more difficult and the pumping action of the heart is less efficient. Chronic back pain and arthritis of the vertebral column may also develop. Treatment options include wearing a back brace, physical therapy, chiropractic care, and surgery (fusion of vertebrae and insertion of metal rods, hooks, and wires to reinforce the surgery).

Kyphosis (kī-FŌ-sis; *kyphoss-* = hump; *-osis* = condition) is an increase in the thoracic curve of the vertebral column that produces a "hunchback" look (**Figure SG7.2b**). In tuberculosis of the spine, vertebral bodies may partially collapse, causing an acute angular bending of the vertebral column. In the elderly, degeneration of the intervertebral discs leads to kyphosis. Kyphosis may also be caused by rickets and poor posture. It is also common in females with advanced osteoporosis.

Lordosis (lor-DŌ-sis; *lord-* = bent backward), sometimes called *hollow back*, is an increase in the lumbar curve of the vertebral column (**Figure SG7.2c**). It may result from increased weight of the abdomen as in pregnancy or extreme obesity, poor posture, rickets, osteoporosis, or tuberculosis of the spine.

Spina Bifida

Spina bifida (SPĪ-na BIF-i-da) is a congenital defect of the vertebral column in which laminae of L5 and/or S1 fail to develop normally and unite at the midline. The least serious form is called *spina bifida occulta*. It occurs in L5 or S1 and produces no symptoms. The only evidence of its presence is a small dimple with a tuft of hair in the overlying skin. Several types of spina bifida involve protrusion of meninges (membranes) and/or spinal cord through the defect in the laminae and are collectively termed *spina bifida cystica* because of the presence of a cystlike sac protruding from the backbone (**Figure SG7.3**). If the sac contains the meninges from the spinal cord and cerebrospinal fluid, the condition is called *spina bifida with meningocele* (me-NING-gō-sēl). If the spinal cord and/or its nerve roots are in the sac, the condition is called *spina*

bifida with meningomyelocele (me-ning-gō-MĪ-ē-lō-sēl). The larger the cyst and the number of neural structures it contains, the more serious the neurological problems. In severe cases, there may be partial or complete paralysis, partial or complete loss of urinary bladder and bowel control, and the absence of reflexes. An increased risk of spina bifida is associated with low levels of a B vitamin called folic acid during pregnancy. Spina bifida may be diagnosed prenatally by a test of the mother's blood for a substance produced by the fetus called alpha-fetoprotein, by sonography, or by amniocentesis (withdrawal of amniotic fluid for analysis).

Fractures of the Vertebral Column

Fractures of the vertebral column often involve C1, C2, C4–T7, and T12–L2. Cervical or lumbar fractures usually result from a flexion–compression type of

Figure SG7.3 Spina bifida with meningomyelocele.

Spina bifida is caused by a failure of laminae to unite at the midline.

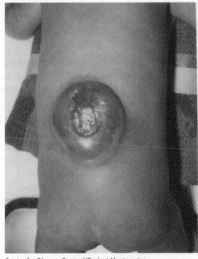

Center for Disease Control/Project Masters, Inc.

Q Deficiency of which B vitamin is linked to spina bifida?

injury such as might be sustained in landing on the feet or buttocks after a fall or having a weight fall on the shoulders. Cervical vertebrae may be fractured or dislodged by a fall on the head with acute flexion of the neck, as might happen on diving into shallow water or being thrown from a horse. Spinal cord or spinal nerve damage may occur as a result of fractures of the vertebral column if the fractures compromise the foramina.

MEDICAL TERMINOLOGY

Chiropractic (kī-rō-PRAK-tik; *cheir-* = hand; *-praktikos* = efficient) A holistic health-care discipline that focuses on nerves, muscles, and bones. A **chiropractor** is a health-care professional who is concerned with the diagnoses, treatment, and prevention of mechanical disorders of the musculoskeletal system and the effects of these disorders on the nervous system and health in general. Treatment involves using the hands to apply specific force to adjust joints of the body (manual adjustment), especially the vertebral column. Chiropractors may also use massage, heat therapy, ultrasound, electrical stimulation, and acupuncture. Chiropractors often provide information about diet, exercise, changes in lifestyle, and stress management. Chiropractors do not prescribe drugs or perform surgery.

Craniostenosis (krā-nē-ō-sten-Ō-sis; *cranio-* = skull; *-stenosis* = narrowing) Premature closure of one or more cranial sutures during the first 18 to 20 months of life, resulting in a distorted skull. Premature closure of the sagittal suture produces a long narrow skull; premature closure of the coronal suture results in a broad skull. Premature closure of all sutures restricts brain growth and development; surgery is necessary to prevent brain damage.

Craniotomy (krā-nē-OT-ō-mē; *cranio-* = skull; *-tome* = cutting) Surgical procedure in which part of the cranium is removed. It may be performed to remove a blood clot, a brain tumor, or a sample of brain tissue for biopsy.

Laminectomy (lam'-i-NEK-tō-mē; *lamina-* = layer) Surgical procedure to remove a vertebral lamina. It may be performed to access the vertebral cavity and relieve the symptoms of a herniated disc.

Lumbar spine stenosis (*sten-* = narrowed) Narrowing of the spinal canal in the lumbar part of the vertebral column, due to hypertrophy of surrounding bone or soft tissues. It may be caused by arthritic changes in the intervertebral discs and is a common cause of back and leg pain.

Spinal fusion (FŪ-zhun) Surgical procedure in which two or more vertebrae of the vertebral column are stabilized with a bone graft or synthetic device. It may be performed to treat a fracture of a vertebra or following removal of a herniated disc.

Whiplash injury Injury to the neck region due to severe hyperextension (backward tilting) of the head followed by severe hyperflexion (forward tilting) of the head, usually associated with a rear-end automobile collision. Symptoms are related to stretching and tearing of ligaments and muscles, vertebral fractures, and herniated vertebral discs.

SELF-QUIZ QUESTIONS

Fill in the blanks in the following statements.

1. Membrane-filled spaces between cranial bones that enable the fetal skull to modify its size and shape for passage through the birth canal are called _____.

2. The hypophyseal fossa of the sella turcica of the sphenoid bone contains the _____.

3. The regions of the vertebral column that consist of fused vertebrae are the _____ and the _____.

Indicate whether the following statements are true or false.

4. The atlanto-occipital joints allow you to rotate the head, as in signifying "no."

5. Ribs that are not attached to the sternum are known as the true ribs.

Choose the one best answer to the following questions.

6. In which of the following bones are paranasal sinuses *not* found?
   (a) frontal bone
   (b) sphenoid bone
   (c) lacrimal bones
   (d) ethmoid bone
   (e) maxillae

7. Which of the following pairs are mismatched?
   (a) mandible: only movable bone in the skull
   (b) hyoid: bone that does not articulate with any other bone
   (c) sacrum: supports lower back
   (d) thoracic vertebrae: articulate with thoracic ribs posteriorly
   (e) inferior nasal conchae: classified as facial bones

8. Which of the following bones are *not* paired?
   (a) vomer
   (b) palatine
   (c) lacrimal
   (d) maxilla
   (e) nasal

9. The suture located between a parietal and temporal bone is the
   (a) lambdoid.
   (b) sagittal.
   (c) coronal.
   (d) anterolateral.
   (e) squamous.

10. The primary vertebral curves that appear during fetal development are the (1) cervical curve, (2) thoracic curve, (3) lumbar curve, (4) coccyx curve, (5) sacral curve.
    (a) 2 and 3
    (b) 1 and 2
    (c) 2 and 4
    (d) 2 and 5
    (e) 1 and 3

11. Which of the following are functions of the cranial bones? (1) protection of the brain; (2) attachment of muscles that move the head; (3) protection of the special sense organs; (4) attachment to the meninges; (5) attachment of muscles that produce facial expressions.
    (a) 1, 2, and 5
    (b) 1, 2, 4, and 5
    (c) 2 and 5
    (d) 1, 2, 3, and 5
    (e) 1, 2, 3, 4, and 5

12. Match the following:
    ____ (a) prominent ridge or elongated projection
    ____ (b) tubelike opening
    ____ (c) large round protuberance at the end of a bone
    ____ (d) smooth, flat articular surface
    ____ (e) sharp, slender projection
    ____ (f) opening for passage of blood vessels, nerves, or ligaments
    ____ (g) large, rounded, rough projection
    ____ (h) shallow depression
    ____ (i) narrow slit between adjacent parts of bones for passage of blood vessels or nerves

    (1) foramen
    (2) tuberosity
    (3) spinous process
    (4) crest
    (5) facet
    (6) fissure
    (7) condyle
    (8) fossa
    (9) meatus

13. Match the following:

_____ (a) supraorbital foramen
_____ (b) temporomandibular joint
_____ (c) external auditory meatus
_____ (d) foramen magnum
_____ (e) optic foramen
_____ (f) cribriform plate
_____ (g) palatine process
_____ (h) ramus, body, and condylar process
_____ (i) transverse foramina, bifid spinous processes
_____ (j) dens
_____ (k) promontory
_____ (l) costal cartilages
_____ (m) xiphoid process

(1) temporal bone
(2) sphenoid bone
(3) cervical vertebrae
(4) ethmoid bone
(5) articulation of mandibular fossa and articular tubercle of the temporal bone to the mandible
(6) occipital bone
(7) frontal bone
(8) maxillae
(9) mandible
(10) axis
(11) sacrum
(12) sternum
(13) ribs

14. Match the following (the same answer may be used more than once):

_____ (a) bones that have greater length than width and consist of a shaft and a variable number of extremities
_____ (b) cube-shaped bones that are nearly equal in length and width
_____ (c) bones that develop in certain tendons where there is considerable friction, tension, and physical stress
_____ (d) small bones located within joints between certain cranial bones
_____ (e) thin bones composed of two nearly parallel plates of compact bone enclosing a layer of spongy bone
_____ (f) bones with complex shapes, including the vertebrae and some facial bones
_____ (g) patella is an example
_____ (h) bones that provide considerable protection and extensive areas for muscle attachment
_____ (i) include femur, tibia, fibula, humerus, ulna, and radius
_____ (j) include cranial bones, sternum, and ribs
_____ (k) include almost all of the carpal (wrist) and tarsal (ankle) bones

(1) irregular bones
(2) long bones
(3) short bones
(4) flat bones
(5) sesamoid bones
(6) sutural bones

15. Match the following:

_____ (a) forms the forehead
_____ (b) form the inferior lateral aspects of the cranium and part of the cranial floor; contain zygomatic process and mastoid process
_____ (c) forms part of the anterior portion of the cranial floor, medial wall of the orbits, superior portions of nasal septum, most of the side walls of the nasal cavity; is a major supporting structure of the nasal cavity
_____ (d) form the prominence of the cheek and part of the lateral wall and floor of each orbit
_____ (e) the largest, strongest facial bone; is the only movable skull bone
_____ (f) a roughly triangular bone on the floor of the nasal cavity; one of the components of the nasal septum
_____ (g) form greater portion of the sides and roof of the cranial cavity
_____ (h) forms the posterior part and most of the base of the cranium; contains the foramen magnum
_____ (i) called the keystone of the cranial floor; contains the sella turcica, optic foramen, and pterygoid processes
_____ (j) form the bridge of the nose
_____ (k) the smallest bones of the face; contain a vertical groove that houses a structure that gathers tears and passes them into the nasal cavity
_____ (l) does not articulate with any other bone
_____ (m) unite to form the upper jawbone and articulate with every bone of the face except the lower jawbone
_____ (n) form the posterior part of the hard palate, part of the floor and lateral wall of the nasal cavity, and a small portion of the floors of the orbits
_____ (o) scroll-like bones that form a part of the lateral walls of the nasal cavity; functions in the turbulent circulation and filtration of air

(1) temporal bones
(2) parietal bones
(3) frontal bone
(4) occipital bone
(5) sphenoid bone
(6) ethmoid bone
(7) nasal bones
(8) maxillae
(9) zygomatic bones
(10) lacrimal bones
(11) palatine bones
(12) vomer
(13) mandible
(14) inferior nasal conchae
(15) hyoid bone

8 THE SKELETAL SYSTEM: THE APPENDICULAR SKELETON

8.1 Pectoral (Shoulder) Girdle

1. Each of the body's two pectoral (shoulder) girdles consists of a clavicle and scapula.
2. Each pectoral girdle attaches an upper limb to the axial skeleton.
3. The clavicle (collarbone) lies horizontally across the anterior part of the thorax superior to the first rib.
4. The medial end of the clavicle articulates with the manubrium of the sternum; the lateral end of the clavicle articulates with the acromion of the scapula.
5. The scapula (shoulder blade) is situated in the superior part of the posterior thorax between the levels of the second and seventh ribs.
6. The scapula articulates with the clavicle and the head of the humerus.

8.2 Upper Limb (Extremity)

1. Each of the two upper limbs (extremities) contains 30 bones.
2. The bones of each upper limb include the humerus, the ulna, the radius, the carpals, the metacarpals, and the phalanges.
3. The humerus (arm bone) is the longest and largest bone of the upper limb.
4. The humerus articulates proximally with the scapula and distally with the ulna and radius.
5. The ulna is located on the medial aspect of the forearm and is longer than the radius.
6. The radius is the smaller bone of the forearm and is located on the lateral aspect of the forearm.
7. The eight carpals are located in the proximal region of the hand.
8. The five metacarpals are located in the intermediate region of the hand.
9. The 14 phalanges are located in the distal part of the hand (fingers).

8.3 Pelvic (Hip) Girdle

1. The pelvic (hip) girdle consists of two hip bones.
2. Each hip bone consists of three parts: the ilium, pubis, and ischium.
3. The hip bones, sacrum, and pubic symphysis form the bony pelvis. It supports the vertebral column and pelvic viscera and attaches the free lower limbs to the axial skeleton.
4. The ilium is the superior portion of the hip bone.
5. The ischium is the inferior, posterior portion of the hip bone.
6. The pubis is the anterior and inferior part of the hip bone.

8.4 False and True Pelves

1. The false pelvis is separated from the true pelvis by the pelvic brim.
2. The true pelvis surrounds the pelvic cavity and houses the rectum and urinary bladder in both genders, the vagina and cervix of the uterus in females, and the prostate in males.
3. The false pelvis is the lower portion of the abdomen that is situated superior to the pelvic brim. It contains the superior portion of the urinary bladder (when full) and the lower intestines in both genders and the uterus, uterine tubes, and ovaries in the female.

8.5 Comparison of Female and Male Pelves

1. Bones of the male skeleton are generally larger and heavier than bones of the female skeleton. They also have more prominent markings for muscle attachments.
2. The female pelvis is adapted for pregnancy and childbirth. Sex-related differences in pelvic structure are listed and illustrated in **Table 8.1**.

8.6 Lower Limb (Extremity)

1. Each of the two lower limbs (extremities) contains 30 bones.
2. The bones of each lower limb include the femur, the patella, the tibia, the fibula, the tarsals, the metatarsals, and the phalanges.
3. The femur (thigh bone) is the longest, heaviest, and strongest bone in the body.
4. The patella (kneecap) is a small, triangular bone located anterior to the knee joint.
5. The tibia (shin bone) is the larger, medial, weight-bearing bone of the leg.
6. The fibula is parallel and lateral to the tibia, but is considerably smaller.
7. The seven tarsal bones are located in the proximal region of the foot.
8. The five metatarsals are located in the intermediate region of the foot.
9. The 14 phalanges are located in the distal part of the foot (toes).
10. The bones of the foot are arranged in two arches, the longitudinal arch and the transverse arch, to provide support and leverage.

8.7 Development of the Skeletal System

1. Most bones form from mesoderm by intramembranous or endochondral ossification; much of the skeleton of the skull arises from ectoderm.
2. Bones of the limbs develop from limb buds, which consist of mesoderm and ectoderm.

CLINICAL CONNECTIONS

Fractured Clavicle (Refer page 201 of Textbook)

The clavicle transmits mechanical force from the upper limb to the trunk. If the force transmitted to the clavicle is excessive, as when you fall on your outstretched arm, a **fractured clavicle** may result. A fractured clavicle may also result from a blow to the superior part of the anterior thorax for example as a result of an impact following an automobile accident. The clavicle is one of the most frequently broken bones in the body. Because the junction of the two curves of the clavicle is its weakest point, the clavicular midregion is the most frequent fracture site. Even in the absence of fracture, compression of the clavicle as a result of automobile accidents involving the use of shoulder harness seatbelts often causes damage to the brachial plexus (the network of nerves that enter the upper limb), which lies between the clavicle and the second rib. A fractured clavicle is usually treated with a figure-eight sling to keep the arm from moving outward.

Boxer's Fracture (Refer page 208 of Textbook)

A **boxer's fracture** is a fracture of the fifth metacarpal, usually near the head of the bone. It frequently occurs after a person punches another person or an object, such as a wall. It is characterized by pain, swelling, and tenderness. There may also be a bump on the side of the hand. Treatment is either by casting or surgery, and the fracture usually heals in about 6 weeks.

Pelvimetry (Refer page 213 of Textbook)

Pelvimetry is the measurement of the size of the inlet and outlet of the birth canal, which may be done by ultrasonography or physical examination. Measurement of the pelvic cavity in pregnant females is important because the fetus must pass through the narrower opening of the pelvis at birth. A cesarean section is usually planned if it is determined that the pelvic cavity is too small to permit passage of the baby.

Patellofemoral Stress Syndrome (Refer page 215 of Textbook)

Patellofemoral stress syndrome (*runner's knee*) is one of the most common problems runners experience. During normal flexion and extension of the knee, the patella tracks (glides) superiorly and inferiorly in the groove between the femoral condyles. In patellofemoral stress syndrome, normal tracking does not occur; instead, the patella tracks laterally as well as superiorly and inferiorly, and the increased pressure on the joint causes aching or tenderness around or under the patella. The pain typically occurs after a person has been sitting for a while, especially after exercise. It is worsened by squatting or walking down stairs. One cause of runner's knee is constantly walking, running, or jogging on the same side of the road. Other predisposing factors include running on hills, running long distances, and an anatomical deformity called **genu valgum**, or *knock-knee* (see the Medical Terminology section at the end of the chapter).

Bone Grafting (Refer page 218 of Textbook)

Bone grafting generally consists of taking a piece of bone, along with its periosteum and nutrient artery, from one part of the body to replace missing bone in another part of the body. The transplanted bone restores the blood supply to the transplanted site, and healing occurs as in a fracture. The fibula is a common source of bone for grafting because even after a piece of the fibula has been removed, walking, running, and jumping can be normal. Recall that the tibia is the weight-bearing bone of the leg.

Fractures of the Metatarsals (Refer page 220 of Textbook)

Fractures of the metatarsals occur when a heavy object falls on the foot or when a heavy object rolls over the foot. Such fractures are also common among dancers, especially ballet dancers. If a ballet dancer is on the tip of her toes and loses her balance, the full body weight is placed on the metatarsals, causing one or more of them to fracture.

Flatfoot and Clawfoot (Refer page 220 of Textbook)

The bones composing the arches of the foot are held in position by ligaments and tendons. If these ligaments and tendons are weakened, the height of the medial longitudinal arch may decrease or "fall." The result is **flatfoot**, the causes of which include excessive weight, postural abnormalities, weakened supporting tissues, and genetic predisposition. Fallen arches may lead to inflammation of the fascia of the sole (plantar fasciitis), Achilles tendinitis, shin splints, stress fractures, bunions, and calluses. A custom-designed arch support often is prescribed to treat flatfoot.

Clawfoot is a condition in which the medial longitudinal arch is abnormally elevated. It is often caused by muscle deformities, such as may occur in diabetics whose neurological lesions lead to atrophy of muscles of the foot.

DISORDERS: HOMEOSTATIC IMBALANCES

Hip Fracture

Although any region of the hip girdle may fracture, the term **hip fracture** most commonly applies to a break in the bones associated with the hip joint—the head, neck, or trochanteric regions of the femur, or the bones that form the acetabulum. In the United States, 300,000 to 500,000 people sustain hip fractures each year. The incidence of hip fractures is increasing, due in part to longer life spans. Decreases in bone mass due to osteoporosis (which occurs more often in females), along with an increased tendency to fall, predispose elderly people to hip fractures.

Hip fractures often require surgical treatment, the goal of which is to repair and stabilize the fracture, increase mobility, and decrease pain. Sometimes the repair is accomplished by using surgical pins, screws, nails, and plates to secure the head of the femur. In severe hip fractures, the femoral head or the acetabulum of the hip bone may be replaced by prostheses (artificial devices). The procedure of replacing either the femoral head or the acetabulum is *hemiarthroplasty* (hem-ē-AR-thrō-plas-tē; *hemi-* = one-half; *-arthro-* = joint; *-plasty* = molding). Replacement of both the femoral head and acetabulum is total *hip arthroplasty*. The acetabular prosthesis is made of plastic, and the femoral prosthesis is metal; both are designed to withstand a high degree of stress. The prostheses are attached to healthy portions of bone with acrylic cement and screws (see **Figure 9.16**).

MEDICAL TERMINOLOGY

Clubfoot or talipes equinovarus (TA-li-pēz ek-wīn-ō-VA-rus; *-pes* = foot; *equino-* = horse) An inherited deformity in which the foot is twisted inferiorly and medially, and the angle of the arch is increased; occurs in 1 of every 1000 births. Treatment consists of manipulating the arch to a normal curvature by casts or adhesive tape, usually soon after birth. Corrective shoes or surgery may also be required.

Genu valgum (JĒ-noo VAL-gum; *genu* = knee; *valgum* = bent outward) A deformity in which the knees are abnormally close together and the space between the ankles is increased due to a lateral angulation of the tibia in relation to the femur. Also called **knock-knee**.

Genu varum (JĒ-noo VAR-um; *varum* = bent toward the midline) A deformity in which the knees are abnormally separated, there is a medial angulation of the tibia in relation to the femur, and the lower limbs are bowed laterally. Also called bowleg.

Hallux valgus (HAL-uks VAL-gus; *hallux* = great toe) Angulation of the great toe away from the midline of the body, typically caused by wearing tightly fitting shoes. When the great toe angles toward the next toe, there is a bony protrusion at the base of the great toe. Also called a **bunion**.

SELF-QUIZ QUESTIONS

Fill in the blanks in the following statements.

1. The bones that make up the palm are the _____ .

2. List the three bones that fuse to form a hip (coxal) bone: _____ , _____ , _____ .

3. The portion of the bony pelvis that is inferior to the pelvic brim is the _____ pelvis; the portion that is superior to the pelvic brim is the _____ pelvis.

Indicate whether the following statements are true or false.

4. The largest carpal bone is the lunate.

5. The anterior joint formed by the two coxal (hip) bones is the pubic symphysis.

Choose the one best answer to the following questions.

6. Which of the following statements are *true*? (1) The pectoral girdle consists of the scapula, the clavicle, and the sternum. (2) Although the joints of the pectoral girdle are not very stable, they allow free movement in many directions. (3) The anterior component of the pectoral girdle is the scapula. (4) The pectoral girdle articulates directly with the vertebral column. (5) The posterior component of the pectoral girdle is the sternum.
   (a) 1, 2, and 3   (b) 2 only   (c) 4 only
   (d) 2, 3, and 5   (e) 3, 4, and 5

7. Which of the following are *true* concerning the elbow joint? (1) When the forearm is extended, the olecranon fossa receives the olecranon. (2) When the forearm is flexed, the radial fossa receives the coronoid process. (3) The head of the radius articulates with the capitulum. (4) The trochlea articulates with the trochlear notch. (5) The head of the ulna articulates with the ulnar notch of the radius.
   (a) 1, 2, 3, 4, and 5   (b) 1, 3, and 4   (c) 1, 3, 4, and 5
   (d) 1, 2, 3, and 4   (e) 2, 3, and 4

8. Which of the following is the most superior of the tarsals and articulates with the distal end of the tibia?
   (a) calcaneus   (b) navicular   (c) cuboid
   (d) cuneiform   (e) talus

9. Which is (are) *not true* concerning the scapula? (1) The lateral border is also known as the axillary border. (2) The scapular notch accommodates the head of the humerus. (3) The scapula is also known as the collarbone. (4) The acromion process articulates with the clavicle. (5) The coracoid process is utilized for muscle attachment.
   (a) 1, 2, and 3   (b) 3 only   (c) 2 and 3
   (d) 3 and 4   (e) 2, 3, and 5

10. Which of the following is *false*?
    (a) A decrease in the height of the medial longitudinal arch creates a condition known as clawfoot.
    (b) The transverse arch is formed by the navicular, cuneiforms, and bases of the five metatarsals.
    (c) The longitudinal arch has medial and lateral parts, both of which originate at the calcaneus.
    (d) Arches help to absorb shocks.
    (e) Arches enable the foot to support the body's weight.

11. Which of the following is (are) involved in the knee joint? Choose the one best answer. (1) fibular notch of the tibia (2) lateral condyle of tibia (3) head of fibula (4) greater trochanter of femur (5) medial condyle of femur.
    (a) 1 and 2   (b) 2 and 5
    (c) 1, 2, 3, and 5   (d) 3 and 4
    (e) 4 only

12. The greater sciatic notch is located on the
    (a) ilium   (b) ischium   (c) femur
    (d) pubis   (e) sacrum

13. Match the following:
    _____ (a) a large, triangular, flat bone found in the posterior part of the thorax
    _____ (b) an S-shaped bone lying horizontally in the superior and anterior part of the thorax
    _____ (c) articulates proximally with the scapula and distally with the radius and ulna
    _____ (d) located on the medial aspect of the forearm
    _____ (e) located on the lateral aspect of the forearm
    _____ (f) the longest, heaviest, and strongest bone of the body
    _____ (g) the larger, medial bone of the leg
    _____ (h) the smaller, lateral bone of the leg
    _____ (i) heel bone
    _____ (j) sesamoid bone that articulates with the femur and tibia

    (1) calcaneus
    (2) scapula
    (3) patella
    (4) radius
    (5) femur
    (6) clavicle
    (7) ulna
    (8) tibia
    (9) humerus
    (10) fibula

14. Match the following:

_____ (a) largest and strongest tarsal bone

_____ (b) most medial bone in the distal row of carpals; has a hook-shaped projection on anterior surface

_____ (c) most medial, pea-shaped bone located in the proximal row of carpals

_____ (d) articulate with metatarsals I–III and cuboid

_____ (e) located in the proximal row of carpals; its name means "moon-shaped"

_____ (f) most lateral bone in the distal row of carpals

_____ (g) largest carpal bone

_____ (h) generally classified as proximal, middle, and distal

_____ (i) most lateral bone in the proximal row of carpals

_____ (j) articulates with the tibia and fibula

_____ (k) located in the proximal row of carpals; its name indicates that it is "three-cornered"

_____ (l) lateral bone that articulates with the calcaneus and metatarsals; IV–V

_____ (m) articulates with metacarpal II

_____ (n) boat-shaped bone that articulates with the talus

(1) cuboid
(2) triquetrum
(3) calcaneus
(4) pisiform
(5) capitate
(6) phalanges
(7) trapezoid
(8) hamate
(9) lunate
(10) scaphoid
(11) cuneiforms
(12) navicular
(13) trapezium
(14) talus

15. Match the following (some answers will be used more than once):

_____ (a) olecranon

_____ (b) olecranon fossa

_____ (c) trochlea

_____ (d) greater trochanter

_____ (e) medial malleolus

_____ (f) acromial end

_____ (g) capitulum

_____ (h) acromion

_____ (i) radial tuberosity

_____ (j) acetabulum

_____ (k) lateral malleolus

_____ (l) glenoid cavity

_____ (m) coronoid process

_____ (n) linea aspera

_____ (o) anterior border

_____ (p) anterior superior iliac spine

_____ (q) fovea capitis

_____ (r) greater tubercle

_____ (s) trochlear notch

_____ (t) obturator foramen

_____ (u) styloid process

(1) clavicle
(2) scapula
(3) humerus
(4) ulna
(5) radius
(6) femur
(7) tibia
(8) fibula
(9) hip bone

9 | JOINTS

Introduction

1. A joint (articulation or arthrosis) is a point of contact between two bones, between bone and cartilage, or between bone and teeth.
2. A joint's structure may permit no movement, slight movement, or free movement.

9.1 Joint Classifications

1. Structural classification is based on the presence or absence of a synovial cavity and the type of connective tissue. Structurally, joints are classified as fibrous, cartilaginous, or synovial.
2. Functional classification of joints is based on the degree of movement permitted. Joints may be synarthroses (immovable), amphiarthroses (slightly movable), or diarthroses (freely movable).

9.2 Fibrous Joints

1. The bones of fibrous joints are held together by dense irregular connective tissue.
2. These joints include immovable or slightly movable sutures (found between skull bones), immovable to slightly movable syndesmoses (such as roots of teeth in the sockets in the mandible and maxilla and the distal tibiofibular joint), and slightly movable interosseous membranes (found between the radius and ulna in the forearm and the tibia and fibula in the leg).

9.3 Cartilaginous joints

1. The bones of cartilaginous joints are held together by cartilage.
2. These joints include slightly movable to immovable hyaline cartilage synchondroses (cartilaginous junction of first rib with manubrium of sternum), slightly movable fibrocartilage symphyses (pubic symphysis), and immovable hyaline cartilage epiphyseal cartilages (epiphyseal or growth plates between the diaphysis and epiphysis and epiphyses of growing bones).

9.4 Synovial Joints

1. Synovial joints contain a space between bones called the synovial cavity. All synovial joints are diarthroses.
2. Other characteristics of synovial joints are the presence of articular cartilage and an articular capsule, made up of a fibrous membrane and a synovial membrane.
3. The synovial membrane secretes synovial fluid, which forms a thin, viscous film over the surfaces within the articular capsule.
4. Many synovial joints also contain accessory ligaments (extracapsular and intracapsular) and articular discs (menisci).

5. Synovial joints contain an extensive nerve and blood supply. The nerves convey information about pain, joint movements, and the degree of stretch at a joint. Blood vessels penetrate the articular capsule and ligaments.
6. Bursae are saclike structures, similar in structure to joint capsules, that alleviate friction in joints such as the shoulder and knee joints.
7. Tendon sheaths are tubelike bursae that wrap around tendons where there is considerable friction.

9.5 Types of Movements at Synovial Joints

1. In a gliding movement, the nearly flat surfaces of bones move back and forth and side to side.
2. In angular movements, a change in the angle between bones occurs. Examples are flexion–extension, lateral flexion, hyperextension, and abduction–adduction. Circumduction refers to the movement of the distal end of a body part in a circle and involves a continuous sequence of flexion, abduction, extension, adduction, and rotation of the joint (or in the opposite direction).
3. In rotation, a bone moves around its own longitudinal axis.
4. Special movements occur at specific synovial joints. Examples are elevation–depression, protraction–retraction, inversion–eversion, dorsiflexion–plantar flexion, supination–pronation, and opposition.
5. **Table 9.1** summarizes the various types of movements at synovial joints.

9.6 Types of Synovial Joints

1. Types of synovial joints are plane, hinge, pivot, condyloid, saddle, and ball-and-socket.
2. In a plane joint the articulating surfaces are flat, and the bones primarily glide back and forth and side to side (many are biaxial); they may also permit rotation (triaxial); examples are joints between carpals and tarsals.
3. In a hinge joint, the convex surface of one bone fits into the concave surface of another, and the motion is angular around one axis (uniaxial); examples are the elbow, knee (a modified hinge joint), and ankle joints.
4. In a pivot joint, a round or pointed surface of one bone fits into a ring formed by another bone and a ligament, and movement is rotational (uniaxial); examples are the atlanto-axial and radioulnar joints.
5. In a condyloid joint, an oval projection of one bone fits into an oval cavity of another, and motion is angular around two axes

(biaxial); examples include the wrist joint and metacarpophalangeal joints of the second through fifth digits.

6. In a saddle joint, the articular surface of one bone is shaped like a saddle and the other bone fits into the saddle like a sitting rider; movement is biaxial. An example is the carpometacarpal joint between the trapezium and the metacarpal of the thumb.

7. In a ball-and-socket joint, the ball-shaped surface of one bone fits into the cuplike depression of another; motion is around three axes (triaxial). Examples include the shoulder and hip joints.

8. **Table 9.2** summarizes the structural and functional categories of joints.

9.7 Factors Affecting Contact and Range of Motion at Synovial Joints

1. The ways that articular surfaces of synovial joints contact one another determine the type of movement that is possible.

2. Factors that contribute to keeping the surfaces in contact and affect range of motion are structure or shape of the articulating bones, strength and tension of the joint ligaments, arrangement and tension of the muscles, apposition of soft parts, hormones, and amount of use.

9.8 Selected Joints of the Body

1. A summary of several selected joints of the body, including articular components, structural and functional classifications, and movements, is presented in **Tables 9.3** and **9.4**.

2. The temporomandibular joint (TMJ), shoulder joint, elbow joint, hip joint, and knee joint are described in Sections 9.9 through 9.13.

9.9 Temporomandibular Joint

1. The temporomandibular joint (TMJ) is between the condyle of the mandible and mandibular fossa and articular tubercle of the temporal bone.

2. The temporomandibular joint is a combined hinge and plane joint.

9.10 Shoulder Joint

1. The shoulder (humeroscapular or glenohumeral) joint is between the head of the humerus and glenoid cavity of the scapula.

2. The shoulder joint is a type of ball-and-socket joint.

9.11 Elbow Joint

1. The elbow joint is between the trochlea of the humerus, the trochlear notch of the ulna, and the head of the radius.

2. The elbow joint is a type of hinge joint.

9.12 Hip Joint

1. The hip (coxal) joint is between the head of the femur and acetabulum of the hip bone.

2. The hip joint is a type of ball-and-socket joint.

9.13 Knee Joint

1. The knee (tibiofemoral) joint is between the patella and patellar surface of the femur; the lateral condyle of the femur, the lateral meniscus, and the lateral condyle of the tibia; and the medial condyle of the femur, the medial meniscus, and the medial condyle of the tibia.

2. The knee joint is a modified hinge joint.

9.14 Aging and Joints

1. With aging, a decrease in synovial fluid, thinning of articular cartilage, and decreased flexibility of ligaments occur.

2. Most individuals experience some degeneration in the knees, elbows, hips, and shoulders due to the aging process.

9.15 Arthroplasty

1. Arthroplasty refers to the surgical replacement of joints.

2. The most commonly replaced joints are the hips, knees, and shoulders.

CLINICAL CONNECTIONS

Autologous Chondrocyte Implantation (Refer page 228 of Textbook)

When there is damage to articular cartilage in the knee joint, especially involving the femur, there is an alternative to partial or total knee replacement (see Section 9.15) called **autologous chondrocyte implantation (ACI)** (aw-TOL-o-gus). Candidates for ACI have cartilage damage due to acute or repetitive trauma, not arthritis. In the procedure, healthy chondrocytes (cartilage cells) are taken from an area of the femoral condyle that is not weight-bearing and sent to a laboratory, where they are cultured for 4 to 5 weeks to generate between 5 million and 10 million cells. When the cultured cells are available, the implantation takes place. The damaged area is prepared by removing dead cartilage from the defect, which is covered by a piece of periosteum, usually taken from the tibia. Then the cultured chondrocytes are injected under the periosteum, where they will grow and mature over time. The patient can put the full weight of the body on the knee in about 10 to 12 weeks.

Torn Cartilage and Arthroscopy (Refer page 229 of Textbook)

The tearing of menisci in the knee, commonly called **torn cartilage**, occurs often among athletes. Such damaged cartilage will begin to wear and may

cause arthritis to develop unless the damaged cartilage is treated surgically. Years ago, if a patient had torn cartilage, the entire meniscus was removed by a procedure called a *meniscectomy* (men′-i-SEK-tō-mē). The problem was that over time the articular cartilage was worn away more quickly. Currently, surgeons perform a partial meniscectomy, in which only the torn segment of the meniscus is removed. Surgical repair of the torn cartilage may be assisted by **arthroscopy** (ar-THROS-kō-pē; -*scopy* = observation). This minimally invasive procedure involves examination of the interior of a joint, usually the knee, with an arthroscope, a lighted, pencil-thin fiber-optic camera used for visualizing the nature and extent of damage. Arthroscopy is also used to monitor the progression of disease and the effects of therapy. The insertion of surgical instruments through other incisions also enables a physician to remove torn cartilage and repair damaged cruciate ligaments in the knee; obtain tissue samples for analysis; and perform surgery on other joints, such as the shoulder, elbow, ankle, and wrist.

Bursitis (Refer page 230 of Textbook)

An acute or chronic inflammation of a bursa, called **bursitis** (bur-SĪ-tis), is usually caused by irritation from repeated, excessive exertion of a joint. The condition may also be caused by trauma, by an acute or chronic infection (including syphilis and tuberculosis), or by rheumatoid arthritis (described in the Disorders: Homeostatic Imbalances section at the end of this chapter). Symptoms include pain, swelling, tenderness, and limited movement. Treatment may include oral anti-inflammatory agents and injections of cortisol-like steroids.

Rotator Cuff Injury, Dislocated and Separated Shoulder, and Torn Glenoid Labrum (Refer page 244 of Textbook)

Rotator cuff injury is a strain or tear in the rotator cuff muscles and is a common injury among baseball pitchers, volleyball players, racket sports players, swimmers, and violinists, due to shoulder movements that involve vigorous circumduction. It also occurs as a result of wear and tear, aging, trauma, poor posture, improper lifting, and repetitive motions in certain jobs, such as placing items on a shelf above your head. Most often, there is tearing of the supraspinatus muscle tendon of the rotator cuff. This tendon is especially predisposed to wear and tear because of its location between the head of the humerus and acromion of the scapula, which compresses the tendon during shoulder movements. Poor posture and poor body mechanics also increase compression of the supraspinatus muscle tendon.

The joint most commonly dislocated in adults is the shoulder joint because its socket is quite shallow and the bones are held together by supporting muscles. Usually in a **dislocated shoulder**, the head of the humerus becomes displaced inferiorly, where the articular capsule is least protected. Dislocations of the mandible, elbow, fingers, knee, or hip are less common. Dislocations are treated with rest, ice, pain relievers, manual manipulation, or surgery followed by use of a sling and physical therapy.

A **separated shoulder** actually refers to an injury not to the shoulder joint but to the acromioclavicular joint, a joint formed by the acromion of the scapula and the acromial end of the clavicle. This condition is usually the result of forceful trauma to the joint, as when the shoulder strikes the ground in a fall. Treatment options are similar to those for treating a dislocated shoulder, although surgery is rarely needed.

In a **torn glenoid labrum**, the fibrocartilaginous labrum may tear away from the glenoid cavity. This causes the joint to catch or feel like it's slipping out of place. The shoulder may indeed become dislocated as a result. A torn labrum is reattached to the glenoid surgically with anchors and sutures. The repaired joint is more stable.

Tennis Elbow, Little League Elbow, and Tommy John Surgery Dislocation of the Radial Head (Refer page 244 of Textbook)

Tennis elbow most commonly refers to pain at or near the lateral epicondyle of the humerus, usually caused by an improperly executed backhand. The extensor muscles strain or sprain, resulting in pain. **Little League elbow,** inflammation of the medial epicondyle, typically develops as a result of a heavy pitching schedule and/or a schedule that involves throwing curve balls, especially among youngsters. In this disorder, the elbow joint may enlarge, fragment, or separate.

A **dislocation of the radial head** is the most common upper limb dislocation in children. In this injury, the head of the radius slides past or ruptures the radial anular ligament, a ligament that forms a collar around the head of the radius at the proximal radioulnar joint. Dislocation is most apt to occur when a strong pull is applied to the forearm while it is extended and supinated, for instance, while swinging a child around with outstretched arms.

Baseball pitchers make more active throws than any other player on the field. As a result of this and the mechanics of pitching, damage to the ulnar collateral ligament is becoming increasingly common. Since 1974, the damaged ligament has been replaced with a tendon taken from the palmaris longus muscle in the wrist or a graft taken from a cadaver. This type of reconstructive surgery for the ulnar collateral ligament is commonly known as **Tommy John surgery**, named for the professional baseball pitcher who first underwent the procedure.

Knee Injuries (Refer page 250 of Textbook)

The knee joint is the joint most vulnerable to damage because it is a mobile, weight-bearing joint and its stability depends almost entirely on its associated ligaments and muscles. Further, there is no complementary fit between the surfaces of the articulating bones. Following are several kinds of **knee injuries.** A **swollen knee** may occur immediately or hours after an injury. The initial swelling is due to escape of blood from damaged blood vessels adjacent to areas of injury, including rupture of the anterior cruciate ligament, damage to synovial membranes, torn menisci, fractures, or collateral ligament sprains. Delayed swelling is due to excessive production of synovial fluid, a condition commonly referred to as "water on the knee."

The firm attachment of the tibial collateral ligament to the medial meniscus is clinically significant because tearing of the ligament typically also results in tearing of the meniscus. Such an injury may occur in sports such as football and rugby when the knee receives a blow from the lateral side while the foot is fixed on the ground. The force of the blow may also tear the anterior cruciate ligament, which is also connected to the medial meniscus. The term **"unhappy triad"** is applied to a knee injury that involves damage to the three components of the knee at the same time: the tibial collateral ligament, medial meniscus, and anterior cruciate ligament.

A **dislocated knee** refers to the displacement of the tibia relative to the femur. The most common type is dislocation anteriorly, resulting from hyperextension of the knee. A frequent consequence of a dislocated knee is damage to the popliteal artery.

If no surgery is required, treatment of knee injuries involves PRICE (protection, rest, ice, compression, and elevation) with some strengthening exercises and perhaps physical therapy.

DISORDERS: HOMEOSTATIC IMBALANCES

Rheumatism and Arthritis

Rheumatism (ROO-ma-tizm) is any painful disorder of the supporting structures of the body—bones, ligaments, tendons, or muscles—that is not caused by infection or injurys. **Arthritis** is a form of rheumatism in which the joints are swollen, stiff, and painful. It afflicts about 45 million people in the United States and is the leading cause of physical disability among adults over age 65.

OSTEOARTHRITIS Osteoarthritis (OA) (os′-tē-ō-ar-THRĪ-tis) is a degenerative joint disease in which joint cartilage is gradually lost. It results from a combination of aging, obesity, irritation of the joints, muscle weakness, and wear and abrasion. Commonly known as "wear-and-tear" arthritis, osteoarthritis is the most common type of arthritis.

Osteoarthritis is a progressive disorder of synovial joints, particularly weight-bearing joints. Articular cartilage deteriorates and new bone forms in the subchondral areas and at the margins of the joint. The cartilage slowly degenerates, and as the bone ends become exposed, spurs (small bumps) of new osseous tissue are deposited on them in a misguided effort by the body to protect against increased friction. These spurs decrease the space of the joint cavity and restrict joint movement. Unlike rheumatoid arthritis (described next), osteoarthritis affects mainly the articular cartilage, although the synovial membrane often becomes inflamed late in the disease.

Two major distinctions between osteoarthritis and rheumatoid arthritis are that osteoarthritis first afflicts the larger joints (knees, hips) and is due to wear and tear, whereas rheumatoid arthritis first strikes smaller joints and is an active attack on the cartilage. Osteoarthritis is the most common reason for hip- and knee-replacement surgery.

RHEUMATOID ARTHRITIS **Rheumatoid arthritis (RA)** is an autoimmune disease in which the immune system of the body attacks its own tissues—in this case, its own cartilage and joint linings. RA is characterized by inflammation of the joint, which causes swelling, pain, and loss of function. Usually, this form of arthritis occurs bilaterally: If one wrist is affected, the other is also likely to be affected, although they are often not affected to the same degree.

The primary symptom of RA is inflammation of the synovial membrane. If untreated, the membrane thickens, and synovial fluid accumulates. The resulting pressure causes pain and tenderness. The membrane then produces an abnormal granulation tissue, called pannus, that adheres to the surface of the articular cartilage and sometimes erodes the cartilage completely. When the cartilage is destroyed, fibrous tissue joins the exposed bone ends. The fibrous tissue ossifies and fuses the joint so that it becomes immovable—the ultimate crippling effect of RA. Growth of the granulation tissue causes the distortion of the fingers that characterizes hands of RA sufferers.

GOUTY ARTHRITIS Uric acid (a substance that gives urine its name) is a waste product produced during the metabolism of nucleic acid (DNA and RNA) subunits. A person who suffers from **gout** (GOWT) either produces excessive amounts of uric acid or is not able to excrete as much as normal. The result is a buildup of uric acid in the blood. This excess acid then reacts with sodium to form a salt called sodium urate. Crystals of this salt accumulate in soft tissues such as the kidneys and in the cartilage of the ears and joints.

In **gouty arthritis**, sodium urate crystals are deposited in the soft tissues of the joints. Gout most often affects the joints of the feet, especially at the base of the big toe. The crystals irritate and erode the cartilage, causing inflammation, swelling, and acute pain. Eventually, the crystals destroy all joint tissues. If the disorder is untreated, the ends of the articulating bones fuse, and the joint becomes immovable. Treatment consists of pain relief (ibuprofen, naproxen, colchicine, and cortisone) followed by administration of allopurinol to keep uric acid levels low so that crystals do not form.

Lyme Disease

A spiral-shaped bacterium called *Borrelia burgdorferi* causes **Lyme disease**, named for the town of Lyme, Connecticut, where it was first reported in 1975. The bacteria are transmitted to humans mainly by deer ticks (*Ixodes dammini*). These ticks are so small that their bites often go unnoticed. Within a few weeks of the tick bite, a rash may appear at the site. Although the rash often resembles a bull's-eye target, there are many variations, and some people never develop a rash. Other symptoms include joint stiffness, fever and chills, headache, stiff neck, nausea, and low back pain. In advanced stages of the disease, arthritis is the main complication. It usually afflicts the larger joints such as the knee, ankle, hip, elbow, or wrist. Antibiotics are generally effective against Lyme disease, especially if they are given promptly. However, some symptoms may linger for years.

Sprain and Strain

A **sprain** is the forcible wrenching or twisting of a joint that stretches or tears its ligaments but does not dislocate the bones. It occurs when the ligaments are stressed beyond their normal capacity. Severe sprains may be so painful that the joint cannot be moved. There is considerable swelling, which results from chemicals released by the damaged cells and hemorrhage of ruptured blood vessels. The lateral ankle joint is most often sprained; the wrist is another area that is frequently sprained. A **strain** is a stretched or partially torn muscle or muscle and tendon. It often occurs when a muscle contracts suddenly and powerfully—such as the leg muscles of sprinters when they spring from the blocks.

Initially sprains should be treated with **PRICE:** protection, rest, ice, compression, and elevation. PRICE therapy may be used on muscle strains, joint inflammation, suspected fractures, and bruises. The five components of PRICE therapy are

- **Protection** means protecting the injury from further damage; for example, stop the activity and use padding and protection, and use splints or a sling, or crutches, if necessary.
- **Rest** the injured area to avoid further damage to the tissues. Stop the activity immediately. Avoid exercise or other activities that cause pain or swelling to the injured area. Rest is needed for repair. Exercising before an injury has healed may increase the probability of re-injury.
- **Ice** the injured area as soon as possible. Applying ice slows blood flow to the area, reduces swelling, and relieves pain. Ice works effectively when applied for 20 minutes, off for 40 minutes, back on for 20 minutes, and so on.
- **Compression** by wrap or bandage helps to reduce swelling. Care must be taken to compress the injured area but not to block blood flow.
- **Elevation** of the injured area above the level of the heart, when possible, will reduce potential swelling.

Tenosynovitis

Tenosynovitis (ten′-o-sīn-ō-VĪ-tis) is an inflammation of the tendons, tendon sheaths, and synovial membranes surrounding certain joints. The tendons most often affected are at the wrists, shoulders, elbows (resulting in *tennis elbow*), finger joints (resulting in *trigger finger*), ankles, and feet. The affected sheaths sometimes become visibly swollen because of fluid accumulation. Tenderness and pain are frequently associated with movement of the body part. The condition often follows trauma, strain, or excessive exercise. Tenosynovitis of the dorsum of the foot may be caused by tying shoelaces too tightly. Gymnasts are prone to developing the condition as a result of chronic, repetitive, and maximum hyperextension at the wrists. Other repetitive movements involving activities such as typing, haircutting, carpentry, and assembly line work can also result in tenosynovitis.

Dislocated Mandible

A **dislocation** (dis′-lō-KĀ-shun; *dis-* = apart) or *luxation* (luks-Ā-shun; *luxatio* = dislocation) is the displacement of a bone from a joint with tearing of ligaments, tendons, and articular capsules. A **dislocated mandible** can occur in several ways. *Anterior displacements* are the most common and occur when the condylar processes of the mandible pass anterior to the articular tubercles. Common causes are extreme mouth opening, as in yawning or taking a large bite, dental procedures, or general anesthesia. *Posterior displacement* can be caused by a direct blow to the chin. *Superior displacements* are typically caused by a direct blow to a partially opened mouth. *Lateral dislocations* are usually associated with mandibular fractures.

MEDICAL TERMINOLOGY

Arthralgia (ar-THRAL-jē-a; *arthr-* = joint; *-algia* = pain) Pain in a joint.
Bursectomy (bur-SEK-tō-mē; *-ectomy* = removal of) Removal of a bursa.
Chondritis (kon-DRĪ-tis; *chondr-* = cartilage) Inflammation of cartilage.

Subluxation (sub-luks-Ā-shun) A partial or incomplete dislocation.
Synovitis (sin′-ō-VĪ-tis) Inflammation of a synovial membrane in a joint.

SELF-QUIZ QUESTIONS

Fill in the blanks in the following statements.

1. A point of contact between two bones, between bone and cartilage, or between bone and teeth is called a(n) _____ .

2. The surgical procedure in which a severely damaged joint is replaced with an artificial joint is known as _____ .

Indicate whether the following statements are true or false.

3. Menisci are fluid-filled sacs located outside of the joint cavity to ease friction between bones and softer tissue.

4. Shrugging your shoulders involves flexion and extension.

5. Synovial fluid becomes more viscous (thicker) as movement at the joint increases.

Choose the one best answer to the following questions.

6. Which of the following are structural classifications of joints? (1) amphiarthrosis, (2) cartilaginous, (3) synovial, (4) synarthrosis, (5) fibrous.
   (a) 1, 2, 3, 4, and 5   (b) 2 and 5   (c) 1 and 4
   (d) 1, 2, 4, and 5   (e) 2, 3, and 5

7. Which of the following joints could be classified functionally as synarthroses? (1) syndesmosis, (2) symphysis, (3) synovial, (4) gomphosis, (5) suture.
   (a) 1 and 2   (b) 3 and 5   (c) 1, 2, and 3
   (d) 4 and 5   (e) 5 only

8. The most common degenerative joint disease in the elderly, often caused by wear-and-tear, is
   (a) rheumatoid arthritis.   (b) osteoarthritis.   (c) rheumatism.
   (d) gouty arthritis.   (e) ankylosing spondylitis.

9. Chewing your food involves (1) flexion, (2) extension, (3) hyperextension, (4) elevation, (5) depression.
   (a) 1 and 2   (b) 1 and 3   (c) 4 and 5
   (d) 3 and 5   (e) 1 and 4

10. Synovial fluid functions to (1) absorb shocks at joints, (2) lubricate joints, (3) form a blood clot in a joint injury, (4) supply oxygen and nutrients to chondrocytes, (5) provide phagocytes to remove debris from joints.
    (a) 1, 2, 4, and 5   (b) 1, 2, 3, 4, and 5   (c) 1, 2, and 4
    (d) 3 and 4   (e) 2, 4, and 5

11. Which of the following statements are *true* concerning a synovial joint? (1) The bones at a synovial joint are covered by a mucous membrane. (2) The articular capsule surrounds a synovial joint, encloses the synovial cavity, and unites the articulating bones. (3) The fibrous membrane of the articular capsule permits considerable movement at a joint. (4) The tensile strength of the fibrous membrane helps prevent bones from disarticulating. (5) All synovial joints contain a fibrous membrane.
    (a) 1, 2, 3, and 4   (b) 2, 3, 4, and 5   (c) 2, 3, and 4
    (d) 1, 2, and 3   (e) 2, 4, and 5

12. Which of the following keep the articular surfaces of synovial joints in contact and affect range of motion? (1) structure or shape of the articulating bones, (2) strength and tension of the joint ligaments, (3) arrangement and tension of muscles, (4) lack of use, (5) contact of soft parts.
    (a) 1, 2, 3, and 5   (b) 2, 3, 4, and 5   (c) 1, 3, 4, and 5
    (d) 1, 3, and 5   (e) 1, 2, 3, 4, and 5

13. Match the following:
    _____ (a) a fibrous joint that unites the bones of the skull; a synarthrosis
    _____ (b) a fibrous joint between the tibia and fibula; an amphiarthrosis
    _____ (c) the articulation between bone and teeth
    _____ (d) the epiphyseal (growth) plate
    _____ (e) joint between the two pubic bones
    _____ (f) joint with a cavity between the bones; diarthrosis
    _____ (g) a bony joint

    (1) synostosis
    (2) synchondrosis
    (3) syndesmosis
    (4) synovial
    (5) suture
    (6) symphysis
    (7) gomphosis

14. Match the following:
    _____ (a) rounded or pointed surface of one bone articulates with a ring formed by another bone and a ligament; allows rotation around its own axis
    _____ (b) articulating bone surfaces are flat or slightly curved; permit gliding movement
    _____ (c) convex, oval projection of one bone fits into oval depression of another bone; permits movement in two axes
    _____ (d) convex surface of one bone articulates with concave surface of another bone; permits flexion and extension
    _____ (e) ball-shaped surface of one bone articulates with cuplike depression of another bone; permits largest degree of movement in three axes
    _____ (f) modified condyloid joint where articulating bones resemble a rider sitting in a saddle

    (1) hinge joint
    (2) saddle joint
    (3) ball-and-socket joint
    (4) plane joint
    (5) condyloid joint
    (6) pivot joint

15. Match the following:

_____ (a) upward movement of a body part

_____ (b) downward movement of a body part

_____ (c) movement of bone toward midline

_____ (d) movement in which relatively flat bone surfaces move back-and-forth and side-to-side with respect to one another

_____ (e) movement of a body part anteriorly in the transverse plane

_____ (f) decrease in angle between bones

_____ (g) movement of an anteriorly projected body part back to the anatomical position

_____ (h) movement of the sole medially

_____ (i) movement of the sole laterally

_____ (j) movement of bone away from midline

_____ (k) action that occurs when you stand on your heels

_____ (l) action that occurs when you stand on your toes

_____ (m) movement of the forearm to turn the palm anteriorly

_____ (n) movement of the forearm to turn the palm posteriorly

_____ (o) movement of thumb across the palm to touch the tips of the fingers of the same hand

_____ (p) increase in angle between bones

_____ (q) movement of distal end of a part of the body in a circle

_____ (r) bone revolves around its own longitudinal axis

(1) pronation

(2) plantar flexion

(3) eversion

(4) abduction

(5) rotation

(6) retraction

(7) opposition

(8) elevation

(9) flexion

(10) adduction

(11) depression

(12) inversion

(13) gliding

(14) extension

(15) protraction

(16) dorsiflexion

(17) circumduction

(18) supination

10 MUSCULAR TISSUE

Introduction

1. Motion results from alternating contraction and relaxation of muscles, which constitute 40–50% of total body weight.
2. The prime function of muscle is changing chemical energy into mechanical energy to perform work.

10.1 Overview of Muscular Tissue

1. The three types of muscular tissue are skeletal, cardiac, and smooth. Skeletal muscle tissue is primarily attached to bones; it is striated and voluntary. Cardiac muscle tissue forms the wall of the heart; it is striated and involuntary. Smooth muscle tissue is located primarily in internal organs; it is nonstriated (smooth) and involuntary.
2. Through contraction and relaxation, muscular tissue performs four important functions, producing body movements, stabilizing body positions, moving substances within the body and regulating organ volume, and producing heat.
3. Four special properties of muscular tissues are (1) electrical excitability, the property of responding to stimuli by producing action potentials; (2) contractility, the ability to generate tension to do work; (3) extensibility, the ability to be extended (stretched); and (4) elasticity, the ability to return to original shape after contraction or extension.

10.2 Structure of Skeletal Muscle Tissue

1. The subcutaneous layer separates skin from muscles, provides a pathway for blood vessels and nerves to enter and exit muscles, and protects muscles from physical trauma. Fascia lines the body wall and limbs that surround and support muscles, allows free movement of muscles, carries nerves and blood vessels, and fills space between muscles.
2. Tendons and aponeuroses are extensions of connective tissue beyond muscle fibers that attach the muscle to bone or to other muscle. A tendon is generally ropelike in shape; an aponeurosis is wide and flat.
3. Skeletal muscles are well supplied with nerves and blood vessels. Generally, an artery and one or two veins accompany each nerve that penetrates a skeletal muscle.
4. Somatic motor neurons provide the nerve impulses that stimulate skeletal muscle to contract.
5. Blood capillaries bring in oxygen and nutrients and remove heat and waste products of muscle metabolism.
6. The major cells of skeletal muscle tissue are termed skeletal muscle fibers. Each muscle fiber has 100 or more nuclei because it arises from the fusion of many myoblasts. Satellite cells are myoblasts that persist after birth. The sarcolemma is a muscle fiber's plasma membrane; it surrounds the sarcoplasm. Transverse tubules are invaginations of the sarcolemma.
7. Each muscle fiber (cell) contains hundreds of myofibrils, the contractile elements of skeletal muscle. Sarcoplasmic reticulum (SR) surrounds each myofibril. Within a myofibril are thin and thick filaments, arranged in compartments called sarcomeres.
8. The overlapping of thick and thin filaments produces striations. Darker A bands alternate with lighter I bands. **Table 10.1** in Textbook summarizes the components of the sarcomere.
9. Myofibrils are composed of three types of proteins: contractile, regulatory, and structural. The contractile proteins are myosin (thick filament) and actin (thin filament). Regulatory proteins are tropomyosin and troponin, both of which are part of the thin filament. Structural proteins include titin (links Z disc to M line and stabilizes thick filament), myomesin (forms M line), nebulin (anchors thin filaments to Z discs and regulates length of thin filaments during development), and dystrophin (links thin filaments to sarcolemma). **Table 10.2** in Textbook summarizes the different types of skeletal muscle fiber proteins. **Table 10.3** in Textbook summarizes the levels of organization within a skeletal muscle.
10. Projecting myosin heads contain actin-binding and ATP-binding sites and are the motor proteins that power muscle contraction.

10.3 Contraction and Relaxation of Skeletal Muscle Fibers

1. Muscle contraction occurs because cross-bridges attach to and "walk" along the thin filaments at both ends of a sarcomere, progressively pulling the thin filaments toward the center of a sarcomere. As the thin filaments slide inward, the Z discs come closer together, and the sarcomere shortens.
2. The contraction cycle is the repeating sequence of events that causes sliding of the filaments: (1) Myosin ATPase hydrolyzes ATP and becomes energized; (2) the myosin head attaches to actin, forming a cross-bridge; (3) the cross-bridge generates force as it rotates toward the center of the sarcomere (power stroke); and (4) binding of ATP to the myosin head detaches it from actin. The myosin head again hydrolyzes the ATP, returns to its original position, and binds to a new site on actin as the cycle continues.
3. An increase in Ca^{2+} concentration in the sarcoplasm starts filament sliding; a decrease turns off the sliding process.

4. The muscle action potential propagating into the T tubule system stimulates voltage-gated Ca^{2+} channels in the T tubule membrane. This causes opening of Ca^{2+} release channels in the SR membrane. Calcium ions diffuse from the SR into the sarcoplasm and combine with troponin. This binding causes tropomyosin to move away from the myosin-binding sites on actin.

5. Ca^{2+} active transport pumps continually remove Ca^{2+} from the sarcoplasm into the SR. When the concentration of calcium ions in the sarcoplasm decreases, tropomyosin slides back over and blocks the myosin-binding sites, and the muscle fiber relaxes.

6. A muscle fiber develops its greatest tension when there is an optimal zone of overlap between thick and thin filaments. This dependency is the length–tension relationship.

7. The neuromuscular junction (NMJ) is the synapse between a somatic motor neuron and a skeletal muscle fiber. The NMJ includes the axon terminals and synaptic end bulbs of a motor neuron, plus the adjacent motor end plate of the muscle fiber sarcolemma.

8. When a nerve impulse reaches the synaptic end bulbs of a somatic motor neuron, it triggers exocytosis of the synaptic vesicles, which releases acetylcholine (ACh). ACh diffuses across the synaptic cleft and binds to ACh receptors, initiating a muscle action potential. Acetylcholinesterase then quickly breaks down ACh into its component parts.

10.4 Muscle Metabolism

1. Muscle fibers have three sources for ATP production: creatine, anaerobic glycolysis, and aerobic respiration.

2. Creatine kinase catalyzes the transfer of a high-energy phosphate group from creatine phosphate to ADP to form new ATP molecules. Together, creatine phosphate and ATP provide enough energy for muscles to contract maximally for about 15 seconds.

3. Glucose is converted to pyruvic acid in the reactions of glycolysis, which yield two ATPs without using oxygen. Anaerobic glycolysis can provide enough energy for 2 minutes of maximal muscle activity.

4. Muscular activity that occurs over a prolonged time depends on aerobic respiration, mitochondrial reactions that require oxygen to produce ATP.

5. The inability of a muscle to contract forcefully after prolonged activity is muscle fatigue.

6. Elevated oxygen use after exercise is called recovery oxygen uptake.

10.5 Control of Muscle Tension

1. A motor neuron and the muscle fibers it stimulates form a motor unit. A single motor unit may contain as few as 2 or as many as 3000 muscle fibers.

2. Recruitment is the process of increasing the number of active motor units.

3. A twitch contraction is a brief contraction of all muscle fibers in a motor unit in response to a single action potential.

4. A record of a contraction is called a myogram. It consists of a latent period, a contraction period, and a relaxation period.

5. Wave summation is the increased strength of a contraction that occurs when a second stimulus arrives before the muscle fiber has relaxed completely following a previous stimulus.

6. Repeated stimuli can produce unfused (incomplete) tetanus, a sustained muscle contraction with partial relaxation between stimuli. More rapidly repeating stimuli produce fused (complete) tetanus, a sustained contraction without partial relaxation between stimuli.

7. Continuous involuntary activation of a small number of motor units produces muscle tone, which is essential for maintaining posture.

8. In a concentric isotonic contraction, the muscle shortens to produce movement and to reduce the angle at a joint. During an eccentric isotonic contraction, the muscle lengthens.

9. Isometric contractions, in which tension is generated without muscle changing its length, are important because they stabilize some joints as others are moved.

10.6 Types of Skeletal Muscle Fibers

1. On the basis of their structure and function, skeletal muscle fibers are classified as slow oxidative (SO), fast oxidative–glycolytic (FOG), and fast glycolytic (FG) fibers.

2. Most skeletal muscles contain a mixture of all three fiber types. Their proportions vary with the typical action of the muscle.

3. The motor units of a muscle are recruited in the following order: first SO fibers, then FOG fibers, and finally FG fibers.

4. **Table 10.4** summarizes the three types of skeletal muscle fibers.

10.7 Exercise and Skeletal Muscle Tissue

1. Various types of exercises can induce changes in the fibers in a skeletal muscle. Endurance-type (aerobic) exercises cause a gradual transformation of some fast glycolytic (FG) fibers into fast oxidative–glycolytic (FOG) fibers.

2. Exercises that require great strength for short periods produce an increase in the size and strength of fast glycolytic (FG) fibers. The increase in size is due to increased synthesis of thick and thin filaments.

10.8 Cardiac Muscle Tissue

1. Cardiac muscle is found only in the heart. Cardiac muscle fibers have the same arrangement of actin and myosin and the same bands, zones, and Z discs as skeletal muscle fibers. The fibers connect to one another through intercalated discs, which contain both desmosomes and gap junctions.

2. Cardiac muscle tissue remains contracted 10 to 15 times longer than skeletal muscle tissue due to prolonged delivery of Ca^{2+} into the sarcoplasm.

3. Cardiac muscle tissue contracts when stimulated by its own autorhythmic fibers. Due to its continuous, rhythmic activity, cardiac muscle depends greatly on aerobic respiration to generate ATP.

10.9 Smooth Muscle Tissue

1. Smooth muscle is nonstriated and involuntary.

2. Smooth muscle fibers contain intermediate filaments and dense bodies; the function of dense bodies is similar to that of the Z discs in striated muscle.

3. Visceral (single-unit) smooth muscle is found in the walls of hollow viscera and of small blood vessels. Many fibers form a network that contracts in unison.

4. Multiunit smooth muscle is found in large blood vessels, large airways to the lungs, arrector pili muscles, and the eye, where it adjusts pupil diameter and lens focus. The fibers operate independently rather than in unison.

5. The duration of contraction and relaxation of smooth muscle is longer than in skeletal muscle since it takes longer for Ca^{2+} to reach the filaments.

6. Smooth muscle fibers contract in response to nerve impulses, hormones, and local factors.

7. Smooth muscle fibers can stretch considerably and still maintain their contractile function.

10.10 Regeneration of Muscular Tissue

1. Skeletal muscle fibers cannot divide and have limited powers of regeneration; cardiac muscle fibers can regenerate under limited circumstances; and smooth muscle fibers have the best capacity for division and regeneration.

2. **Table 10.5** summarizes the major characteristics of the three types of muscular tissue.

10.11 Development of Muscle

1. With few exceptions, muscles develop from mesoderm.

2. Skeletal muscles of the head and limbs develop from general mesoderm. Other skeletal muscles develop from the mesoderm of somites.

10.12 Aging and Muscular Tissue

1. With aging, there is a slow, progressive loss of skeletal muscle mass, which is replaced by fibrous connective tissue and fat.

2. Aging also results in a decrease in muscle strength, slower muscle reflexes, and loss of flexibility.

CLINICAL CONNECTIONS

Fibromyalgia (Refer page 256 of Textbook)

Fibromyalgia (fī-brō-mī-AL-jē-a; -*algia* = painful condition) is a chronic, painful, nonarticular rheumatic disorder that affects the fibrous connective tissue components of muscles, tendons, and ligaments. A striking sign is pain that results from gentle pressure at specific "tender points." Even without pressure, there is pain, tenderness, and stiffness of muscles, tendons, and surrounding soft tissues. Besides muscle pain, those with fibromyalgia report severe fatigue, poor sleep, headaches, depression, irritable bowel syndrome, and inability to carry out their daily activities. There is no specific identifiable cause. Treatment consists of stress reduction, regular exercise, application of heat, gentle massage, physical therapy, medication for pain, and a low-dose antidepressant to help improve sleep.

Muscular Hypertrophy, Fibrosis, and
Muscular Atrophy (Refer page 256 of Textbook)

The muscle growth that occurs after birth occurs by enlargement of existing muscle fibers, called **muscular hypertrophy** (hī-PER-trō-fē; *hyper-* = above or excessive; -*trophy* = nourishment). Muscular hypertrophy is due to increased production of myofibrils, mitochondria, sarcoplasmic reticulum, and other organelles. It results from very forceful, repetitive muscular activity, such as strength training. Because hypertrophied muscles contain more myofibrils, they are capable of more forceful contractions. During childhood, human growth hormone and other hormones stimulate an increase in the size of skeletal muscle fibers. The hormone testosterone promotes further enlargement of muscle fibers.

A few myoblasts do persist in mature skeletal muscle as *satellite cells* (see **Figure 10.2a, b**). Satellite cells retain the capacity to fuse with one another or with damaged muscle fibers to regenerate functional muscle fibers. However, when the number of new skeletal muscle fibers that can be formed by satellite cells is not enough to compensate for significant skeletal muscle damage or degeneration, the muscular tissue undergoes **fibrosis**, the replacement of muscle fibers by fibrous scar tissue.

Muscular atrophy (AT-rō-fē; *a-* = without, -*trophy* = nourishment) is a decrease in size of individual muscle fibers as a result of progressive loss of myofibrils. Atrophy that occurs because muscles are not used is termed *disuse atrophy*. Bedridden individuals and people with casts experience disuse atrophy because the flow of nerve impulses to inactive skeletal muscle is greatly reduced, but the condition is reversible. If instead its nerve supply is disrupted or cut, the muscle undergoes *denervation atrophy*. Over a period of 6 months to 2 years, the muscle shrinks to about one-fourth its original size, and its fibers are irreversibly replaced by fibrous connective tissue.

Rigor Mortis (Refer page 266 of Textbook)

After death, cellular membranes become leaky. Calcium ions leak out of the sarcoplasmic reticulum into the sarcoplasm and allow myosin heads to bind to actin. ATP synthesis ceases shortly after breathing stops, however, so the cross-bridges cannot detach from actin. The resulting condition, in which muscles are in a state of rigidity (cannot contract or stretch), is called **rigor mortis** (rigidity of death). Rigor mortis begins 3–4 hours after death and lasts about 24 hours; then it disappears as proteolytic enzymes from lysosomes digest the cross-bridges.

Electromyography (Refer page 271 of Textbook)

Electromyography (EMG) (e-lek′-trō-mī-OG-ra-fē; *electro-* = electricity; -*myo-* = muscle; -*graph* = to write) is a test that measures the electrical activity (muscle action potentials) in resting and contracting muscles. Normally, resting muscle produces no electrical activity; a slight contraction produces some electrical activity; and a more forceful contraction produces increased electrical activity. In the procedure, a ground electrode is placed over the muscle to be tested to eliminate background electrical activity. Then, a fine needle attached by wires to a recording instrument is inserted into the muscle. The electrical activity of the muscle is displayed as waves on an oscilloscope and heard through a loudspeaker.

EMG helps to determine if muscle weakness or paralysis is due to a malfunction of the muscle itself or the nerves supplying the muscle. EMG is also used to diagnose certain muscle disorders, such as muscular dystrophy, and to understand which muscles function during complex movements.

Creatine Supplementation (Refer page 271 of Textbook)

Creatine is both synthesized in the body and derived from foods such as milk, red meat, and some fish. Adults need to synthesize and ingest a total of about 2 grams of creatine daily to make up for the urinary loss of creatinine, the breakdown product of creatine. Some studies have demonstrated improved performance from **creatine supplementation** during explosive movements, such as sprinting. Other studies, however, have failed to find a performance-enhancing effect of creatine supplementation. Moreover, ingesting extra creatine decreases the body's own synthesis of creatine, and it is not known whether natural synthesis recovers after long-term creatine supplementation. In addition, creatine supplementation can cause dehydration and may cause kidney dysfunction. Further research is needed to determine both the long-term safety and the value of creatine supplementation.

Anaerobic Training versus Aerobic Training (Refer page 275 of Textbook)

Regular, repeated activities such as jogging or aerobic dancing increase the supply of oxygen-rich blood available to skeletal muscles for aerobic respiration. By contrast, activities such as weight lifting rely more on anaerobic production of ATP through glycolysis. Such **anaerobic training** activities stimulate synthesis of muscle proteins and result, over time, in increased muscle size (muscle hypertrophy). Athletes who engage in anaerobic training should have a diet that includes an adequate amount of proteins. This protein intake will allow the body to synthesize muscle proteins and to increase muscle mass. As a result, **aerobic training** builds endurance for prolonged activities; in contrast, anaerobic training builds muscle strength for short-term feats. **Interval training** is a workout regimen that incorporates both types of training—for example, alternating sprints with jogging.

Hypotonia and Hypertonia (Refer page 275 of Textbook)

Hypotonia (hī´-pō-TŌ-nē-a; *hypo-* = below) refers to decreased or lost muscle tone. Such muscles are said to be flaccid. Flaccid muscles are loose and appear flattened rather than rounded. Certain disorders of the nervous system and disruptions in the balance of electrolytes (especially sodium, calcium, and, to a lesser extent, magnesium) may result in **flaccid paralysis** (pa-RAL-i-sis), which

is characterized by loss of muscle tone, loss or reduction of tendon reflexes, and atrophy (wasting away) and degeneration of muscles.

Hypertonia (hī´-per-TŌ-nē-a; *hyper-* = above) refers to increased muscle tone and is expressed in two ways: spasticity or rigidity. **Spasticity** (spas-TIS-i-tē) is characterized by increased muscle tone (stiffness) associated with an increase in tendon reflexes and pathological reflexes (such as the Babinski sign, in which the great toe extends with or without fanning of the other toes in response to stroking the outer margin of the sole). Certain disorders of the nervous system and electrolyte disturbances such as those previously noted may result in **spastic paralysis**, partial paralysis in which the muscles exhibit spasticity. **Rigidity** refers to increased muscle tone in which reflexes are not affected, as occurs in tetanus. Tetanus is a disease caused by a bacterium, *Clostridium tetani,* that enters the body through exposed wounds. It leads to muscle stiffness and spasms that can make breathing difficult and can become life-threatening as a result. The bacteria produce a toxin that interferes with the nerves controlling the muscles. The first signs are typically spasms and stiffness in the muscles of the face and jaws.

Anabolic Steroids (Refer page 279 of Textbook)

The use of **anabolic steroids** (an-a-BOL-ik = to build up proteins), or "roids," by athletes has received widespread attention. These steroid hormones, similar to testosterone, are taken to increase muscle size by increasing the synthesis of proteins in muscle and thus increasing strength during athletic contests. However, the large doses needed to produce an effect have damaging, sometimes even devastating side effects, including liver cancer, kidney damage, increased risk of heart disease, stunted growth, wide mood swings, increased acne, and increased irritability and aggression. Additionally, females who take anabolic steroids may experience atrophy of the breasts and uterus, menstrual irregularities, sterility, facial hair growth, and deepening of the voice. Males may experience diminished testosterone secretion, atrophy of the testes, sterility, and baldness.

DISORDERS: HOMEOSTATIC IMBALANCES

Abnormalities of skeletal muscle function may be due to disease or damage of any of the components of a motor unit: somatic motor neurons, neuromuscular junctions, or muscle fibers. The term **neuromuscular disease** encompasses problems at all three sites; the term **myopathy** (mī-OP-a-thē; *-pathy* = disease) signifies a disease or disorder of the skeletal muscle tissue itself.

Myasthenia Gravis

Myasthenia gravis (mī-as-THĒ-nē-a GRAV-is; *mys-* = muscle; *-aisthesis* = sensation) is an autoimmune disease that causes chronic, progressive damage of the neuromuscular junction. The immune system inappropriately produces antibodies that bind to and block some ACh receptors, thereby decreasing the number of functional ACh receptors at the motor end plates of skeletal muscles (see **Figure 10.9**). Because 75% of patients with myasthenia gravis have hyperplasia or tumors of the thymus, it is thought that thymic abnormalities cause the disorder. As the disease progresses, more ACh receptors are lost. Thus, muscles become increasingly weaker, fatigue more easily, and may eventually cease to function.

Myasthenia gravis occurs in about 1 in 10,000 people and is more common in women, typically ages 20 to 40 at onset; men usually are ages 50 to 60 at onset. The muscles of the face and neck are most often affected. Initial symptoms include weakness of the eye muscles, which may produce double vision, and weakness of the throat muscles that may produce difficulty in swallowing. Later, the person has difficulty chewing and talking. Eventually the muscles of the limbs may become involved. Death may result from paralysis of the respiratory muscles, but often the disorder does not progress to that stage.

Anticholinesterase drugs such as pyridostigmine (Mestinon) or neostigmine, the first line of treatment, act as inhibitors of acetylcholinesterase, the enzyme that breaks down ACh. Thus, the inhibitors raise the level of ACh that is available to bind with still-functional receptors. More recently, steroid drugs such as prednisone have been used with success to reduce antibody levels. Another treatment is plasmapheresis, a procedure that removes the antibodies from the blood. Often, surgical removal of the thymus (thymectomy) is helpful.

Muscular Dystrophy

The term **muscular dystrophy** (DIS-trō-fē´; *dys-* = difficult; *-trophy* = nourishment) refers to a group of inherited muscle-destroying diseases that cause progressive degeneration of skeletal muscle fibers. The most common form of muscular dystrophy is *Duchenne muscular dystrophy* (DMD) (doo-SHAN). Because the mutated gene is on the X chromosome, and because males have only one X chromosome, DMD strikes boys almost exclusively. (Sex-linked inheritance is described in Chapter 29.) Worldwide, about 1 in every 3500 male babies—about 21,000 in all—are born with DMD each year. The disorder usually becomes apparent between the ages of 2 and 5, when parents notice the child falls often and has difficulty running, jumping, and hopping. By age 12 most boys with DMD are unable to walk. Respiratory or cardiac failure usually causes death by age 20.

In DMD, the gene that codes for the protein dystrophin is mutated, so little or no dystrophin is present in the sarcolemma. Without the reinforcing effect of dystrophin, the sarcolemma tears easily during muscle contraction, causing muscle fibers to rupture and die. The dystrophin gene was discovered in 1987, and by 1990 the first attempts were made to treat DMD patients with gene therapy. The muscles of three boys with DMD were injected with myoblasts bearing functional dystrophin genes, but only a few muscle fibers gained the ability to produce dystrophin. Similar clinical trials with additional patients have also failed. An alternative approach to the problem is to find a way to induce muscle fibers to produce the protein utrophin, which is similar to dystrophin. Experiments with dystrophin-deficient mice suggest this approach may work.

Abnormal Contractions of Skeletal Muscle

One kind of abnormal muscular contraction is a **spasm**, a sudden involuntary contraction of a single muscle in a large group of muscles. A painful spasmodic contraction is known as a **cramp**. Cramps may be caused by inadequate blood flow to muscles, overuse of a muscle, dehydration, injury, holding a position for prolonged periods, and low blood levels of electrolytes, such as potassium. A **tic** is a spasmodic twitching made involuntarily by muscles that are ordinarily under voluntary control. Twitching of the eyelid and facial muscles are examples of tics. A **tremor** is a rhythmic, involuntary, purposeless contraction that produces a quivering or shaking movement. A **fasciculation** (fa-sik-ū-LĀ-shun) is an involuntary, brief twitch of an entire motor unit that is visible under the skin; it occurs irregularly and is not associated with movement of the affected muscle. Fasciculations may be seen in multiple sclerosis (see Disorders: Homeostatic Imbalances in Chapter 12) or in amyotrophic lateral sclerosis (Lou Gehrig's disease; see Clinical Connection: Amyotrophic Lateral Sclerosis in Chapter 16). A **fibrillation** (fi-bri-LĀ-shun) is a spontaneous contraction of a single muscle fiber that is not visible under the skin but can be recorded by electromyography. Fibrillations may signal destruction of motor neurons.

Exercise-induced Muscle Damage

Comparison of electron micrographs of muscle tissue taken from athletes before and after intense exercise reveals considerable **exercise-induced muscle damage**, including torn sarcolemmas in some muscle fibers, damaged myofibrils, and disrupted Z discs. Microscopic muscle damage after exercise also is indicated by increases in blood levels of proteins, such as myoglobin and the enzyme creatine kinase, which are normally confined within muscle fibers. From 12 to 48 hours after a period of strenuous exercise, skeletal muscles often become sore. Such **delayed onset muscle soreness (DOMS)** is accompanied by stiffness, tenderness, and swelling. Although the causes of DOMS are not completely understood, microscopic muscle damage appears to be a major factor. In response to exercise-induced muscle damage, muscle fibers undergo repair: new regions of sarcolemma are formed to replace torn sarcolemmas, and more muscle proteins (including those of the myofibrils) are synthesized in the sarcoplasm of the muscle fibers.

MEDICAL TERMINOLOGY

Myalgia (mī-AL-jē-a; -*algia* = painful condition) Pain in or associated with muscles.

Myoma (mī-Ō-ma; -*oma* = tumor) A tumor consisting of muscle tissue.

Myomalacia (mī′-ō-ma-LĀ-shē-a; -*malacia* = soft) Pathological softening of muscle tissue.

Myositis (mī′-ō-SĪ-tis; -*itis* = inflammation of) Inflammation of muscle fibers (cells).

Myotonia (mī′-ō-TO-nē-a; -*tonia* = tension) Increased muscular excitability and contractility, with decreased power of relaxation; tonic spasm of the muscle.

Volkmann's contracture (FŌLK-manz kon-TRAK-chur; *contra-* = against) Permanent shortening (contracture) of a muscle due to replacement of destroyed muscle fibers by fibrous connective tissue, which lacks extensibility. Typically occurs in forearm flexor muscles. Destruction of muscle fibers may occur from interference with circulation caused by a tight bandage, a piece of elastic, or a cast.

SELF-QUIZ QUESTIONS

Fill in the blanks in the following statements.

1. A single somatic motor neuron and all of the muscle fibers it stimulates is known as a _____.

2. The wasting away of muscle due to lack of use is known as _____ while the replacement of skeletal muscle fibers with scar tissue is known as

   _____.

3. The synaptic end bulbs of somatic motor neurons contain synaptic vesicles filled with the neurotransmitter _____.

Indicate whether the following statements are *True* or *False*.

4. The ability of muscle cells to respond to stimuli and produce electrical signals is known as excitability.

5. The sequence of events resulting in skeletal muscle contraction are
   (a) generation of a nerve impulse;
   (b) release of the neurotransmitter acetylcholine;
   (c) generation of a muscle action potential;
   (d) release of calcium ions from the sarcoplasmic reticulum;
   (e) calcium ion binding to troponin;
   (f) power stroke with actin and myosin binding and release.

Choose the one best answer to the following questions.

6. In muscle physiology, the latent period refers to
   (a) the period of lost excitability that occurs when two stimuli are applied immediately one after the other.
   (b) the brief contraction of a motor unit.
   (c) the period of elevated oxygen use after exercise.
   (d) an inability of a muscle to contract forcefully after prolonged activity.
   (e) a brief delay that occurs between application of a stimulus and the beginning of contraction.

7. Which of the following muscle proteins and their descriptions are mismatched?
   (a) titin: regulatory protein that holds troponin in place
   (b) myosin: contractile motor protein
   (c) tropomyosin: regulatory protein that blocks myosin-binding sites
   (d) actin: contractile protein that contains myosin-binding sites
   (e) calsequestrin: calcium-binding protein

8. During muscle contraction all of the following occur *except*
   (a) cross-bridges are formed when the energized myosin head attaches to actin's myosin-binding site.
   (b) ATP undergoes hydrolysis.
   (c) the thick filaments slide inward toward the M line.
   (d) calcium concentration in the cytosol increases.
   (e) the Z discs are drawn toward each other.

9. Which of the following is *not* true concerning sarcomeres (before contraction begins) and muscle fiber length–tension relationships?
   (a) If sarcomeres are stretched, the tension in the fiber decreases.
   (b) If a muscle cell is stretched so that there is no overlap of the filaments, no tension is generated.
   (c) Extremely compressed sarcomeres result in less muscle tension.
   (d) Maximum tension occurs when the zone of overlap between a thick and thin filament extends from the edge of the H zone to one end of a thick filament.
   (e) If sarcomeres shorten, the tension in them increases.

10. Which of the following are sources of ATP for muscle contraction? (1) creatine phosphate, (2) glycolysis, (3) anaerobic cellular respiration, (4) aerobic cellular respiration, (5) acetylcholine.
    (a) 1, 2, and 3   (b) 2, 3, and 4   (c) 2, 3, and 5
    (d) 1, 2, 3, and 4   (e) 2, 3, 4, and 5

11. What would happen if ATP were suddenly unavailable after the sarcomere had begun to shorten?
    (a) Nothing. The contraction would proceed normally.
    (b) The myosin heads would be unable to detach from actin.
    (c) Troponin would bind with the myosin heads.
    (d) Actin and myosin filaments would separate completely and be unable to recombine.
    (e) The myosin heads would detach completely from actin.

12. Match the following:
    _____ (a) a sheath of areolar connective tissue that wraps around individual skeletal muscle fibers
    _____ (b) dense irregular connective tissue that separates a muscle into groups of individual muscle fibers
    _____ (c) bundles of muscle fibers
    _____ (d) the outermost connective tissue layer that encircles an entire skeletal muscle
    _____ (e) dense irregular connective tissue that lines the body wall and limbs and holds functional muscle units together
    _____ (f) a cord of dense regular connective tissue that attaches muscle to the periosteum of bone
    _____ (g) muscle cell
    _____ (h) areolar and adipose connective tissue that separates muscle from skin
    _____ (i) connective tissue elements extended as a broad, flat layer
    _____ (j) a two-layer tube of fibrous connective tissue enclosing certain tendons

    (1) aponeurosis
    (2) fascia
    (3) subcutaneous layer
    (4) tendon
    (5) endomysium
    (6) perimysium
    (7) epimysium
    (8) tendon (synovial) sheath
    (9) fascicles
    (10) muscle fiber

13. Match the following:
    _____ (a) synapse between a motor neuron and a muscle fiber
    _____ (b) invaginations of the sarcolemma from the surface toward the center of the muscle fiber
    _____ (c) myoblasts that persist in mature skeletal muscle
    _____ (d) plasma membrane of a muscle fiber
    _____ (e) oxygen-binding protein found only in muscle fibers
    _____ (f) Ca^{2+}-storing tubular system similar to smooth endoplasmic reticulum
    _____ (g) contracting unit of a skeletal muscle fiber
    _____ (h) middle area in the sarcomere where thick and thin filaments are found
    _____ (i) area in the sarcomere where only thin filaments are present but thick filaments are not
    _____ (j) separates the sarcomeres from each other
    _____ (k) area of only thick filaments
    _____ (l) cytoplasm of a muscle fiber
    _____ (m) composed of supporting proteins holding thick filaments together at the H zone

    (1) A band
    (2) I band
    (3) Z disc
    (4) H zone
    (5) M line
    (6) sarcomere
    (7) neuromuscular junction
    (8) myoglobin
    (9) satellite cells
    (10) transverse tubules
    (11) sarcoplasmic reticulum
    (12) sarcolemma
    (13) sarcoplasm

14. Match the following (some questions will have more than one answer):
    _____ (a) has fibers joined by intercalated discs
    _____ (b) thick and thin filaments are not arranged as orderly sarcomeres
    _____ (c) uses satellite cells to repair damaged muscle fibers
    _____ (d) striated
    _____ (e) contraction begins slowly but lasts for long periods
    _____ (f) has an extended contraction due to prolonged calcium delivery from both the sarcoplasmic reticulum and the interstitial fluid
    _____ (g) does not exhibit autorhythmicity
    _____ (h) uses pericytes to repair damaged muscle fibers
    _____ (i) uses troponin as a regulatory protein
    _____ (j) can be classified as single-unit or multiunit
    _____ (k) can be autorhythmic
    _____ (l) uses calmodulin as a regulatory protein

    (1) skeletal muscle
    (2) cardiac muscle
    (3) smooth muscle

15. Match the following:

_____ (a) the smooth muscle action that allows the fibers to maintain their contractile function even when stretched

_____ (b) a brief contraction of all the muscle fibers in a motor unit of a muscle in response to a single action potential in its motor neuron

_____ (c) sustained contraction of a muscle, with no relaxation between stimuli

_____ (d) larger contractions resulting from stimuli arriving at different times

_____ (e) process of increasing the number of activated motor units

_____ (f) contraction in which the muscle shortens

_____ (g) inability of a muscle to maintain its strength of contraction or tension during prolonged activity

_____ (h) sustained, but wavering contraction with partial relaxation between stimuli

_____ (i) produced by the continual involuntary activation of a small number of skeletal muscle motor units; results in firmness in skeletal muscle

_____ (j) contraction in which muscle tension is generated without shortening of the muscle

_____ (k) amount of oxygen needed to restore the body's metabolic conditions back to resting levels after exercise

_____ (l) contraction in which a muscle lengthens

(1) muscle fatigue

(2) twitch contraction

(3) wave summation

(4) fused (complete) tetanus

(5) concentric isotonic contraction

(6) motor unit recruitment

(7) muscle tone

(8) eccentric isotonic contraction

(9) isometric contraction

(10) stress–relaxation response

(11) recovery oxygen uptake

(12) unfused (incomplete) tetanus

11 | THE MUSCULAR SYSTEM

CHAPTER REVIEW

11.1 How Skeletal Muscles Produce Movements

1. Skeletal muscles that produce movement do so by pulling on bones.
2. The attachment to the more stationary bone is the origin; the attachment to the more movable bone is the insertion.
3. Bones serve as levers, and joints serve as fulcrums. Two different forces act on the lever: load (resistance) and effort.
4. Levers are categorized into three types—first-class, second-class, and third-class (most common)—according to the positions of the fulcrum, the effort, and the load on the lever.
5. Fascicular arrangements include parallel, fusiform, circular, triangular, and pinnate (see **Table 11.1**). Fascicular arrangement affects a muscle's power and range of motion.
6. A prime mover produces the desired action; an antagonist produces an opposite action. Synergists assist a prime mover by reducing unnecessary movement. Fixators stabilize the origin of a prime mover so that it can act more efficiently.

11.2 How Skeletal Muscles Are Named

1. Distinctive features of different skeletal muscles include direction of muscle fascicles; size, shape, action, number of origins (or heads), and location of the muscle; and sites of origin and insertion of the muscle (see **Table 11.2**).
2. Most skeletal muscles are named based on combinations of features.

11.3 Overview of the Principal Skeletal Muscles

1. In Sections 11.4 through 11.23, you will learn about the principal skeletal muscles in various regions of the body.
2. Each of these sections contains several features that will help you understand the importance of the principal skeletal muscles of the body.

11.4 Muscles of the Head That Produce Facial Expressions

1. Muscles of the head that produce facial expressions move the skin rather than a joint when they contract.
2. These muscles permit us to express a wide variety of emotions.

11.5 Muscles of the Head That Move the Eyeballs (Extrinsic Eye Muscles) and Upper Eyelids

1. The muscles of the head that move the eyeballs are among the fastest contracting and most precisely controlled skeletal muscles in the body; they permit us to elevate, depress, abduct, adduct, and medially and laterally rotate the eyeballs.
2. The muscles that move the upper eyelids open the eyes.

11.6 Muscles of the Head That Move the Mandible and Assist in Mastication and Speech

1. The muscles that move the mandible at the temporomandibular joint are known as the muscles of mastication (chewing).
2. Muscles that move the mandible not only play a role in mastication but also in speech.

11.7 Muscles of the Head That Move the Tongue and Assist in Mastication and Speech

1. The muscles of the head that move the tongue are important in mastication and speech.
2. These muscles are also involved in deglutition (swallowing).

11.8 Muscles of the Anterior Neck That Assist in Deglutition and Speech

1. Muscles of the anterior neck that assist in deglutition and speech, called suprahyoid muscles, are located above the hyoid bone.
2. The anterior neck also contains infrahyoid muscles, which along with suprahyoid muscles, help stabilize the hyoid bone.

11.9 Muscles of the Neck That Move the Head

1. Muscles of the neck that move the head alter the position of the head.
2. These muscles also help balance the head on the vertebral column.

11.10 Muscles of the Abdomen That Protect Abdominal Viscera and Move the Vertebral Column

1. Muscles of the abdomen help contain and protect the abdominal viscera and move the vertebral column.
2. These muscles also compress the abdomen and help produce the force required for defecation, urination, vomiting, and childbirth.

11.11 Muscles of the Thorax That Assist in Breathing

1. Muscles of the thorax used in breathing alter the size of the thoracic cavity so that inhalation and exhalation can occur.
2. These muscles also assist in venous return of blood to the heart.

11.12 Muscles of the Pelvic Floor That Support Pelvic Viscera and Function as Sphincters

1. Muscles of the pelvic floor support the pelvic viscera and resist the thrust that accompanies increases in intra-abdominal pressure.
2. These muscles also function as sphincters at the anorectal junction, urethra, and vagina.

11.13 Muscles of the Perineum

1. The perineum is the region of the truck inferior to the pelvic diaphragm.
2. Muscles of the perineum assist in urination, erection of the penis and clitoris, ejaculation, and defecation.

11.14 Muscles of the Thorax That Move the Pectoral Girdle

1. Muscles of the thorax that move the pectoral girdle stabilize the scapula so it can function as a stable point of origin for most of the muscles that move the humerus.
2. These muscles also move the scapula to increase the range of motion of the humerus.

11.15 Muscles of the Thorax and Shoulder That Move the Humerus

1. Muscles of the thorax that move the humerus originate for the most part on the scapula (scapular muscles).
2. The remaining muscles originate on the axial skeleton (axial muscles).

11.16 Muscles of the Arm That Move the Radius and Ulna

1. Muscles of the arm that move the radius and ulna are involved in flexion and extension at the elbow joint.
2. These muscles are organized into flexor and extensor compartments.

11.17 Muscles of the Forearm That Move the Wrist, Hand, Thumb, and Digits

1. Muscles of the forearm that move the wrist, hand, thumb, and digits are many and varied.
2. Those muscles that act on the digits are called extrinsic muscles.

11.18 Muscles of the Palm That Move the Digits—Intrinsic Muscles of the Hand

1. The muscles of the palm that move the digits (intrinsic muscles) are important in skilled.

CLINICAL CONNECTIONS

Intramuscular Injections (Refer page 288 of Textbook)

An **intramuscular** (*IM*) **injection** penetrates the skin and subcutaneous layer to enter the muscle itself. Intramuscular injections are preferred when prompt absorption is desired, when larger doses than can be given subcutaneously are indicated, or when the drug is too irritating to give subcutaneously. The common sites for intramuscular injections include the gluteus medius muscle of the buttock (see **Figure 11.3b**), lateral side of the thigh in the midportion of the vastus lateralis muscle (see **Figure 11.3a**), and the deltoid muscle of the shoulder (see **Figure 11.3b**). Muscles in these areas, especially the gluteal muscles in the buttock, are fairly thick, and absorption is promoted by their extensive blood supply. To avoid injury, intramuscular injections are given deep within the muscle, away from major nerves and blood vessels. Intramuscular injections have a faster speed of delivery than oral medications but are slower than intravenous infusions.

Benefits of Stretching (Refer page 289 of Textbook)

The overall goal of **stretching** is to achieve normal range of motion of joints and mobility of soft tissues surrounding the joints. For most individuals, the best stretching routine involves *static stretching,* that is, slow sustained stretching that holds a muscle in a lengthened position. The muscles should be stretched to the point of slight discomfort (not pain) and held for about 30 seconds. Stretching should be done after warming up to increase the range of motion most effectively.

1. *Improved physical performance.* A flexible joint has the ability to move through a greater range of motion, which improves performance.
2. *Decreased risk of injury.* Stretching decreases resistance in various soft tissues so there is less likelihood of exceeding maximum tissue extensibility during an activity (i.e., injuring the soft tissues).
3. *Reduced muscle soreness.* Stretching can reduce some of the muscle soreness that results after exercise.
4. *Improved posture.* Poor posture results from improper position of various parts of the body and the effects of gravity over a number of years. Stretching can help realign soft tissues to improve and maintain good posture.

Bell's Palsy (Refer page 294 of Textbook)

Bell's palsy, also known as *facial paralysis,* is a unilateral paralysis of the muscles of facial expression. It is due to damage or disease of the facial (VII) nerve. Possible causes include inflammation of the facial nerve due to an ear infection, ear surgery that damages the facial nerve, or infection by the herpes simplex virus. The paralysis causes the entire side of the face to droop in severe cases. The person cannot wrinkle the forehead, close the eye, or pucker the lips on the affected side. Drooling and difficulty in swallowing also occur. Eighty percent of patients recover completely within a few weeks to a few months. For others, paralysis is permanent. The symptoms of Bell's palsy mimic those of a stroke.

Strabismus (Refer page 297 of Textbook)

Strabismus (stra-BIZ-mus; *strabismos* = squinting) is a condition in which the two eyeballs are not properly aligned. This can be hereditary or it can be due to birth injuries, poor attachments of the muscles, problems with the brain's control center, or localized disease. Strabismus can be constant or intermittent. In strabismus, each eye sends an image to a different area of the brain and because the brain usually ignores the messages sent by one of the eyes, the ignored eye becomes weaker; hence "lazy eye," or *amblyopia,* develops. *External strabismus* results when a lesion in the oculomotor (III) nerve causes the eyeball to move laterally when at rest, and results in an inability to move the eyeball medially and inferiorly. A lesion in the abducens (VI) nerve results in *internal strabismus,* a condition in which the eyeball moves medially when at rest and cannot move laterally.

Treatment options for strabismus depend on the specific type of problem and include surgery, visual therapy (retraining the brain's control center), and orthoptics (eye muscle training to straighten the eyes).

Gravity and the Mandible (Refer page 299 of Textbook)

As just noted, three of the four muscles of mastication close the mandible and only the lateral pterygoid opens the mouth. The force of **gravity on the mandible** offsets this imbalance. When the masseter, temporalis, and medial pterygoid muscles relax, the mandible drops. Now you know why the mouth of many persons, particularly the elderly, is open while the person is asleep in a chair. In contrast, astronauts in zero gravity must work hard to open their mouths.

Intubation during Anesthesia (Refer page 299 of Textbook)

When general anesthesia is administered during surgery, a total relaxation of the muscles results. Once the various types of drugs for anesthesia have been given (especially the paralytic agents), the patient's airway must be protected and the lungs ventilated because the muscles involved with respiration are among those paralyzed. Paralysis of the genioglossus muscle causes the tongue to fall posteriorly, which may obstruct the airway to the lungs. To avoid this, the mandible is either manually thrust forward and held in place (known as the "sniffing position"), or a tube is inserted from the lips through the laryngopharynx (inferior portion of the throat) into the trachea (**endotracheal intubation**). People can also be intubated nasally (through the nose).

Dysphagia (Refer page 302 of Textbook)

Dysphagia (dis-FĀ-jē-a; *dys-* = abnormal; *-phagia* = to eat) is a clinical term for difficulty in swallowing. Some individuals are unable to swallow while others have difficulty swallowing liquids, foods, or saliva. Causes include nervous system disorders that weaken or damage muscles of deglutition (stroke, Parkinson's disease, cerebral palsy); infections; cancer of the head, neck, or esophagus; and injuries to the head, neck, or chest.

Inguinal Hernia and Sports Hernia (Refer page 306 of Textbook)

A **hernia** (HER-nē-a) is a protrusion of an organ through a structure that normally contains it, which creates a lump that can be seen or felt through the skin's surface. The inguinal region is a weak area in the abdominal wall. It is often the site of an **inguinal hernia**, a rupture or separation of a portion of the inguinal area of the abdominal wall resulting in the protrusion of a part of the small intestine. A hernia is much more common in males than in females because the inguinal canals in males are larger to accommodate the spermatic cord and ilioinguinal nerve. Treatment of hernias most often involves surgery. The organ that protrudes is "tucked" back into the abdominal cavity and the defect in the abdominal muscles is repaired. In addition, a mesh is often applied to reinforce the area of weakness.

A **sports hernia** is a painful strain (tear) in the soft tissues (muscles, tendons, and ligaments) in the lower abdomen or groin. Unlike an inguinal hernia, a sports hernia does not cause a visible lump. It occurs more frequently in males and is due to simultaneous contraction of the abdominal and adductor muscles that attach to the hip bone and pull in different directions. This occurs during activities that involve rapid acceleration and changes in direction, kicking, and side-to-side motions such as those that occur in ice hockey, soccer, football, rugby, tennis, and high jumping. Treatment of sports hernia includes rest, ice, anti-inflammatory medications, physical therapy, and surgery.

Injury of Levator Ani and Urinary Stress Incontinence
(Refer page 310 of Textbook)

During childbirth, the levator ani muscle supports the head of the fetus, and the muscle may be injured during a difficult childbirth or traumatized during an *episiotomy* (a cut made with surgical scissors to prevent or direct tearing of the perineum during the birth of a baby). The consequence of such injuries may be **urinary stress incontinence**, that is, the leakage of urine whenever intra-abdominal pressure is increased—for example, during coughing. One way to treat urinary stress incontinence is to strengthen and tighten the muscles that support the pelvic viscera. This is accomplished by *Kegel exercises*, the alternate contraction and relaxation of muscles of the pelvic floor. To find the correct muscles, the person imagines that she is urinating and then contracts the muscles as if stopping in midstream. The muscles should be held for a count of three, then relaxed for a count of three. This should be done 5–10 times each hour—sitting, standing, and lying down. Kegel exercises are also encouraged during pregnancy to strengthen the muscles for delivery.

Impingement Syndrome (Refer page 316 of Textbook)

One of the most common causes of shoulder pain and dysfunction in athletes is known as **impingement syndrome**, which is sometimes confused with another common complaint, compartment syndrome, discussed in Disorders: Homeostatic Imbalances at the end of this chapter. The repetitive movement of the arm over the head that is common in baseball, overhead racquet sports, lifting weights over the head, spiking a volleyball, and swimming puts these athletes at risk. Impingement syndrome may also be caused by a direct blow or stretch injury. Continual pinching of the supraspinatus tendon as a result of overhead motions causes it to become inflamed and results in pain. If movement is continued despite the pain, the tendon may degenerate near the attachment to the humerus and ultimately may tear away from the bone (rotator cuff injury). Treatment consists of resting the injured tendons, strengthening the shoulder through exercise, massage therapy, and surgery if the injury is particularly severe. During surgery, an inflamed bursa may be removed, bone may be trimmed, and/or the coracoacromial ligament may be detached. Torn rotator cuff tendons may be trimmed and then reattached with sutures, anchors, or surgical tacks. These steps make more space, thus relieving pressure and allowing the arm to move freely.

Rotator Cuff Injury (Refer page 318 of Textbook)

Rotator cuff injury is a strain or tear in the rotator cuff muscles and is common among baseball pitchers, volleyball players, racquet sports players, and swimmers due to shoulder movements that involve vigorous circumduction. It also occurs as a result of wear and tear, aging, trauma, poor posture, improper lifting, and repetitive motions in certain jobs, such as placing items on a shelf above your head. Most often, there is tearing of the supraspinatus muscle tendon or the rotator cuff. This tendon is especially predisposed to wear and tear because of its location between the head of the humerus and acromion of the scapula, which compresses the tendon during shoulder movements. Poor posture and poor body mechanics also increase compression of the supraspinatus muscle tendon.

Golfer's Elbow (Refer page 323 of Textbook)

Golfer's elbow is a condition that can be caused by strain of the flexor muscles, especially the flexor carpi radialis, as a result of repetitive movements such as swinging a golf club. Strain can, however, be caused by many actions. Pianists, violinists, movers, weight lifters, bikers, and those who use computers are among those who may develop pain near the medial epicondyle (*medial epicondylitis*).

Carpal Tunnel Syndrome (Refer page 330 of Textbook)

Structures within the carpal tunnel (see **Figure 11.17f**), especially the median nerve, are vulnerable to compression, and the resulting condition is called **carpal tunnel syndrome**. Compression of the median nerve leads to sensory changes over the lateral side of the hand and muscle weakness in the thenar eminence. This results in pain, numbness, and tingling of the fingers. The condition may be caused by inflammation of the digital tendon sheaths, fluid retention, excessive exercise, infection, trauma, and/or repetitive activities that involve flexion of the wrist, such as keyboarding, cutting hair, or playing the piano. Treatment may involve the use of nonsteroidal anti-inflammatory drugs (such as ibuprofen or aspirin), wearing a wrist splint, corticosteroid injections, or surgery to cut the flexor retinaculum and release pressure on the median nerve.

Back Injuries and Heavy Lifting (Refer page 334 of Textbook)

The four factors associated with increased risk of **back injury** are amount of force, repetition, posture, and stress applied to the backbone. Poor physical condition, poor posture, lack of exercise, and excessive body weight contribute to the number and severity of sprains and strains. Back pain caused by a muscle strain or ligament sprain will normally heal within a short time and may never cause further problems. However, if ligaments and muscles are weak, discs in the lower back can become weakened and may herniate (rupture) with excessive lifting or a sudden fall, causing considerable pain.

Full flexion at the waist, as in touching your toes, overstretches the erector spinae muscles. Muscles that are overstretched cannot contract effectively. Straightening up from such a position is therefore initiated by the hamstring muscles on the back of the thigh and the gluteus maximus muscles of the buttocks. The erector spinae muscles join in as the degree of flexion decreases. Improperly lifting a heavy weight, however, can strain the erector spinae muscles. The result can be painful muscle spasms, tearing of tendons and ligaments of the lower back, and herniating of intervertebral discs. The lumbar muscles are adapted for maintaining posture, not for lifting. This is why it is important to bend at the knees and use the powerful extensor muscles of the thighs and buttocks while lifting a heavy load.

Groin Pull (Refer page 336 of Textbook)

The five major muscles of the inner thigh function to move the legs medially. This muscle group is important in activities such as sprinting, hurdling, and horseback riding. A rupture or tear of one or more of these muscles can cause a **groin pull**. Groin pulls most often occur during sprinting or twisting, or from kicking a solid, perhaps stationary object. Symptoms of a groin pull may be sudden or may not surface until the day after the injury; they include sharp pain in the inguinal region, swelling, bruising, or inability to contract the muscles. As with most strain injuries, treatment involves PRICE (*P*rotection, *R*est, *I*ce, *C*ompression, and *E*levation). After the injured part is protected from further damage, ice should be applied immediately, and the injured part should be elevated and rested. An elastic bandage should be applied, if possible, to compress the injured tissue.

Pulled Hamstrings (Refer page 341 of Textbook)

A strain or partial tear of the proximal hamstring muscles is referred to as **pulled hamstrings** or *hamstring strains*. Like pulled groins (see Section 11.20), they are common sports injuries in individuals who run very hard and/or are required to perform quick starts and stops. Sometimes the violent muscular exertion required to perform a feat tears away a part of the tendinous origins of the hamstrings, especially the biceps femoris, from the ischial tuberosity. This is usually accompanied by a contusion (bruising), tearing of some of the muscle fibers, and rupture of blood vessels, producing a hematoma (collection of blood) and sharp pain. Adequate training with good balance between the quadriceps femoris and hamstrings and stretching exercises before running or competing are important in preventing this injury.

Shin Splint Syndrome (Refer page 343 of Textbook)

Shin splint syndrome, or simply *shin splints*, refers to pain or soreness along the tibia, specifically the medial, distal two-thirds. It may be caused by tendinitis of the anterior compartment muscles, especially the tibialis anterior muscle, inflammation of the periosteum (periostitis) around the tibia, or stress fractures of the tibia. The tendinitis usually occurs when poorly conditioned runners run on hard or banked surfaces with poorly supportive running shoes. The condition may also occur with vigorous activity of the legs following a period of relative inactivity or running in cold weather without proper warmup. The muscles in the anterior compartment (mainly the tibialis anterior) can be strengthened to balance the stronger posterior compartment muscles.

DISORDERS: HOMEOSTATIC IMBALANCES

Running-Related Injuries

Many individuals who jog or run sustain some type of **running-related injury**. Although such injuries may be minor, some can be quite serious. Untreated or inappropriately treated minor injuries may become chronic. Among runners, common sites of injury include the ankle, knee, calcaneal (Achilles) tendon, hip, groin, foot, and back. Of these, the knee often is the most severely injured area.

Running injuries are frequently related to faulty training techniques. This may involve improper or lack of sufficient warm-up routines, running too much, or running too soon after an injury. Or it might involve extended running on hard and/or uneven surfaces. Poorly constructed or worn-out running shoes can also contribute to injury, as can any biomechanical problem (such as a fallen arch) aggravated by running.

Most sports injuries should be treated initially with PRICE (*P*rotection, *R*est, *I*ce, *C*ompression, and *E*levation). Immediately protect the injured part, rest, and apply ice immediately after the injury, and elevate the injured part. Then apply an elastic bandage, if possible, to compress the injured tissue. Continue using PRICE for 2 to 3 days, and resist the temptation to apply heat, which may worsen the swelling. Follow-up treatment may include alternating moist heat and ice massage to enhance blood flow in the injured area. Sometimes it is helpful to take nonsteroidal anti-inflammatory drugs (NSAIDs) or to have local injections of corticosteroids. During the recovery period, it is important to keep active, using an alternative fitness program that does not worsen the original injury. This activity should be determined in consultation with a physician. Finally, careful exercise is needed to rehabilitate the injured area itself. Massage therapy may also be used to prevent or treat many sports injuries.

Compartment Syndrome

As noted earlier in this chapter, skeletal muscles in the limbs are organized into functional units called *compartments*. In a disorder called **compartment syndrome**, some external or internal pressure constricts the structures within a compartment, resulting in damaged blood vessels and subsequent reduction of the blood supply (ischemia) to the structures within the compartment. Symptoms include pain, burning, pressure, pale skin, and paralysis. Common causes of compartment syndrome include crushing and penetrating injuries, contusion (damage to subcutaneous tissues without the skin being broken), muscle strain (overstretching of a muscle), or an improperly fitted cast. The pressure increase in the compartment can have serious consequences, such as hemorrhage, tissue injury, and edema (buildup of interstitial fluid). Because deep fasciae (connective tissue coverings) that enclose the compartments are very strong, accumulated blood and interstitial fluid cannot escape, and the increased pressure can literally choke off the blood flow and deprive nearby muscles and nerves of oxygen. One treatment option is **fasciotomy** (fash-ē-OT-ō-mē), a surgical procedure in which muscle fascia is cut to relieve the pressure. Without intervention, nerves can suffer damage, and muscles can develop scar tissue that results in permanent shortening of the muscles, a condition called *contracture.* If left untreated, tissues may die and the limb may no longer be able to function. Once the syndrome has reached this stage, amputation may be the only treatment option.

Plantar Fasciitis

Plantar fasciitis (fas-ē-ī-tis) or *painful heel syndrome* is an inflammatory reaction due to chronic irritation of the plantar aponeurosis (fascia) at its origin on the calcaneus (heel bone). The aponeurosis becomes less elastic with age. This condition is also related to weight-bearing activities (walking, jogging, lifting heavy objects), improperly constructed or fitting shoes, excess weight (which puts pressure on the feet), and poor biomechanics (flat feet, high arches, and abnormalities in gait that may cause uneven distribution of weight on the feet). Plantar fasciitis is the most common cause of heel pain in runners and arises in response to the repeated impact of running. Treatments include ice, deep heat, stretching exercises, weight loss, prosthetics (such as shoe inserts or heel lifts), steroid injections, and surgery.

MEDICAL TERMINOLOGY

Charley horse (CHAR-lē HŌRS) A popular name for a cramp or stiffness of muscles due to tearing of the muscle, followed by bleeding into the area. It is a common sports injury due to trauma or excessive activity and frequently occurs in the quadriceps femoris muscle, especially among football players.

Muscle strain Tearing of fibers in a skeletal muscle or its tendon that attaches the muscle to bone. The tearing can also damage small blood vessels, causing local bleeding (bruising) and pain (caused by irritation of nerve endings in the region). Muscle strains usually occur when a muscle is stretched beyond its limit, for example, in response to sudden, quick heavy lifting; during sports activities; or while performing work tasks. Also called **muscle pull** or **muscle tear**.

Paralysis (pa-RAL-i-sis; *para-* = departure from normal; *-lysis* = loosening) Loss of muscle function (voluntary movement) through injury, disease, or damage to its nerve supply. Most paralysis is due to stroke or spinal cord injury.

Repetitive strain injuries (RSIs) Conditions resulting from overuse of equipment, poor posture, poor body mechanics, or activity that requires repeated movements; for example, various conditions of assembly line workers. Examples of overuse of equipment include overuse of a computer, hammer, guitar, or piano. Also called repetitive motion injuries.

Rhabdomyosarcoma (rab′-dō-mī′-ō-sar-KŌ-ma; *rhab-* = rod-shaped; *-myo-* = muscle; *-sarc-* = flesh; *-oma* = tumor) A tumor of skeletal muscle. Usually occurs in children and is highly malignant, with rapid metastasis.

Torticollis (tor-ti-KOL-is; *tortus-* = twisted; *-column* = neck) A contraction or shortening of the sternocleidomastoid muscle that causes the head to tilt toward the affected side and the chin to rotate toward the opposite side. It may be acquired or congenital. Also called **wryneck**.

Tic Spasmodic twitching made involuntarily by muscles that are usually under conscious control, for example, twitching of an eyelid.

SELF-QUIZ QUESTIONS

Fill in the blanks in the following statements.

1. The muscle that forms the major portion of the cheek is the _____.

2. The three superficial posterior plantar flexors of the leg are the _____, _____, and _____. All three of these muscles insert on the _____ by way of the Achilles tendon.

Indicate whether the following statements are true or false.

3. Longer fibers in a muscle result in a greater range of motion.

4. When flexing the forearm, the biceps brachii acts as the prime mover and the triceps brachii acts as the antagonist.

Choose the one best answer to the following questions.

5. Which of the following muscles does *not* flex the thigh?
   (a) rectus femoris
   (b) gracilis
   (c) sartorius
   (d) iliacus
   (e) tensor fascia latae

6. The iliotibial tract is composed of the tendon of the gluteus maximus muscle, the deep fascia that encircles the thigh, and the tendon of which of the following muscles?
   (a) iliacus
   (b) gluteus minimus
   (c) tensor fascia latae
   (d) adductor longus
   (e) vastus lateralis

7. In order for movement to occur, (1) muscles generally need to cross a joint, (2) contraction of the muscle will pull on the origin, (3) muscles that move a body part cannot cover the moving part, (4) muscles need to exert force on tendons that pull on bones, (5) the insertion must act to stabilize the joint.
   (a) 1, 2, 3, 4, and 5
   (b) 1, 2, 3, and 4
   (c) 1, 2, and 4
   (d) 1, 3, and 4
   (e) 3 and 4

8. Because you did not do well on your recent anatomy and physiology exam, you leave the classroom pouting. Which one of these muscles are you using?

   (a) mentalis
   (b) orbicularis oris
   (c) risorius
   (d) levator labii superioris
   (e) zygomaticus minor

9. The rectus femoris has fascicles arranged on both sides of a centrally positioned tendon. This pattern of fascicle arrangement is
   (a) unipennate.
   (b) fusiform.
   (c) multipennate.
   (d) parallel.
   (e) bipennate.

10. Which of the following muscle names and their naming descriptors are mismatched?
    (a) adductor brevis: short muscle that moves a bone closer to the midline
    (b) rectus abdominis: muscle with fibers parallel to the midline of the abdomen
    (c) levator scapula: muscle that raises the scapula
    (d) sternohyoid: muscle attached to the sternum and hyoid
    (e) serratus anterior: comblike muscle located on the body's anterior surface

11. Match the following:
    _____ (a) muscle that stabilizes the origin of the prime mover
    _____ (b) site of muscle attachment to a stationary bone
    _____ (c) muscle that stretches to allow desired motion
    _____ (d) muscle that contracts to stabilize intermediate joints
    _____ (e) site of muscle attachment to a movable bone
    _____ (f) group of muscles, along with their blood and nerves, that have a common function
    _____ (g) contracting muscle that produces the desired motion
    _____ (h) fleshy part of the muscle

    (1) compartment
    (2) origin
    (3) insertion
    (4) belly
    (5) synergist
    (6) fixator
    (7) prime mover (agonist)
    (8) antagonist

12. Match the following:

_____ (a) compression of median nerve resulting in pain and numbness and tingling in the fingers

_____ (b) tendinitis of the anterior compartment muscles of the leg; inflammation of the tibial periosteum

_____ (c) improperly aligned eyeballs due to lesions in either the oculomotor or abducens nerves

_____ (d) stretching or tearing of distal attachments of adductor muscles

_____ (e) rupture of a portion of the inguinal area of the abdominal wall resulting in protrusion of part of the small intestine

_____ (f) caused by repetitive movement of the arm over the head that results in inflammation of the supraspinatus tendon

_____ (g) inflammation due to chronic irritation of the plantar aponeurosis at its origin on the calcaneus; most common cause of heel pain in runners

_____ (h) painful inflammation of tendons, tendon sheaths, and synovial membranes of joints

_____ (i) paralysis of facial muscles as a result of damage to the facial nerve

_____ (j) common in individuals who perform quick starts and stops; tearing away of part of the tendinous origins from the ischial tuberosity

_____ (k) permanent shortening of a muscle due to nerve damage and scar tissue development

_____ (l) may occur as a result of injury to levator ani muscle

_____ (m) external or internal pressure constricts structures in a compartment, causing a reduction of blood supply to the structures

(1) tenosynovitis
(2) Bell's palsy
(3) inguinal hernia
(4) urinary stress incontinence
(5) compartment syndrome
(6) groin pull
(7) pulled hamstrings
(8) strabismus
(9) shin splints
(10) plantar fasciitis
(11) impingement syndrome
(12) contracture
(13) carpal tunnel syndrome

13. Match the following:

_____ (a) rectus femoris, vastus lateralis, vastus medialis, vastus intermedius

_____ (b) biceps femoris, semitendinosus, semimembranosus

_____ (c) erector spinae; includes iliocostalis, longissimus, and spinalis groups

_____ (d) thenar, hypothenar, intermediate

_____ (e) biceps brachii, brachialis, coracobrachialis

_____ (f) latissimus dorsi

_____ (g) subscapularis, supraspinatus, infraspinatus, teres minor

_____ (h) diaphragm, external intercostals, internal intercostals

_____ (i) trapezius, levator scapulae, rhomboid major, rhomboid minor

(1) breathing muscles
(2) constitute flexor compartment of the arm
(3) hamstrings
(4) intrinsic muscle groups of the hand
(5) muscles that strengthen and stabilize the shoulder joint; the rotator cuff
(6) quadriceps femoris muscle
(7) largest muscle mass of the back
(8) posterior thoracic muscles
(9) swimmer's muscle

14. Match the following (some answers may be used more than once):

_____ (a) trapezius
_____ (b) orbicularis oculi
_____ (c) levator ani
_____ (d) rectus abdominis
_____ (e) triceps brachii
_____ (f) gastrocnemius
_____ (g) temporalis
_____ (h) external anal sphincter
_____ (i) external oblique
_____ (j) iliocostalis thoracis
_____ (k) digastric
_____ (l) styloglossus
_____ (m) masseter
_____ (n) adductor longus
_____ (o) zygomaticus major
_____ (p) latissimus dorsi
_____ (q) flexor carpi radialis
_____ (r) pronator teres
_____ (s) sternocleidomastoid
_____ (t) quadriceps femoris
_____ (u) deltoid
_____ (v) tibialis anterior
_____ (w) sartorius
_____ (x) gluteus maximus
_____ (y) superior rectus
_____ (z) trapezius

(1) muscle of facial expression
(2) muscle of mastication
(3) muscle that moves the eyeballs
(4) extrinsic muscle that moves the tongue
(5) suprahyoid muscle
(6) muscle of the perineum
(7) muscle that moves the head
(8) abdominal wall muscle
(9) pelvic floor muscle
(10) pectoral girdle muscle
(11) muscle that moves the humerus
(12) muscle that moves the radius and ulna
(13) muscle that moves the wrist, hand, and digits
(14) muscle that moves the vertebral column
(15) muscle that moves the femur
(16) muscle that acts on the femur, tibia, and fibula
(17) muscle that moves the foot and toes

15. Match the following (some answers may be used more than once):

_____ (a) most common lever in the body
_____ (b) lever formed by the head resting on the vertebral column
_____ (c) always produces a mechanical advantage
_____ (d) EFL
_____ (e) FLE
_____ (f) FEL
_____ (g) adduction of the thigh

(1) first-class lever
(2) second-class lever
(3) third-class lever

12 NERVOUS TISSUE

12.1 Overview of the Nervous system

1. The central nervous system (CNS) consists of the brain and spinal cord.
2. The peripheral nervous system (PNS) consists of all nervous tissue outside the CNS. Components of the PNS include nerves and sensory receptors.
3. The PNS is divided into a sensory (afferent) division and a motor (efferent) division.
4. The sensory division conveys sensory input into the CNS from sensory receptors.
5. The motor division conveys motor output from the CNS to effectors (muscles and glands).
6. The efferent division of the PNS is further subdivided into a somatic nervous system (conveys motor output from the CNS to skeletal muscles only) and an autonomic nervous system (conveys motor output from the CNS to smooth muscle, cardiac muscle, and glands). The autonomic nervous system in turn is divided into a sympathetic nervous system, parasympathetic nervous system, and an enteric nervous system.
7. The nervous system helps maintain homeostasis and integrates all body activities by sensing changes (sensory function), interpreting them (integrative function), and reacting to them (motor function).

12.2 Histology of Nervous Tissue

1. Nervous tissue consists of neurons (nerve cells) and neuroglia. Neurons have the property of electrical excitability and are responsible for most unique functions of the nervous system: sensing, thinking, remembering, controlling muscle activity, and regulating glandular secretions.
2. Most neurons have three parts. The dendrites are the main receiving or input region. Integration occurs in the cell body, which includes typical cellular organelles. The output part typically is a single axon, which propagates nerve impulses toward another neuron, a muscle fiber, or a gland cell.
3. Synapses are the site of functional contact between two excitable cells. Axon terminals contain synaptic vesicles filled with neurotransmitter molecules.
4. Slow axonal transport and fast axonal transport are systems for conveying materials to and from the cell body and axon terminals.
5. On the basis of their structure, neurons are classified as multipolar, bipolar, or unipolar.

6. Neurons are functionally classified as sensory (afferent) neurons, motor (efferent) neurons, and interneurons. Sensory neurons carry sensory information into the CNS. Motor neurons carry information out of the CNS to effectors (muscles and glands). Interneurons are located within the CNS between sensory and motor neurons.
7. Neuroglia support, nurture, and protect neurons and maintain the interstitial fluid that bathes them. Neuroglia in the CNS include astrocytes, oligodendrocytes, microglial cells, and ependymal cells. Neuroglia in the PNS include Schwann cells and satellite cells.
8. Two types of neuroglia produce myelin sheaths: Oligodendrocytes myelinate axons in the CNS, and Schwann cells myelinate axons in the PNS.
9. White matter consists of aggregates of myelinated axons; gray matter contains cell bodies, dendrites, and axon terminals of neurons, unmyelinated axons, and neuroglia.
10. In the spinal cord, gray matter forms an H-shaped inner core that is surrounded by white matter. In the brain, a thin, superficial shell of gray matter covers the cerebral and cerebellar hemispheres.

12.3 Electrical Signals in Neurons: An Overview

1. Neurons communicate with one another using graded potentials, which are used for short-distance communication only, and action potentials, which allow communication over long distances within the body.
2. The electrical signals produced by neurons and muscle fibers rely on four kinds of ion channels: leak channels, ligand-gated channels, mechanically-gated channels, and voltage-gated channels. **Table 12.1** summarizes the different types of ion channels in neurons.

12.4 Resting Membrane Potential

1. A resting membrane potential exists across the plasma membrane of excitable cells that are unstimulated (at rest). The resting membrane potential exists because of a small buildup of negative ions in the cytosol along the inside surface of the membrane, and an equal buildup of positive ions in the extracellular fluid along the outside surface of the membrane.
2. A typical value for the resting membrane potential of a neuron is −70 mV. A cell that exhibits a membrane potential is polarized.
3. The resting membrane potential is determined by three major factors: (1) unequal distribution of ions in the ECF and cytosol; (2) inability of most cytosolic anions to leave the cell; and (3) electrogenic nature of the Na^+/K^+ ATPases.

12.5 Graded Potentials

1. A graded potential is a small deviation from the resting membrane potential that occurs because ligand-gated or mechanically-gated channels open or close.
2. A hyperpolarizing graded potential makes the membrane potential more negative (more polarized).
3. A depolarizing graded potential makes the membrane potential less negative (less polarized).
4. The amplitude of a graded potential varies, depending on the strength of the stimulus.

12.6 Action Potentials

1. According to the all-or-none principle, if a stimulus is strong enough to generate an action potential, the impulse generated is of a constant size. A stronger stimulus does not generate a larger action potential. Instead, the greater the stimulus strength above threshold, the greater the frequency of the action potentials.
2. During an action potential, voltage-gated Na^+ and K^+ channels open and close in sequence. This results first in depolarization, the reversal of membrane polarization (from -70 mV to $+30$ mV). Then repolarization, the recovery of the resting membrane potential (from $+30$ mV to -70 mV), occurs.
3. During the first part of the refractory period (RP), another impulse cannot be generated at all (absolute RP); a little later, it can be triggered only by a larger-than-normal stimulus (relative RP).
4. Because an action potential travels from point to point along the membrane without getting smaller, it is useful for long-distance communication.
5. Nerve impulse propagation in which the impulse "leaps" from one node of Ranvier to the next along a myelinated axon is saltatory conduction. Saltatory conduction is faster than continuous conduction.
6. Axons with larger diameters conduct impulses at higher speeds than do axons with smaller diameters.
7. The intensity of a stimulus is encoded in the frequency of action potentials and in the number of sensory neurons that are recruited.
8. **Table 12.2** compares graded potentials and action potentials.

12.7 Signal Transmission at Synapses

1. A synapse is the functional junction between one neuron and another, or between a neuron and an effector such as a muscle or a gland. The two types of synapses are electrical and chemical.
2. A chemical synapse produces one-way information transfer—from a presynaptic neuron to a postsynaptic neuron.
3. An excitatory neurotransmitter is one that can depolarize the postsynaptic neuron's membrane, bringing the membrane potential closer to threshold. An inhibitory neurotransmitter hyperpolarizes the membrane of the postsynaptic neuron, moving it further from threshold.
4. There are two major types of neurotransmitter receptors: ionotropic receptors and metabotropic receptors. An ionotropic receptor contains a neurotransmitter binding site and an ion channel. A metabotropic receptor contains a neurotransmitter binding site and is coupled to a separate ion channel by a G protein.
5. Neurotransmitter is removed from the synaptic cleft in three ways: diffusion, enzymatic degradation, and uptake by cells (neurons and neuroglia).
6. If several presynaptic end bulbs release their neurotransmitter at about the same time, the combined effect may generate a nerve impulse, due to summation. Summation may be spatial or temporal.
7. The postsynaptic neuron is an integrator. It receives excitatory and inhibitory signals, integrates them, and then responds accordingly.
8. **Table 12.3** summarizes the structural and functional elements of a neuron.

12.8 Neurotransmitters

1. Both excitatory and inhibitory neurotransmitters are present in the CNS and the PNS. A given neurotransmitter may be excitatory in some locations and inhibitory in others.
2. Neurotransmitters can be divided into two classes based on size: (1) small-molecule neurotransmitters (acetylcholine, amino acids, biogenic amines, ATP and other purines, nitric oxide, and carbon monoxide), and (2) neuropeptides, which are composed of 3 to 40 amino acids.
3. Chemical synaptic transmission may be modified by affecting synthesis, release, or removal of a neurotransmitter or by blocking or stimulating neurotransmitter receptors.
4. **Table 12.4** describes several important neuropeptides.

12.9 Neural Circuits

1. Neurons in the central nervous system are organized into networks called neural circuits.
2. Neural circuits include simple series, diverging, converging, reverberating, and parallel after-discharge circuits.

12.10 Regeneration and Repair of Nervous Tissue

1. The nervous system exhibits plasticity (the capability to change based on experience), but it has very limited powers of regeneration (the capability to replicate or repair damaged neurons).
2. Neurogenesis, the birth of new neurons from undifferentiated stem cells, is normally very limited. Repair of damaged axons does not occur in most regions of the CNS.
3. Axons and dendrites that are associated with a neurolemma in the PNS may undergo repair if the cell body is intact, the Schwann cells are functional, and scar tissue formation does not occur too rapidly.

CLINICAL CONNECTIONS

Neurotoxins and Local Anesthetics (Refer page 375 of Textbook)

Certain shellfish and other organisms contain **neurotoxins** (noo'-rō-TOK-sins), substances that produce their poisonous effects by acting on the nervous system. One particularly lethal neurotoxin is tetrodotoxin (TTX), present in the viscera of Japanese puffer fish. TTX effectively blocks action potentials by inserting itself into voltage-gated Na^+ channels so they cannot open.

Local anesthetics are drugs that block pain and other somatic sensations. Examples include procaine (Novocaine®) and lidocaine, which may be used to produce anesthesia in the skin during suturing of a gash, in the mouth during dental work, or in the lower body during childbirth. Like TTX, these drugs act by blocking the opening of voltage-gated Na^+ channels. Action potentials cannot propagate past the obstructed region, so pain signals do not reach the CNS.

Localized cooling of a nerve can also produce an anesthetic effect because axons propagate action potentials at lower speeds when cooled. The application of ice to injured tissue can reduce pain because propagation of the pain sensations along axons is partially blocked.

Strychnine Poisoning (Refer page 383 of Textbook)

The importance of inhibitory neurons can be appreciated by observing what happens when their activity is blocked. Normally, inhibitory neurons in the spinal cord called *Renshaw cells* release the neurotransmitter glycine at inhibitory synapses with somatic motor neurons. This inhibitory input to their motor neurons prevents excessive contraction of skeletal muscles. **Strychnine** (STRIK-nīn) is a lethal poison that is mainly used as a pesticide to control rats, moles, gophers, and coyotes. When ingested, it binds to and blocks glycine receptors. The normal, delicate balance between excitation and inhibition in the CNS is disturbed, and motor neurons generate nerve impulses without restraint. All skeletal muscles, including the diaphragm, contract fully and remain contracted. Because the diaphragm cannot relax, the victim cannot inhale, and suffocation results.

Modifying the Effects of Neurotransmitters (Refer page 386 of Textbook)

Substances naturally present in the body as well as drugs and toxins can modify the effects of neurotransmitters in several ways:

1. Neurotransmitter synthesis can be stimulated or inhibited. For instance, many patients with Parkinson's disease (see Disorders: Homeostatic Imbalances in Chapter 16) receive benefit from the drug L-dopa because it is a precursor of dopamine. For a limited period of time, taking L-dopa boosts dopamine production in affected brain areas.

2. Neurotransmitter release can be enhanced or blocked. Amphetamines promote release of dopamine and norepinephrine. Botulinum toxin causes paralysis by blocking release of acetylcholine from somatic motor neurons.

3. The neurotransmitter receptors can be activated or blocked. An agent that binds to receptors and enhances or mimics the effect of a natural neurotransmitter is an **agonist** (AG-ōn-ist). Isoproterenol (Isuprel®) is a powerful agonist of epinephrine and norepinephrine. It can be used to dilate the airways during an asthma attack. An agent that binds to and blocks neurotransmitter receptors is an **antagonist** (an-TAG-ō-nist). Zyprexa®, a drug prescribed for schizophrenia, is an antagonist of serotonin and dopamine.

4. Neurotransmitter removal can be stimulated or inhibited. For example, cocaine produces euphoria—intensely pleasurable feelings—by blocking transporters for dopamine reuptake. This action allows dopamine to linger longer in synaptic clefts, producing excessive stimulation of certain brain regions.

DISORDERS: HOMEOSTATIC IMBALANCES

Multiple Sclerosis

Multiple sclerosis (MS) is a disease that causes a progressive destruction of myelin sheaths surrounding neurons in the CNS. It afflicts about 350,000 people in the United States and 2 million people worldwide. It usually appears between the ages of 20 and 40, affecting females twice as often as males. MS is most common in whites, less common in blacks, and rare in Asians. MS is an autoimmune disease—the body's own immune system spearheads the attack. The condition's name describes the anatomical pathology: In multiple regions the myelin sheaths deteriorate to scleroses, which are hardened scars or plaques. Magnetic resonance imaging (MRI) studies reveal numerous plaques in the white matter of the brain and spinal cord. The destruction of myelin sheaths slows and then short-circuits propagation of nerve impulses.

The most common form of the condition is relapsing–remitting MS, which usually appears in early adulthood. The first symptoms may include a feeling of heaviness or weakness in the muscles, abnormal sensations, or double vision. An attack is followed by a period of remission during which the symptoms temporarily disappear. One attack follows another over the years, usually every year or two. The result is a progressive loss of function interspersed with remission periods, during which symptoms abate.

Although the cause of MS is unclear, both genetic susceptibility and exposure to some environmental factor (perhaps a herpes virus) appear to contribute. Since 1993, many patients with relapsing–remitting MS have been treated with injections of beta-interferon. This treatment lengthens the time between relapses, decreases the severity of relapses, and slows formation of new lesions in some cases. Unfortunately, not all MS patients can tolerate beta-interferon, and therapy becomes less effective as the disease progresses.

Epilepsy

Epilepsy (ep′-i-LEP-sē) is characterized by short, recurrent attacks of motor, sensory, or psychological malfunction, although it almost never affects intelligence. The attacks, called *epileptic seizures*, afflict about 1% of the world's

population. They are initiated by abnormal, synchronous electrical discharges from millions of neurons in the brain, perhaps resulting from abnormal reverberating circuits. The discharges stimulate many of the neurons to send nerve impulses over their conduction pathways. As a result, lights, noise, or smells may be sensed when the eyes, ears, and nose have not been stimulated. Moreover, the skeletal muscles of a person having a seizure may contract involuntarily. *Partial seizures* begin in a small area on one side of the brain and produce milder symptoms; *generalized seizures* involve larger areas on both sides of the brain and loss of consciousness.

Epilepsy has many causes, including brain damage at birth (the most common cause); metabolic disturbances (hypoglycemia, hypocalcemia, uremia, hypoxia); infections (encephalitis or meningitis); toxins (alcohol, tranquilizers, hallucinogens); vascular disturbances (hemorrhage, hypotension); head injuries; and tumors and abscesses of the brain. Seizures associated with fever are most common in children under the age of two. However, most epileptic seizures have no demonstrable cause.

Epileptic seizures often can be eliminated or alleviated by antiepileptic drugs, such as phenytoin, carbamazepine, and valproate sodium. An implantable device that stimulates the vagus (X) nerve has produced dramatic results in reducing seizures in some patients whose epilepsy was not well controlled by drugs. In very severe cases, surgical intervention may be an option.

Excitotoxicity

A high level of glutamate in the interstitial fluid of the CNS causes **excitotoxicity** (ek-sī′-tō-tok-SIS-i-tē)—destruction of neurons through prolonged activation of excitatory synaptic transmission. The most common cause of excitotoxicity is oxygen deprivation of the brain due to ischemia (inadequate blood flow), as happens during a stroke. Lack of oxygen causes the glutamate transporters to fail, and glutamate accumulates in the interstitial spaces between neurons and neuroglia, literally stimulating the neurons to death. Clinical trials are underway to see whether antiglutamate drugs administered after a stroke can offer some protection from excitotoxicity.

Depression

Depression is a disorder that affects over 18 million people each year in the United States. People who are depressed feel sad and helpless, have a lack of interest in activities that they once enjoyed, and experience suicidal thoughts. There are several types of depression. A person with **major depression** experiences symptoms of depression that last for more than two weeks. A person with **dysthymia** (dis-THĬ-mē-a) experiences episodes of depression that alternate with periods of feeling normal. A person with **bipolar disorder**, or *manic-depressive illness*, experiences recurrent episodes of depression and extreme elation (mania). A person with **seasonal affective disorder (SAD)** experiences depression during the winter months, when day length is short

(see Clinical Connection: Seasonal Affective Disorder and Jet Lag in Chapter 18). Although the exact cause is unknown, research suggests that depression is linked to an imbalance of the neurotransmitters serotonin, norepinephrine, and dopamine in the brain. Factors that may contribute to depression include heredity, stress, chronic illnesses, certain personality traits (such as low self-esteem), and hormonal changes. Medication is the most common treatment for depression. For example, **selective serotonin reuptake inhibitors (SSRIs)** are drugs that provide relief from some forms of depression. By inhibiting reuptake of serotonin by serotonin transporters, SSRIs prolong the activity of this neurotransmitter at synapses in the brain. SSRIs include fluoxetine (Prozac®), paroxetine (Paxil®), and sertraline (Zoloft®).

MEDICAL TERMINOLOGY

Guillain-Barré syndrome (GBS) (GHE-an ba-RĀ) An acute demyelinating disorder in which macrophages strip myelin from axons in the PNS. It is the most common cause of acute paralysis in North America and Europe and may result from the immune system's response to a bacterial infection. Most patients recover completely or partially, but about 15% remain paralyzed.

Neuroblastoma (noor-ō-blas-TŌ-ma) A malignant tumor that consists of immature nerve cells (neuroblasts); occurs most commonly in the abdomen and most frequently in the adrenal glands. Although rare, it is the most common tumor in infants.

Neuropathy (noo-ROP-a-thē; *neuro-* = a nerve; *-pathy* = disease) Any disorder that affects the nervous system but particularly a disorder of a cranial or spinal nerve. An example is facial *neuropathy* (Bell's palsy), a disorder of the facial (VII) nerve.

Rabies (RĀ-bēz; *rabi-* = mad, raving) A fatal disease caused by a virus that reaches the CNS via fast axonal transport. It is usually transmitted by the bite of an infected dog or other meat-eating animal. The symptoms are excitement, aggressiveness, and madness, followed by paralysis and death.

SELF-QUIZ QUESTIONS

Fill in the blanks in the following statements.

1. The subdivisions of the PNS are the _____, _____, and _____.

2. The two divisions of the autonomic nervous system are the _____ division and the _____ division.

Indicate whether the following statements are true or false.

3. At a chemical synapse between two neurons, the neuron receiving the signal is called the presynaptic neuron, and the neuron sending the signal is called the postsynaptic neuron.

4. Neurons in the PNS are always capable of repair while those in the CNS are not.

Choose the one best answer to the following questions.

5. Which of the following statements are *true*? (1) The sensory function of the nervous system involves sensory receptors sensing certain changes in the internal and external environments. (2) Sensory neurons receive electrical signals from sensory receptors. (3) The integrative function of the nervous system involves analyzing sensory information, storing some of it, and making decisions regarding appropriate responses. (4) Interneurons are located primarily in the PNS. (5) Motor function involves the activation of effectors (muscles and glands).
   (a) 1, 2, 3, and 4   (b) 2, 4, and 5   (c) 1, 2, 3, and 5
   (d) 1, 2, and 4   (e) 2, 3, 4, and 5

6. A neuron's resting membrane potential is established and maintained by (1) a high concentration of K^+ in the extracellular fluid and a high concentration of Na^+ in the cytosol, (2) the plasma membrane's higher permeability to Na^+ because of the presence of numerous Na^+ leakage channels, (3) differences in both ion concentrations and electrical gradients, (4) the fact that there are numerous large, nondiffusible anions in the cytosol, (5) sodium–potassium pumps that help to maintain the proper distribution of sodium and potassium.

   (a) 1, 2, and 5   (b) 1, 2, and 3   (c) 2, 3, and 4
   (d) 3, 4, and 5   (e) 1, 2, 3, 4, and 5

7. Place the following events in a chemical synapse in the correct order: (1) release of neurotransmitters into the synaptic cleft, (2) arrival of nerve impulse at the presynaptic neuron's synaptic end bulb (or varicosity), (3) either depolarization or hyperpolarization of postsynaptic membrane, (4) inward flow of Ca^{2+} through activated voltage-gated Ca^{2+} channels in the synaptic end bulb membrane, (5) exocytosis of synaptic vesicles, (6) opening of ligand-gated channels on the postsynaptic plasma membrane, (7) binding of neurotransmitters to receptors in the postsynaptic neuron's plasma membrane.
   (a) 2, 1, 5, 4, 7, 6, 3   (b) 1, 2, 4, 5, 7, 6, 3
   (c) 2, 4, 5, 1, 7, 6, 3   (d) 4, 5, 1, 7, 6, 3, 2
   (e) 2, 5, 1, 4, 6, 7, 3

8. Several neurons in the brain sending impulses to a single motor neuron that terminates at a neuromuscular junction is an example of a _____ circuit.
   (a) reverberating   (b) simple series
   (c) parallel after-discharge   (d) diverging
   (e) converging

9. Which of the following statements are *true*? (1) If the excitatory effect is greater than the inhibitory effect but less than the threshold of stimulation, the result is a subthreshold EPSP. (2) If the excitatory effect is greater than the inhibitory effect and reaches or surpasses the threshold level of stimulation, the result is a threshold or suprathreshold EPSP and one or more nerve impulses. (3) If the inhibitory effect is greater than the excitatory effect, the membrane hyperpolarizes, resulting in inhibition of the postsynaptic neuron and the inability of the neuron to generate a nerve impulse. (4) The greater the summation of hyperpolarizations, the more likely a nerve impulse will be initiated.
   (a) 1 and 4   (b) 2 and 4   (c) 1, 3, and 4
   (d) 2, 3, and 4   (e) 1, 2, and 3

10. Which of the following statements are *true*? (1) The basic types of ion channels are gated, leakage, and electrical. (2) Ion channels allow for the development of graded potentials and action potentials. (3) Voltage-gated channels open in response to changes in membrane potential. (4) Ligand-gated channels open due to the presence of specific chemicals. (5) A graded potential is useful for communication over long distances.
    (a) 1, 2, and 3   (b) 2, 3, and 4   (c) 2, 3, and 5
    (d) 2, 3, 4, and 5   (e) 1, 3, and 5

11. Which of the following statements are *true*? (1) The frequency of impulses and number of activated sensory neurons encode differences in stimuli intensity. (2) Larger-diameter axons conduct nerve impulses faster than smaller-diameter ones. (3) Continuous conduction is faster than saltatory conduction. (4) The presence or absence of a myelin sheath is an important factor that determines the speed of nerve impulse propagation. (5) Action potentials are localized, but graded potentials are propagated.
    (a) 1, 3, and 5   (b) 3 and 4   (c) 2, 4, and 5
    (d) 2 and 4   (e) 1, 2, and 4

12. Neurotransmitters are removed from the synaptic cleft by (1) axonal transport, (2) diffusion away from the cleft, (3) neurosecretory cells, (4) enzymatic breakdown, (5) cellular uptake.
    (a) 1, 2, 3, and 4   (b) 2, 4, and 5   (c) 2, 3, and 4
    (d) 1, 4, and 5   (e) 1, 2, 3, 4, and 5

13. Match the following:
    ____ (a) neurons with just one process extending from the cell body; are always sensory neurons
    ____ (b) small phagocytic neuroglia
    ____ (c) help maintain an appropriate chemical environment for generation of action potentials by neurons; part of the blood–brain barrier
    ____ (d) provide myelin sheath for CNS axons
    ____ (e) contains neuronal cell bodies, dendrites, axon terminals, unmyelinated axons, and neuroglia
    ____ (f) a cluster of cell bodies within the CNS
    ____ (g) form CSF and assist in its circulation; form blood–cerebrospinal barrier
    ____ (h) neurons having several dendrites and one axon; most common neuronal type
    ____ (i) neurons with one main dendrite and one axon; found in the retina of the eye
    ____ (j) provide myelin sheath for PNS axons
    ____ (k) support neurons in PNS ganglia

    (1) astrocytes
    (2) oligodendrocytes
    (3) ganglion
    (4) ependymal cells
    (5) satellite cells
    (6) unipolar neurons
    (7) bipolar neurons
    (8) multipolar neurons
    (9) gray matter
    (10) white matter
    (11) enteric plexuses
    (12) microglia
    (13) Schwann cells
    (14) nucleus
    (15) nerves

    ____ (l) a cluster of neuronal cell bodies located outside the brain and spinal cord
    ____ (m) composed primarily of myelinated axons
    ____ (n) bundles of axons and associated connective tissue and blood vessels in the PNS
    ____ (o) extensive neuronal networks that help regulate the digestive system

14. Match the following:
    ____ (a) a sequence of rapidly occurring events that decreases and eventually reverses the membrane potential and then restores it to the resting state; a nerve impulse
    ____ (b) a small deviation from the resting membrane potential that makes the membrane either more or less polarized
    ____ (c) period of time when a second action potential can be initiated with a very strong stimulus
    ____ (d) the minimum level of depolarization required for a nerve impulse to be generated
    ____ (e) the recovery of the resting membrane potential
    ____ (f) a neurotransmitter-caused depolarization of the postsynaptic membrane
    ____ (g) a neurotransmitter-caused hyperpolarization of the postsynaptic membrane
    ____ (h) time during which a neuron cannot produce an action potential even with a very strong stimulus
    ____ (i) polarization that is less negative than the resting level
    ____ (j) results from the buildup of neurotransmitter released simultaneously by several presynaptic end bulbs
    ____ (k) the hyperpolarization that occurs after the repolarizing phase of an action potential
    ____ (l) polarization that is more negative than the resting level
    ____ (m) results from the buildup of neurotransmitter from the rapid, successive release by a single presynaptic end bulb

    (1) graded potential
    (2) action potential
    (3) excitatory postsynaptic potential
    (4) inhibitory postsynaptic potential
    (5) absolute refractory period
    (6) repolarization
    (7) after-hyperpolarizing phase
    (8) spatial summation
    (9) threshold
    (10) relative refractory period
    (11) temporal summation
    (12) depolarizing graded potential
    (13) hyperpolarizing graded potential

15. Match the following:

—— (a) the part of the neuron that contains the nucleus and organelles

—— (b) rough endoplasmic reticulum in neurons; site of protein synthesis

—— (c) store neurotransmitter

—— (d) the process that propagates nerve impulses toward another neuron, muscle fiber, or gland cell

—— (e) the highly branched receiving or input portions of a neuron

—— (f) a multilayered lipid and protein covering for axons produced by neuroglia

—— (g) the outer nucleated cytoplasmic layer of the Schwann cell

—— (h) first portion of the axon, closest to the axon hillock

—— (i) site of communication between two neurons or between a neuron and an effector cell

—— (j) form the cytoskeleton of a neuron

—— (k) gaps in the myelin sheath of an axon

—— (l) general term for any neuronal process

—— (m) area where the axon joins the cell body

—— (n) area where nerve impulses arise

—— (o) the numerous fine processes at the ends of an axon and its collaterals

—— (p) interstitial fluid-filled space separating two neurons

(1) myelin sheath

(2) neurolemma

(3) nodes of Ranvier

(4) cell body

(5) Nissl bodies

(6) neurofibrils

(7) dendrites

(8) axon

(9) axon hillock

(10) initial segment

(11) trigger zone

(12) synaptic cleft

(13) nerve fiber

(14) axon terminals

(15) synapse

(16) synaptic vesicles

13 THE SPINAL CORD AND SPINAL NERVES

CHAPTER REVIEW

13.1 Spinal Cord Anatomy

1. The spinal cord is protected by the vertebral column, the meninges, cerebrospinal fluid, and denticulate ligaments.
2. The three meninges are coverings that run continuously around the spinal cord and brain. They are the dura mater, arachnoid mater, and pia mater.
3. The spinal cord begins as a continuation of the medulla oblongata and ends at about the second lumbar vertebra in an adult.
4. The spinal cord contains cervical and lumbar enlargements that serve as points of origin for nerves to the limbs.
5. The tapered inferior portion of the spinal cord is the conus medullaris, from which arise the filum terminale and cauda equina.
6. Spinal nerves connect to each segment of the spinal cord by two roots. The posterior or dorsal root contains sensory axons, and the anterior or ventral root contains motor neuron axons.
7. The anterior median fissure and the posterior median sulcus partially divide the spinal cord into right and left sides.
8. The gray matter in the spinal cord is divided into horns, and the white matter into columns. In the center of the spinal cord is the central canal, which runs the length of the spinal cord.
9. Parts of the spinal cord observed in transverse section are the gray commissure; central canal; anterior, posterior, and lateral gray horns; and anterior, posterior, and lateral white columns, which contain ascending and descending tracts. Each part has specific functions.
10. The spinal cord conveys sensory and motor information by way of ascending and descending tracts, respectively.

13.2 Spinal Nerves

1. The 31 pairs of spinal nerves are named and numbered according to the region and level of the spinal cord from which they emerge. There are 8 pairs of cervical, 12 pairs of thoracic, 5 pairs of lumbar, 5 pairs of sacral, and 1 pair of coccygeal nerves.
2. Spinal nerves typically are connected with the spinal cord by a posterior root and an anterior root. All spinal nerves contain both sensory and motor axons (they are mixed nerves).
3. Three connective tissue coverings associated with spinal nerves are the endoneurium, perineurium, and epineurium.
4. Branches of a spinal nerve include the posterior ramus, anterior ramus, meningeal branch, and rami communicantes.
5. The anterior rami of spinal nerves, except for T2–T12, form networks of nerves called plexuses.

6. Emerging from the plexuses are nerves bearing names that typically describe the general regions they supply or the route they follow.
7. Anterior rami of nerves T2–T12 do not form plexuses and are called intercostal (thoracic) nerves. They are distributed directly to the structures they supply in intercostal spaces.
8. Sensory neurons within spinal nerves and the trigeminal (V) nerve serve specific, constant segments of the skin called dermatomes.
9. Knowledge of dermatomes helps a physician determine which segment of the spinal cord or which spinal nerve is damaged.

13.3 Cervical Plexus

1. The cervical plexus is formed by the roots (anterior rami) of the first four cervical nerves (C1–C4), with contributions from C5.
2. Nerves of the cervical plexus supply the skin and muscles of the head, neck, and upper part of the shoulders; they connect with some cranial nerves and innervate the diaphragm.

13.4 Brachial Plexus

1. The roots (anterior rami) of spinal nerves C5–C8 and T1 form the brachial plexus.
2. Nerves of the brachial plexus supply the upper limbs and several neck and shoulder muscles.

13.5 Lumbar Plexus

1. The roots (anterior rami) of spinal nerves L1–L4 form the lumbar plexus.
2. Nerves of the lumbar plexus supply the anterolateral abdominal wall, external genitals, and part of the lower limbs.

13.6 Sacral and Coccygeal Plexuses

1. The roots (anterior rami) of spinal nerves L1–L5 and S1–S4 form the sacral plexus.
2. Nerves of the sacral plexus supply the buttocks, perineum, and part of the lower limbs.
3. The roots (anterior rami) of the spinal nerves S4–S5 and the coccygeal nerves form the coccygeal plexus.
4. Nerves of the coccygeal plexus supply the skin of the coccygeal region.

13.7 Spinal Cord Physiology

1. The white matter tracts in the spinal cord are highways for nerve impulse propagation. Along these tracts, sensory input travels toward the brain, and motor output travels from the brain toward

skeletal muscles and other effector tissues. Sensory input travels along two main routes in the white matter of the spinal cord: the posterior column and the spinothalamic tract. Motor output travels along two main routes in the white matter of the spinal cord: direct pathways and indirect pathways.

2. A second major function of the spinal cord is to serve as an integrating center for spinal reflexes. This integration occurs in the gray matter.

3. A reflex is a fast, predictable sequence of involuntary actions, such as muscle contractions or glandular secretions, which occurs in response to certain changes in the environment. Reflexes may be spinal or cranial and somatic or autonomic (visceral).

4. The components of a reflex arc are sensory receptor, sensory neuron, integrating center, motor neuron, and effector.

5. Somatic spinal reflexes include the stretch reflex, the tendon reflex, the flexor (withdrawal) reflex, and the crossed extensor reflex; all exhibit reciprocal innervation.

6. A two-neuron or monosynaptic reflex arc consists of one sensory neuron and one motor neuron. A stretch reflex, such as the patellar reflex, is an example.

7. The stretch reflex is ipsilateral and is important in maintaining muscle tone.

8. A polysynaptic reflex arc contains sensory neurons, interneurons, and motor neurons. The tendon reflex, flexor (withdrawal) reflex, and crossed extensor reflexes are examples.

9. The tendon reflex is ipsilateral and prevents damage to muscles and tendons when muscle force becomes too extreme. The flexor reflex is ipsilateral and moves a limb away from the source of a painful stimulus. The crossed extensor reflex extends the limb contralateral to a painfully stimulated limb, allowing the weight of the body to shift when a supporting limb is withdrawn.

10. Several important somatic reflexes are used to diagnose various disorders. These include the patellar reflex, Achilles reflex, Babinski sign, and abdominal reflex.

CLINICAL CONNECTIONS

Spinal Tap (Refer page 393 of Textbook)

In a **spinal tap** (*lumbar puncture*), a local anesthetic is given, and a long hollow needle is inserted into the subarachnoid space to withdraw cerebrospinal fluid (CSF) for diagnostic purposes; to introduce antibiotics, contrast media for myelography, or anesthetics; to administer chemotherapy; to measure CSF pressure; and/or to evaluate the effects of treatment for diseases such as meningitis. During this procedure, the patient lies on his or her side with the vertebral column flexed. Flexion of the vertebral column increases the distance between the spinous processes of the vertebrae, which allows easy access to the subarachnoid space. The spinal cord ends around the second lumbar vertebra (L2); however, the spinal meninges and circulating cerebrospinal fluid extend to the second sacral vertebra (S2). Between vertebrae L2 and S2 the spinal meninges are present, but the spinal cord is absent. Consequently, a spinal tap is normally performed in adults between the L3 and L4 or L4 and L5 lumbar vertebrae because this region provides safe access to the subarachnoid space without the risk of damaging the spinal cord. (A line drawn across the highest points of the iliac crests, called the supracristal line, passes through the spinous process of the fourth lumbar vertebra and is used as a landmark for administering a spinal tap.)

Injuries to the Phrenic Nerves (Refer page 401 of Textbook)

The phrenic nerves originate from C3, C4, and C5 and supply the diaphragm. Complete severing of the spinal cord above the origin of the phrenic nerves (C3, C4, and C5) causes respiratory arrest. In **injuries to the phrenic nerves**, breathing stops because the phrenic nerves no longer send nerve impulses to the diaphragm. The phrenic nerves may also be damaged due to pressure from malignant tracheal or esophageal tumors in the mediastinum.

Injuries to Nerves Emerging from the Brachial Plexus

(Refer page 404 of Textbook)

Injury to the superior roots of the brachial plexus (C5–C6) may result from forceful pulling away of the head from the shoulder, as might occur from a heavy fall on the shoulder or excessive stretching of an infant's neck during childbirth. The presentation of this injury is characterized by an upper limb in which the shoulder is adducted, the arm is medially rotated, the elbow is extended, the forearm is pronated, and the wrist is flexed (**Figure 13.9c**). This condition is called **Erb-Duchenne palsy** or *waiter's tip position*. There is loss of sensation along the lateral side of the arm.

Injury to the radial (and axillary) **nerve** can be caused by improperly administered intramuscular injections into the deltoid muscle. The radial nerve may also be injured when a cast is applied too tightly around the mid-humerus. Radial nerve injury is indicated by **wrist drop**, the inability to extend the wrist and fingers (**Figure 13.9c**). Sensory loss is minimal due to the overlap of sensory innervation by adjacent nerves.

Injury to the median nerve may result in **median nerve palsy**, which is indicated by numbness, tingling, and pain in the palm and fingers. There is also inability to pronate the forearm and flex the proximal interphalangeal joints of all digits and the distal interphalangeal joints of the second and third digits (**Figure 13.9c**). In addition, wrist flexion is weak and is accompanied by adduction, and thumb movements are weak.

Injury to the ulnar nerve may result in **ulnar nerve palsy**, which is indicated by an inability to abduct or adduct the fingers, atrophy of the interosseous muscles of the hand, hyperextension of the metacarpophalangeal joints, and flexion of the interphalangeal joints, a condition called **clawhand** (**Figure 13.9c**). There is also loss of sensation over the little finger.

Injury to the long thoracic nerve results in paralysis of the serratus anterior muscle. The medial border of the scapula protrudes, giving it the appearance of a wing. When the arm is raised, the vertebral border and inferior angle of the scapula pull away from the thoracic wall and protrude outward, causing the medial border of the scapula to protrude; because the scapula looks like a wing, this condition is called **winged scapula** (**Figure 13.9c**). The arm cannot be abducted beyond the horizontal position.

Compression of the brachial plexus on one or more of its nerves is sometimes known as **thoracic outlet syndrome**. The subclavian artery and subclavian vein may also be compressed. The compression may result from spasm of the scalene or pectoralis minor muscles, the presence of a cervical rib (an embryological anomaly), or misaligned ribs. The patient may experience pain, numbness, weakness, or tingling in the upper limb, across the upper thoracic area, and over the scapula on the affected side. The symptoms of thoracic outlet syndrome are exaggerated during physical or emotional stress because the added stress increases the contraction of the involved muscles.

Injuries to the Lumbar Plexus (Refer page 405 of Textbook)

The largest nerve arising from the lumbar plexus is the femoral nerve. **Injuries to the femoral nerve**, which can occur in stab or gunshot wounds, are indicated by an inability to extend the leg and by loss of sensation in the skin over the anteromedial aspect of the thigh.

Injuries to the obturator nerve result in paralysis of the adductor muscles of the thigh and loss of sensation over the medial aspect of the thigh. It may result from pressure on the nerve by the fetal head during pregnancy.

Injury to the Sciatic Nerve (Refer page 406 of Textbook)

The most common form of back pain is caused by compression or irritation of the sciatic nerve, the longest nerve in the human body. The sciatic nerve is actually two nerves—tibial and common fibular—bound together by a common sheath of connective tissue. It splits into its two divisions, usually at the knee. **Injury to the sciatic nerve** results in **sciatica** (sī-AT-i-ka), pain that may extend from the buttock down the posterior and lateral aspect of the leg and the lateral aspect of the foot. The sciatic nerve may be injured because of a herniated (slipped) disc, dislocated hip, osteoarthritis of the lumbosacral spine, pathological shortening of the lateral rotator muscles of the thigh (especially piriformis), pressure from the uterus during pregnancy, inflammation, irritation, or an improperly administered gluteal intramuscular injection. In addition, sitting on a wallet or other object for a long period of time can compress the nerve and induce pain.

In many sciatic nerve injuries, the common fibular portion is the most affected, frequently from fractures of the fibula or by pressure from casts or splints over the thigh or leg. Damage to the common fibular nerve causes the foot to be plantar flexed, a condition called **foot drop**, and inverted, a condition called **equinovarus** (e-KWĪ-nō-va-rus). There is also loss of function along the anterolateral aspects of the leg and dorsum of the foot and toes. Injury to the tibial portion of the sciatic nerve results in dorsiflexion of the foot plus eversion, a condition called **calcaneovalgus** (kal-KĀ-nē-ō-val'-gus). Loss of sensation on the sole also occurs. Treatments for sciatica are similar to those for a herniated (slipped) disc—rest, pain medications, exercises, ice or heat, and massage.

Reflexes and Diagnosis (Refer page 413 of Textbook)

Reflexes are often used for diagnosing disorders of the nervous system and locating injured tissue. If a reflex ceases to function or functions abnormally, the physician may suspect that the damage lies somewhere along a particular conduction pathway. Many somatic reflexes can be tested simply by tapping or stroking the body. Among the somatic reflexes of clinical significance are the following:

- **Patellar reflex** (knee jerk). This stretch reflex involves extension of the leg at the knee joint by contraction of the quadriceps femoris muscle in response to tapping the patellar ligament (see **Figure 13.14**). This reflex is blocked by damage to the sensory or motor nerves supplying the muscle or to the integrating centers in the second, third, or fourth lumbar segments of the spinal cord. It is often absent in people with chronic diabetes mellitus or neurosyphilis, both of which cause degeneration of nerves. It is exaggerated in disease or injury involving certain motor tracts descending from the higher centers of the brain to the spinal cord.

- **Achilles reflex** (a-KIL-ēz) (ankle jerk). This stretch reflex involves plantar flexion of the foot by contraction of the gastrocnemius and soleus muscles in response to tapping the calcaneal (Achilles) tendon. Absence of the Achilles reflex indicates damage to the nerves supplying the posterior leg muscles or to neurons in the lumbosacral region of the spinal cord. This reflex may also disappear in people with chronic diabetes, neurosyphilis, alcoholism, and subarachnoid hemorrhages. An exaggerated Achilles reflex indicates cervical cord compression or a lesion of the motor tracts of the first or second sacral segments of the cord.

- **Babinski sign** (ba-BIN-skē). This reflex results from gentle stroking of the lateral outer margin of the sole. The great toe extends, with or without a lateral fanning of the other toes. This phenomenon normally occurs in children under 1½ years of age and is due to incomplete myelination of fibers in the corticospinal tract. A positive Babinski sign after age 1½ is abnormal and indicates an interruption of the corticospinal tract as the result of a lesion of the tract, usually in the upper portion. The normal response after age 1½ is the **plantar flexion reflex**, or negative Babinski—a curling under of all the toes.

- **Abdominal reflex.** This reflex involves contraction of the muscles that compress the abdominal wall in response to stroking the side of the abdomen. The response is an abdominal muscle contraction that causes the umbilicus to move in the direction of the stimulus. Absence of this reflex is associated with lesions of the corticospinal tracts. It may also be absent because of lesions of the peripheral nerves, lesions of integrating centers in the thoracic part of the cord, or multiple sclerosis.

Most autonomic reflexes are not practical diagnostic tools because it is difficult to stimulate visceral effectors, which are deep inside the body. An exception is the pupillary light reflex, in which the pupils of both eyes decrease in diameter when either eye is exposed to light. Because the reflex arc includes synapses in lower parts of the brain, the **absence of a normal pupillary light reflex** may indicate brain damage or injury.

DISORDERS: HOMEOSTATIC IMBALANCES

The spinal cord can be damaged in several ways. Outcomes range from little or no long-term neurological deficits to severe deficits and even death.

Traumatic Injuries

Most **spinal cord injuries** are due to trauma as a result of factors such as automobile accidents, falls, contact sports, diving, and acts of violence. The effects of the injury depend on the extent of direct trauma to the spinal cord or compression of the cord by fractured or displaced vertebrae or blood clots. Although any segment of the spinal cord may be involved, the most common sites of injury are in the cervical, lower thoracic, and upper lumbar regions. Depending on the location and extent of spinal cord damage, paralysis may occur. **Monoplegia** (mon'-ō-PLĒ-jē-a; mono- = one; -plegia = blow or strike) is paralysis of one limb only. **Diplegia** (di- = two) is paralysis of both upper limbs or both lower limbs. **Paraplegia** (para- = beyond) is paralysis of both lower limbs. **Hemiplegia** (hemi- = half) is paralysis of the upper limb, trunk, and lower limb on one side of the body, and **quadriplegia** (quad- = four) is paralysis of all four limbs.

Complete transection (tran-SEK-shun; trans- = across; -section = a cut) of the spinal cord means that the cord is severed from one side to the other, thus cutting all sensory and motor tracts. It results in a loss of all sensations and voluntary movement below the level of the transection. A person will have permanent loss of all sensations in dermatomes below the injury because ascending nerve impulses cannot propagate past the transection to reach the brain. At the same time, all voluntary muscle contractions will be lost below the transection because nerve impulses descending from the brain also cannot pass. The extent of paralysis of skeletal muscles depends on the level of injury. The closer the injury is to the head, the greater the area of the body that may be affected. The following list outlines which muscle functions may be retained at progressively lower levels of spinal cord transection. (These are spinal cord levels and not vertebral column levels. Recall that spinal cord levels differ from vertebral column levels because of the differential growth of the cord versus the column, especially as you progress inferiorly.)

- C1–C3: no function maintained from the neck down; ventilator needed for breathing; electric wheelchair with breath, head, or shoulder-controlled device required (see **Figure A**)
- C4–C5: diaphragm, which allows breathing
- C6–C7: some arm and chest muscles, which allows feeding, some dressing, and manual wheelchair required (see **Figure B**)
- T1–T3: intact arm function
- T4–T9: control of trunk above the umbilicus
- T10–L1: most thigh muscles, which allows walking with long leg braces (see **Figure C**)
- L1–L2: most leg muscles, which allows walking with short leg braces (see **Figure D**)

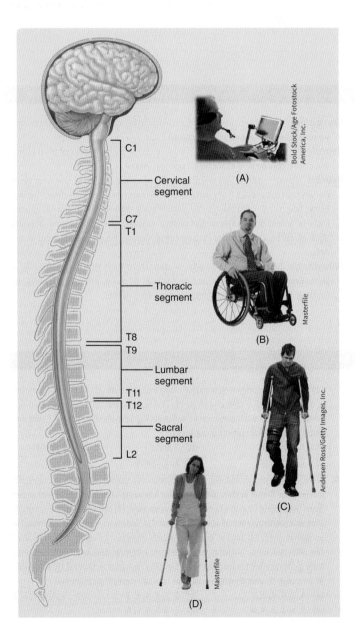

(A)

Bold Stock/Age Fotostock America, Inc.

C1

Cervical segment

C7
T1

Thoracic segment

(B)

Masterfile

T8
T9

Lumbar segment

T11
T12

Sacral segment

L2

(C)

Andersen Ross/Getty Images, Inc.

(D)

Masterfile

Hemisection is a partial transection of the cord on either the right or the left side. After hemisection, three main symptoms, known together as *Brown-Séquard syndrome* (sē-KAR), occur below the level of the injury: (1) Damage of the posterior column (sensory tracts) causes loss of proprioception and fine touch sensations on the *ipsilateral* (same) side as the injury. (2) Damage of the lateral corticospinal tract (motor tract) causes ipsilateral paralysis. (3) Damage of the spinothalamic tracts (sensory tracts) causes loss of pain and temperature sensations on the *contralateral* (opposite) side.

Following complete transection, and to varying degrees after hemisection, spinal shock occurs. **Spinal shock** is an immediate response to spinal cord injury characterized by temporary **areflexia** (a′-re-FLEK-sē-a), loss of reflex function. The areflexia occurs in parts of the body served by spinal nerves below the level of the injury. Signs of acute spinal shock include slow heart rate, low blood pressure, flaccid paralysis of skeletal muscles, loss of somatic sensations, and urinary bladder dysfunction. Spinal shock may begin within 1 hour after injury and may last from several minutes to several months, after which reflex activity gradually returns.

In many cases of traumatic injury of the spinal cord, the patient may have an improved outcome if an anti-inflammatory corticosteroid drug called methylprednisolone is given within 8 hours of the injury. This is because the degree of neurologic deficit is greatest immediately following traumatic injury as a result of *edema* (collection of fluid within tissues) as the immune system responds to the injury.

Spinal Cord Compression

Although the spinal cord is normally protected by the vertebral column, certain disorders may put pressure on it and disrupt its normal functions. Spinal cord compression may result from fractured vertebrae, herniated intervertebral discs, tumors, osteoporosis, or infections. If the source of the compression is determined before neural tissue is destroyed, spinal cord function usually returns to normal. Depending on the location and degree of compression, symptoms include pain, weakness or paralysis, and either decreased or complete loss of sensation below the level of the injury.

Degenerative Diseases

A number of **degenerative diseases** affect the functions of the spinal cord. One of these is multiple sclerosis, the details of which were presented in Disorders: Homeostatic Imbalances at the end of Chapter 12. Another progressive degenerative disease is amyotrophic lateral sclerosis (Lou Gehrig's disease), which affects motor neurons of the brain and spinal cord and results in muscle weakness and atrophy. Details are presented in Clinical Connection: Amyotrophic Lateral Sclerosis in Chapter 16.

Shingles

Shingles is an acute infection of the peripheral nervous system caused by herpes zoster (HER-pēz ZOS-ter), the virus that also causes chickenpox. After a person recovers from chickenpox, the virus retreats to a posterior root ganglion. If the virus is reactivated, the immune system usually prevents it from spreading. From time to time, however, the reactivated virus overcomes a weakened immune system, leaves the ganglion, and travels down sensory neurons of the skin by fast axonal transport (described in Section 12.2). The result is pain, discoloration of the skin, and a characteristic line of skin blisters. The line of blisters marks the distribution (dermatome) of the particular cutaneous sensory nerve belonging to the infected posterior root ganglion.

Poliomyelitis

Poliomyelitis (pō′-lē-ō-mī-e-LĪ-tis), or simply *polio*, is caused by a virus called poliovirus. The onset of the disease is marked by fever, severe headache, a stiff neck and back, deep muscle pain and weakness, and loss of certain somatic reflexes. In its most serious form, the virus produces paralysis by destroying cell bodies of motor neurons, specifically those in the anterior horns of the spinal cord and in the nuclei of the cranial nerves. Polio can cause death from respiratory or heart failure if the virus invades neurons in vital centers that control breathing and heart functions in the brainstem. Even though polio vaccines have virtually eradicated polio in the United States, outbreaks of polio continue throughout the world. Due to international travel, polio could be easily reintroduced into North America if individuals are not vaccinated appropriately.

Several decades after suffering a severe attack of polio and following their recovery from it, some individuals develop a condition called **post-polio syndrome**. This neurological disorder is characterized by progressive muscle weakness, extreme fatigue, loss of function, and pain, especially in muscles and joints. Post-polio syndrome seems to involve a slow degeneration of motor neurons that innervate muscle fibers. Triggering factors appear to be a fall, a minor accident, surgery, or prolonged bed rest. Possible causes include overuse of surviving motor neurons over time, smaller motor neurons because of the initial infection by the virus, reactivation of dormant polio viral particles, immune-mediated responses, hormone deficiencies, and environmental toxins. Treatment consists of muscle-strengthening exercises, administration of pyridostigmine to enhance the action of acetylcholine in stimulating muscle contraction, and administration of nerve growth factors to stimulate both nerve and muscle growth.

MEDICAL TERMINOLOGY

Epidural block (ep′-i-DOO-ral) Injection of an anesthetic drug into the epidural space, the space between the dura mater and the vertebral column, in order to cause a temporary loss of sensation. Such injections in the lower lumbar region are used to control pain during childbirth.

Meningitis (men-in-JĪ-tis; *-itis* = inflammation) Inflammation of the meninges due to an infection, usually caused by a bacterium or virus. Symptoms include fever, headache, stiff neck, vomiting, confusion, lethargy, and drowsiness. Bacterial meningitis is much more serious and is treated with antibiotics. Viral meningitis has no specific treatment. Bacterial meningitis may be fatal if not treated promptly; viral meningitis usually resolves on its own in 1–2 weeks. A vaccine is available to help protect against some types of bacterial meningitis.

Myelitis (mī-e-LĪ-tis; *myel-* = spinal cord) Inflammation of the spinal cord.

Nerve block Loss of sensation in a region due to injection of a local anesthetic; an example is local dental anesthesia.

Neuralgia (noo-RAL-jē-a; *neur-* = nerve; *-algia* = pain) Attacks of pain along the entire course or a branch of a sensory nerve.

Neuritis (*neur-* = nerve; *-itis* = inflammation) Inflammation of one or several nerves that may result from irritation to the nerve produced by direct blows, bone fractures, contusions, or penetrating injuries. Additional causes include infections, vitamin deficiency (usually thiamine), and poisons such as carbon monoxide, carbon tetrachloride, heavy metals, and some drugs.

Paresthesia (par-es-THĒ-zē-a; *par-* = departure from normal; *-esthesia* = sensation) An abnormal sensation such as burning, pricking, tickling, or tingling resulting from a disorder of a sensory nerve.

SELF-QUIZ QUESTIONS

Fill in the blanks in the following statements.

1. Because they contain both sensory and motor axons, spinal nerves are considered to be _____ nerves.

2. The five components of a reflex arc, in order from the beginning to the end, are (1) _____, (2) _____, (3) _____, (4) _____, and (5) _____.

Indicate whether the following statements are true or false.

3. Gray matter of the spinal cord contains somatic motor and sensory nuclei, autonomic motor and sensory nuclei, and functions to receive and integrate both incoming and outgoing information.

4. The epidural space is located between the wall of the vertebral canal and the pia mater.

Choose the one best answer to the following questions.

5. Which of the following is not true? (1) Dermatomes are areas of the body that are stimulated by motor neurons exiting specific spinal nerves. (2) The stretch reflex helps to maintain muscle tone. (3) The Achilles reflex is an example of a stretch reflex. (4) The abdominal reflex is used to diagnose problems with autonomic reflexes. (5) Spinal nerves T2–T12 do not enter into the formation of a plexus.
   (a) 1, 2, and 4   (b) 2 and 5   (c) 1 and 4
   (d) 1, 3, and 5   (e) 1, 3, and 4

6. While identifying and labeling cadaver muscles, your lab partner accidentally pokes your finger with a pin. Place the following steps in the correct order from beginning to end of your body's response. (1) Impulses travel through anterior (ventral) root of spinal nerve(s). (2) Sensory neuron relays impulse to spinal cord. (3) Motor impulses reach muscles, causing withdrawal of the affected limb. (4) Integrating centers interpret sensory impulses, and then generate motor impulses. (5) Sensory receptor activated by stimulus. (6) Impulse travels through posterior (dorsal) root of spinal nerve.
   (a) 5, 3, 6, 4, 1, 2   (b) 5, 2, 1, 4, 6, 3   (c) 5, 2, 6, 4, 1, 3
   (d) 3, 5, 1, 2, 4, 6   (e) 2, 1, 5, 4, 6, 3

7. The connective tissue surrounding each individual axon is
   (a) endoneurium.   (b) epineurium.   (c) perineurium.
   (d) fascicle.   (e) arachnoid mater.

8. The tracts of the posterior column are involved in (1) conscious proprioception. (2) touch, (3) pain, (4) thermal sensations, (5) pressure, (6) vibration.
   (a) 1, 2, 4, and 5   (b) 2, 4, and 6   (c) 1, 2, 5, and 6
   (d) 3, 4, 5, and 6   (e) 1, 3, 5, and 6

9. Which of the following is a motor tract?
   (a) posterior spinocerebellar   (b) lateral spinothalamic
   (c) anterior spinocerebellar   (d) lateral corticospinal
   (e) posterior column

10. Cutting the posterior root of a spinal nerve would
   (a) interfere with the circulation of cerebrospinal fluid.
   (b) impair motor control of skeletal muscles.
   (c) interfere with the ability of the brain to transmit motor impulses.
   (d) impair motor control of organs.
   (e) interfere with the flow of sensory impulses.

11. Which of the following statements is *false*?
   (a) The two main spinal cord sensory paths are the spinothalamic and anterior columns.
   (b) The spinothalamic tracts convey impulses for sensing pain, temperature, itching, and tickling.
   (c) Direct pathways convey nerve impulses destined to cause precise, voluntary movements of skeletal muscles.
   (d) Indirect pathways convey nerve impulses that program automatic movements, help coordinate body movements with visual stimuli, maintain skeletal muscle tone and posture, and contribute to equilibrium.
   (e) The direct pathways are motor pathways.

12. Which of the following are *true*?
   (1) The anterior (ventral) gray horns contain cell bodies of neurons that cause skeletal muscle contraction. (2) The gray commissure connects the white matter of the right and left sides of the spinal cord. (3) Cell bodies of autonomic motor neurons are located in the lateral gray horns. (4) Sensory (ascending) tracts conduct motor impulses down the spinal cord. (5) Gray matter in the spinal cord consists of cell bodies of neurons, neuroglia, unmyelinated axons, and dendrites of interneurons and motor neurons.
   (a) 1, 2, 3, and 5   (b) 2 and 4   (c) 2, 3, 4, and 5
   (d) 1, 3, and 5   (e) 1, 2, 3, and 4

13. Match the following (some answers may be used more than once):
   ____ (a) a reflex resulting in the contraction of a skeletal muscle when it is stretched
   ____ (b) receptors that monitor changes in muscle length
   ____ (c) a balance-maintaining reflex
   ____ (d) operates as a feedback mechanism to control muscle tension by causing muscle relaxation when muscle force becomes too extreme
   ____ (e) reflex arc that consists of one sensory and one motor neuron
   ____ (f) acts as a feedback mechanism to control muscle length by causing muscle contraction
   ____ (g) sensory impulses enter on one side of the spinal cord and motor impulses exit on the opposite side
   ____ (h) occurs when sensory nerve impulse travels up and down the spinal cord, thereby activating several motor neurons and more than one effector
   ____ (i) polysynaptic reflex initiated in response to a painful stimulus
   ____ (j) receptors that monitor changes in muscle tension
   ____ (k) maintains proper muscle tone
   ____ (l) reflex pathway that contains sensory neurons, interneurons, and motor neurons
   ____ (m) motor nerve impulses exit the spinal cord on the same side that sensory impulses entered the spinal cord
   ____ (n) protects the tendon and muscle from damage due to excessive tension
   ____ (o) a neural circuit that coordinates body movements by causing contraction of one muscle and relaxation of antagonistic muscles or relaxation of a muscle and contraction of the antagonists

   (1) stretch reflex
   (2) tendon reflex
   (3) flexor (withdrawal) reflex
   (4) crossed extensor reflex
   (5) intersegmental reflex arc
   (6) contralateral reflex arc
   (7) ipsilateral reflex arc
   (8) muscle spindles
   (9) tendon (Golgi tendon) organs
   (10) reciprocal innervation
   (11) monosynaptic reflex
   (12) polysynaptic reflex

14. Match the following:

_____ (a) the joining together of the anterior rami of adjacent nerves

_____ (b) spinal nerve branches that serve the deep muscles and skin of the posterior surface of the trunk

_____ (c) spinal nerve branches that serve the muscles and structures of the upper and lower limbs and the lateral and ventral trunk

_____ (d) area of the spinal cord from which nerves to and from the upper limbs arise

_____ (e) area of the spinal cord from which nerves to and from the lower limbs arise

_____ (f) the roots form the nerves that arise from the inferior part of the spinal cord but do not leave the vertebral column at the same level as they exit the cord

_____ (g) contains motor neuron axons and conducts impulses from the spinal cord to the peripheral organs and cells

_____ (h) avascular covering of spinal cord composed of delicate collagen fibers and some elastic fibers

_____ (i) contains sensory neuron axons and conducts impulses from the peripheral receptors into the spinal cord

_____ (j) superficial spinal cord covering of dense, irregular connective tissue

_____ (k) an extension of the pia mater that anchors the spinal cord to the coccyx

_____ (l) extending the length of the spinal cord, these pia mater thickenings fuse with the arachnoid mater and dura mater and help to protect the spinal cord from shock and sudden displacement

_____ (m) thin transparent connective tissue composed of interlacing bundles of collagen fibers and some elastic fibers adhering to the spinal cord's surface

_____ (n) space within the spinal cord filled with cerebrospinal fluid

_____ (o) spinal nerve branch that supplies vertebrae, vertebral ligaments, blood vessels of the spinal cord, and meninges

(1) cervical enlargement
(2) lumbar enlargement
(3) central canal
(4) denticulate ligaments
(5) cauda equina
(6) meningeal branch
(7) pia mater
(8) arachnoid mater
(9) dura mater
(10) posterior (dorsal) root
(11) anterior (ventral) root
(12) posterior (dorsal) ramus
(13) anterior (ventral) ramus
(14) plexus
(15) filum terminale

15. Match the following:

_____ (a) provides the entire nerve supply of the shoulders and upper limbs

_____ (b) provides the nerve supply of the skin and muscles of the head, neck, and superior part of the shoulders and chest

_____ (c) provides the nerve supply of the anterolateral abdominal wall, external genitals, and part of the lower limbs

_____ (d) supplies the buttocks, perineum, and lower limbs

_____ (e) formed by the anterior rami of C1–C4 with some contribution by C5

_____ (f) formed by anterior rami of S4–S5 and coccygeal nerves

_____ (g) formed by the anterior rami of L1–L4

_____ (h) formed by the anterior rami of C5–C8 and T1

_____ (i) formed by the anterior rami of L4–L5 and S1–S4

_____ (j) phrenic nerve arises from this plexus

_____ (k) median nerve arises from this plexus

_____ (l) sciatic nerve arises from this plexus

_____ (m) femoral nerve arises from this plexus

_____ (n) supplies a small area of skin in coccygeal region

_____ (o) injury to this plexus can affect breathing

(1) cervical plexus
(2) brachial plexus
(3) lumbar plexus
(4) sacral plexus
(5) coccygeal plexus

14 | THE BRAIN AND CRANIAL NERVES

14.1 Brain Organization, Protection, and Blood Supply

1. The major parts of the brain are the brainstem, cerebellum, diencephalon, and cerebrum.
2. The brain is protected by cranial bones and the cranial meninges.
3. The cranial meninges are continuous with the spinal meninges. From superficial to deep, they are the dura mater, arachnoid mater, and pia mater.
4. Blood flow to the brain is mainly via the internal carotid and vertebral arteries.
5. Any interruption of the oxygen or glucose supply to the brain can result in weakening of, permanent damage to, or death of brain cells.
6. The blood–brain barrier (BBB) causes different substances to move between the blood and the brain tissue at different rates and prevents the movement of some substances from blood into the brain.

14.2 Cerebrospinal Fluid

1. Cerebrospinal fluid (CSF) is formed in the choroid plexuses and circulates through the lateral ventricles, third ventricle, fourth ventricle, subarachnoid space, and central canal. Most of the fluid is absorbed into the blood across the arachnoid villi of the superior sagittal sinus.
2. Cerebrospinal fluid provides mechanical protection, chemical protection, and circulation of nutrients.

14.3 The Brainstem and Reticular Formation

1. The medulla oblongata is continuous with the superior part of the spinal cord and contains both sensory tracts and motor tracts. It contains a cardiovascular center, which regulates heart rate and blood vessel diameter (cardiovascular center), and a medullary respiratory center, which helps control breathing. It also contains the gracile nucleus, cuneate nucleus, gustatory nucleus, cochlear nuclei, and vestibular nuclei, which are components of sensory pathways to the brain. Also present in the medulla is the inferior olivary nucleus, which provides instructions that the cerebellum uses to adjust muscle activity when you learn new motor skills. Other nuclei of the medulla coordinate vomiting, swallowing, sneezing, coughing, and hiccupping. The medulla also contains nuclei associated with the vestibulocochlear (VIII), glossopharyngeal (IX), vagus (X), accessory (XI), and hypoglossal (XII) nerves.

2. The pons is superior to the medulla. It contains both sensory tracts and motor tracts. Pontine nuclei relay nerve impulses related to voluntary skeletal movements from the cerebral cortex to the cerebellum. The pons also contains the pontine respiratory group, which helps control breathing. Vestibular nuclei, which are present in the pons and medulla, are part of the equilibrium pathway to the brain. Also present in the pons are nuclei associated with the trigeminal (V), abducens (VI), and facial (VII) nerves and the vestibular branch of the vestibulocochlear (VIII) nerve.
3. The midbrain connects the pons and diencephalon and surrounds the cerebral aqueduct. It contains both sensory tracts and motor tracts. The superior colliculi coordinate movements of the head, eye, and trunk in response to visual stimuli; the inferior colliculi coordinate movements of the head, eyes, and trunk in response to auditory stimuli. The midbrain also contains nuclei associated with the oculomotor (III) and trochlear (IV) nerves.
4. A large part of the brainstem consists of small areas of gray matter and white matter called the reticular formation, which helps maintain consciousness, causes awakening from sleep, and contributes to regulating muscle tone.

14.4 The Cerebellum

1. The cerebellum occupies the inferior and posterior aspects of the cranial cavity. It consists of two lateral hemispheres and a medial, constricted vermis.
2. It connects to the brainstem by three pairs of cerebellar peduncles.
3. The cerebellum smooths and coordinates the contractions of skeletal muscles. It also maintains posture and balance.

14.5 The Diencephalon

1. The diencephalon surrounds the third ventricle and consists of the thalamus, hypothalamus, and epithalamus.
2. The thalamus is superior to the midbrain and contains nuclei that serve as relay stations for most sensory input to the cerebral cortex. It also contributes to motor functions by transmitting information from the cerebellum and basal nuclei to the primary motor area of the cerebral cortex. In addition, the thalamus plays a role in maintenance of consciousness.
3. The hypothalamus is inferior to the thalamus. It controls the autonomic nervous system, produces hormones, and regulates emotional and behavioral patterns (along with the limbic system). The hypothalamus also contains a feeding center and

satiety center, which regulate eating, and a thirst center, which regulates drinking. In addition, the hypothalamus controls body temperature by serving as the body's thermostat. Also present in the hypothalamus is the suprachiasmatic nucleus, which regulates circadian rhythms and functions as the body's internal biological clock.

4. The epithalamus consists of the pineal gland and the habenular nuclei. The pineal gland secretes melatonin, which is thought to promote sleep and to help set the body's biological clock.

5. Circumventricular organs (CVOs) can monitor chemical changes in the blood because they lack the blood–brain barrier.

14.6 The Cerebrum

1. The cerebrum is the largest part of the brain. Its cortex contains gyri (convolutions), fissures, and sulci.

2. The cerebral hemispheres are divided into four lobes: frontal, parietal, temporal, and occipital.

3. The white matter of the cerebrum is deep to the cortex and consists primarily of myelinated axons extending to other regions as association, commissural, and projection fibers.

4. The basal nuclei are several groups of nuclei in each cerebral hemisphere. They help initiate and terminate movements, suppress unwanted movements, and regulate muscle tone.

5. The limbic system encircles the upper part of the brainstem and the corpus callosum. It functions in emotional aspects of behavior and memory.

6. **Table 14.2** summarizes the functions of various parts of the brain.

14.7 Functional Organization of the Cerebral Cortex

1. The sensory areas of the cerebral cortex allow perception of sensory information. The motor areas control the execution of voluntary movements. The association areas are concerned with more complex integrative functions such as memory, personality traits, and intelligence.

2. The primary somatosensory area (areas 1, 2, and 3) receives nerve impulses from somatic sensory receptors for touch, pressure, vibration, itch, tickle, temperature, pain, and proprioception and is involved in the perception of these sensations. Each point within the area receives impulses from a specific part of the face or body. The primary visual area (area 17) receives visual information and is involved in visual perception. The primary auditory area (areas 41 and 42) receives information for sound and is involved in auditory perception. The primary gustatory area (area 43) receives impulses for taste and is involved in gustatory perception and taste discrimination. The primary olfactory area (area 28) receives impulses for smell and is involved in olfactory perception.

3. Motor areas include the primary motor area (area 4), which controls voluntary contractions of specific muscles or groups of muscles, and Broca's speech area (areas 44 and 45), which controls production of speech.

4. The somatosensory association area (areas 5 and 7) permits you to determine the exact shape and texture of an object simply by touching it and to sense the relationship of one body part to another. It also stores memories of past somatic sensory experiences.

5. The visual association area (areas 18 and 19) relates present to past visual experiences and is essential for recognizing and evaluating what is seen. The facial recognition area (areas 20, 21, and 37) stores information about faces and allows you to recognize people by their faces. The auditory association area (area 22) allows you to recognize a particular sound as speech, music, or noise.

6. The orbitofrontal cortex (area 11) allows you to identify odors and discriminate among different odors. Wernicke's area (area 22 and possibly 39 and 40) interprets the meaning of speech by translating words into thoughts. The common integrative area (areas 5, 7, 39, and 40) integrates sensory interpretations from the association areas and impulses from other areas, allowing thoughts based on sensory inputs.

7. The prefrontal cortex (areas 9, 10, 11, and 12) is concerned with personality, intellect, complex learning abilities, judgment, reasoning, conscience, intuition, and development of abstract ideas. The premotor area (area 6) generates nerve impulses that cause specific groups of muscles to contract in specific sequences. It also serves as a memory bank for complex movements. The frontal eye field area (area 8) controls voluntary scanning movements of the eyes.

8. Subtle anatomical differences exist between the two hemispheres, and each has unique functions. Each hemisphere receives sensory signals from and controls movements of the opposite side of the body. The left hemisphere is more important for language, numerical and scientific skills, and reasoning. The right hemisphere is more important for musical and artistic awareness, spatial and pattern perception, recognition of faces, emotional content of language, identifying odors, and generating mental images of sight, sound, touch, taste, and smell.

9. Brain waves generated by the cerebral cortex are recorded from the surface of the head in an electroencephalogram (EEG). The EEG may be used to diagnose epilepsy, infections, and tumors.

14.8 Cranial Nerves: An Overview

1. Twelve pairs of cranial nerves originate from the nose, eyes, inner ear, brainstem, and spinal cord.

2. They are named primarily based on their distribution and are numbered I–XII in order of attachment to the brain.

14.9 Olfactory (I) Nerve

1. The olfactory (I) nerve is entirely sensory.

2. It contains axons that conduct nerve impulses for olfaction (sense of smell).

14.10 Optic (II) Nerve

1. The optic (II) nerve is purely sensory.

2. It contains axons that conduct nerve impulses for vision.

14.11 Oculomotor (III), Trochlear (IV), and Abducens (VI) Nerves

1. The oculomotor (III), trochlear (IV), and abducens (VI) nerves are the cranial nerves that control the muscles that move the eyeballs.

2. They are all motor nerves.

14.12 Trigeminal (V) Nerve

1. The trigeminal (V) nerve is a mixed cranial nerve and the largest of the cranial nerves.
2. It conveys touch, pain, and thermal sensations from the scalp, face, and oral cavity and controls chewing muscles and middle ear muscle.

14.13 Facial (VII) Nerve

1. The facial (VII) nerve is a mixed cranial nerve.
2. It conveys taste from anterior two-thirds of the tongue as well as touch, pain, and thermal sensations from skin in the external ear canal; it also controls muscles of facial expression and middle ear muscle; promotes secretion of tears; and promotes secretion of saliva.

14.14 Vestibulocochlear (VIII) Nerve

1. The vestibulocochlear (VIII) nerve is a sensory cranial nerve.
2. It conveys sensory information for audition (hearing) and equilibrium (balance).

14.15 Glossopharyngeal (IX) Nerve

1. The glossopharyngeal (IX) nerve is a mixed cranial nerve.
2. It conveys taste from posterior one-third of tongue, proprioception from some swallowing muscles, and touch, pain, and thermal sensations from the skin of external ear and upper pharynx; monitors blood pressure and oxygen and carbon dioxide levels in blood; assists in swallowing; and promotes secretion of saliva.

14.16 Vagus (X) Nerve

1. The vagus (X) nerve is a mixed cranial nerve.
2. It conveys taste from the epiglottis, proprioception from throat and voice box muscles, touch, pain, and thermal sensations from skin of external ear, and sensations from thoracic and abdominal organs; monitors blood pressure and oxygen and carbon dioxide levels in blood; promotes swallowing, vocalization, and coughing, motility and excretion of gastrointestinal organs, and constriction of respiratory passageways; and decreases heart rate.

14.17 Accessory (XI) Nerve

1. The accessory (XI) nerve is a motor cranial nerve.
2. It controls movements of the head.

14.18 Hypoglossal (XII) Nerve

1. The hypoglossal (XII) nerve is a motor cranial nerve.
2. It promotes speech and swallowing.

14.19 Development of the Nervous System

1. The development of the nervous system begins with a thickening of a region of the ectoderm called the neural plate.
2. During embryological development, primary brain vesicles form from the neural tube and serve as forerunners of various parts of the brain.
3. The telencephalon forms the cerebrum, the diencephalon develops into the thalamus and hypothalamus, the mesencephalon develops into the midbrain, the metencephalon develops into the pons and cerebellum, and the myelencephalon forms the medulla.

14.20 Aging and the Nervous System

1. The brain grows rapidly during the first few years of life.
2. Age-related effects involve loss of brain mass and decreased capacity for sending nerve impulses.

CLINICAL CONNECTIONS

Breaching the Blood–Brain Barrier (Refer page 420 of Textbook)

Because it is so effective, the blood–brain barrier prevents the passage of helpful substances as well as those that are potentially harmful. Researchers are exploring ways to move drugs that could be therapeutic for brain cancer or other CNS disorders past the BBB. In one method, the drug is injected in a concentrated sugar solution. The high osmotic pressure of the sugar solution causes the endothelial cells of the capillaries to shrink, which opens gaps between their tight junctions, making the BBB more leaky and allowing the drug to enter the brain tissue.

Hydrocephalus (Refer page 422 of Textbook)

Abnormalities in the brain—tumors, inflammation, or developmental malformations—can interfere with the circulation of CSF from the ventricles into the subarachnoid space. When excess CSF accumulates in the ventricles, the CSF pressure rises. Elevated CSF pressure causes a condition called **hydrocephalus** (hī′-drō-SEF-a-lus; *hydro-* = water; *-cephal-* = head). The abnormal accumulation of CSF may be due to an obstruction to CSF flow or an abnormal rate of CSF production and/or reabsorption. In a baby whose fontanels have not yet closed, the head bulges due to the increased pressure. If the condition persists, the fluid buildup compresses and damages the delicate nervous tissue. Hydrocephalus is relieved by draining the excess CSF. In one procedure, called *endoscopic third ventriculostomy* (*ETV*), a neurosurgeon makes a hole in the floor of the third ventricle and the CSF drains directly into the subarachnoid space. In adults, hydrocephalus may occur after head injury, meningitis, or subarachnoid hemorrhage. Because the adult skull bones are fused together, this condition can quickly become life-threatening and requires immediate intervention.

Injury to the Medulla (Refer page 427 of Textbook)

Given the vital activities controlled by the medulla, it is not surprising that **injury to the medulla** from a hard blow to the back of the head or upper neck such as falling back on ice can be fatal. Damage to the medullary respiratory center is particularly serious and can rapidly lead to death. Symptoms of nonfatal injury to the medulla may include cranial nerve malfunctions on the same side of the body as the injury, paralysis and loss of sensation on the opposite side of the body, and irregularities in breathing or heart rhythm. Alcohol overdose also suppresses the medullary rhythmicity center and may result in death.

Ataxia (Refer page 431 of Textbook)

Damage to the cerebellum can result in a loss of ability to coordinate muscular movements, a condition called **ataxia** (a-TAK-sē-a; *a-* = without; *-taxia* = order). Blindfolded people with ataxia cannot touch the tip of their nose with a finger because they cannot coordinate movement with their sense of where a body part is located. Another sign of ataxia is a changed speech pattern due to uncoordinated speech muscles. Cerebellar damage may also result in staggering or abnormal walking movements. People who consume too much alcohol show signs of ataxia because alcohol inhibits activity of the cerebellum. Such individuals have difficulty in passing sobriety tests. Ataxia can also occur as a result of degenerative diseases (multiple sclerosis and Parkinson's disease), trauma, brain tumors, and genetic factors, and as a side effect of medication prescribed for bipolar disorder.

Brain Injuries (Refer page 444 of Textbook)

Brain injuries are commonly associated with head trauma and result in part from displacement and distortion of neural tissue at the moment of impact. Additional tissue damage may occur when normal blood flow is restored after a period of ischemia (reduced blood flow). The sudden increase in oxygen level produces large numbers of oxygen free radicals (charged oxygen molecules with an unpaired electron). Brain cells recovering from the effects of a stroke or cardiac arrest also release free radicals. Free radicals cause damage by disrupting cellular DNA and enzymes and by altering plasma membrane permeability. Brain injuries can also result from hypoxia (cellular oxygen deficiency).

Various degrees of brain injury are described by specific terms. A **concussion** (kon-KUSH-un) is an injury characterized by an abrupt, but temporary, loss of consciousness (from seconds to hours), disturbances of vision, and problems with equilibrium. It is caused by a blow to the head or the sudden stopping of a moving head (as in an automobile accident) and is the most common brain injury. A concussion produces no obvious bruising of the brain. Signs of a concussion are headache, drowsiness, nausea and/or vomiting, lack of concentration, confusion, or post-traumatic amnesia (memory loss).

There has been a tremendous amount of public interest and concern about a condition called **chronic traumatic encephalopathy (CTE)**. It is a progressive, degenerative brain disorder caused by concussions and other repeated head injuries and occurs primarily among athletes who participate in contact sports such as football, ice hockey, and boxing as well as combat veterans and individuals with a history of repetitive brain trauma. Within the axons of neurons are microtubules that act as scaffolds to support the axon and serve as tracks for axonal transport (see Section 12.2). The assembly of microtubules into structural and functional unit in axons is promoted by a protein in brain tissue called *tau (TOW)*. Repeated brain injuries can cause a buildup of tau, causing it to tangle and clump together. The clumps initially kill affected brain cells and then spread to nearby cells, killing them as well. These changes in the brain can begin months, years, or even decades after the last brain trauma. This is CTE. Possible symptoms of CTE include memory loss, confusion, impulsive or erratic behavior, impaired judgment, depression, paranoia, aggression, difficulty with balance and motor skills, and eventually dementia. At present, there is no treatment or cure for CTE and a definitive diagnosis can only be made after death by brain tissue analysis when an autopsy is performed.

A brain **contusion** (kon-TOO-zhun) is bruising due to trauma and includes the leakage of blood from microscopic vessels. It is usually associated with a concussion. In a contusion, the pia mater may be torn, allowing blood to enter the subarachnoid space. The area most commonly affected is the frontal lobe. A contusion usually results in an immediate loss of consciousness (generally lasting no longer than 5 minutes), loss of reflexes, transient cessation of respiration, and decreased blood pressure. Vital signs typically stabilize in a few seconds.

A **laceration** (las-er-Ā-shun) is a tear of the brain, usually from a skull fracture or a gunshot wound. A laceration results in rupture of large blood vessels, with bleeding into the brain and subarachnoid space. Consequences include cerebral hematoma (localized pool of blood, usually clotted, that swells against the brain tissue), edema, and increased intracranial pressure. If the blood clot is small enough, it may pose no major threat and may be absorbed. If the blood clot is large, it may require surgical removal. Swelling infringes on the limited space that the brain occupies in the cranial cavity. Swelling causes excruciating headaches. Brain tissue can also undergo *necrosis* (cellular death) due to the swelling; if the swelling is severe enough, the brain can herniate through the foramen magnum, resulting in death.

Dental Anesthesia (Refer page 445 of Textbook)

The inferior alveolar nerve, a branch of the mandibular nerve, supplies all of the teeth in one half of the mandible; it is often anesthetized in dental procedures. The same procedure will anesthetize the lower lip because the mental nerve is a branch of the inferior alveolar nerve. Because the lingual nerve runs very close to the inferior alveolar nerve near the mental foramen, it too is often anesthetized at the same time. For anesthesia to the upper teeth, the superior alveolar nerve endings, which are branches of the maxillary nerve, are blocked by inserting the needle beneath the mucous membrane. The anesthetic solution is then infiltrated slowly throughout the area of the roots of the teeth to be treated.

Anosmia (Refer page 445 of Textbook)

Loss of the sense of smell, called **anosmia** (an-OZ-mē-a), may result from infections of the nasal mucosa, head injuries in which the cribriform plate of the ethmoid bone is fractured, lesions along the olfactory pathway or in the brain, meningitis, smoking, or cocaine use.

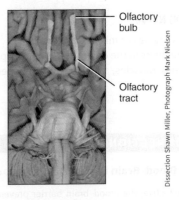

Olfactory bulb

Olfactory tract

Dissection Shawn Miller, Photograph Mark Nielsen

Anopia (Refer page 446 of Textbook)

Fractures in the orbit, brain lesions, damage along the visual pathway, diseases of the nervous system (such as multiple sclerosis), pituitary gland tumors, or cerebral aneurysms (enlargements of blood vessels due to weakening of their walls) may result in visual field defects and loss of visual acuity. Blindness due to a defect in or loss of one or both eyes is called **anopia** (an-Ō-pē-a).

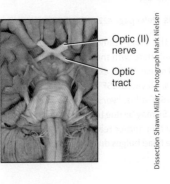

Optic (II) nerve

Optic tract

Dissection Shawn Miller, Photograph Mark Nielsen

Strabismus, Ptosis, and Diplopia (Refer page 448 of Textbook)

Damage to the oculomotor (III) nerve causes **strabismus** (stra-BIZ-mus) (a condition in which both eyes do not fix on the same object, since one or both eyes may turn inward or outward), **ptosis** (TŌ-sis) (drooping) of the upper eyelid, dilation of the pupil, movement of the eyeball downward and outward on the damaged side, loss of accommodation for near vision, and **diplopia** (di-PLŌ-pē-a) (double vision).

Trochlear (IV) nerve damage can also result in strabismus and diplopia.

With damage to the abducens (VI) nerve, the affected eyeball cannot move laterally beyond the midpoint, and the eyeball usually is directed medially. This leads to strabismus and diplopia.

Causes of damage to the oculomotor, trochlear, and abducens nerves include trauma to the skull or brain, compression resulting from aneurysms, and lesions of the superior orbital fissure. Individuals with damage to these nerves are forced to tilt their heads in various directions to help bring the affected eyeball into the correct frontal plane.

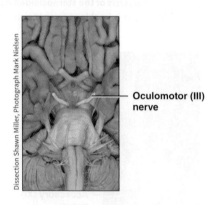

Oculomotor (III) nerve

Dissection Shawn Miller, Photograph Mark Nielsen

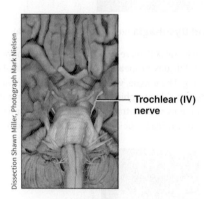

Trochlear (IV) nerve

Dissection Shawn Miller, Photograph Mark Nielsen

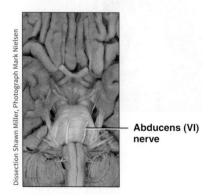

Abducens (VI) nerve

Dissection Shawn Miller, Photograph Mark Nielsen

Trigeminal Neuralgia (Refer page 449 of Textbook)

Neuralgia (pain) relayed via one or more branches of the trigeminal (V) nerve, caused by conditions such as inflammation or lesions, is called **trigeminal neuralgia** (*tic douloureux*). This is a sharp cutting or tearing pain that lasts for a few seconds to a minute and is caused by anything that presses on the trigeminal nerve or its branches. It occurs almost exclusively in people over 60 and can be the first sign of a disease, such as multiple sclerosis or diabetes, or lack of vitamin B$_{12}$, which damage the nerves. Injury of the mandibular nerve may cause paralysis of the chewing muscles and a loss of the sensations of touch, temperature, and proprioception in the lower part of the face.

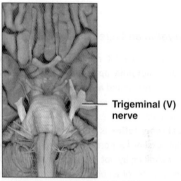

Trigeminal (V) nerve

Dissection Shawn Miller, Photograph Mark Nielsen

Bell's Palsy (Refer page 450 of Textbook)

Damage to the facial (VII) nerve due to conditions such as viral infection (shingles) or a bacterial infection (Lyme disease) produces **Bell's palsy** (paralysis of the facial muscles), loss of taste, decreased salivation, and loss of ability to close the eyes, even during sleep. The nerve can also be damaged by trauma, tumors, and stroke.

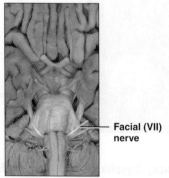

Facial (VII) nerve

Dissection Shawn Miller, Photograph Mark Nielsen

Vertigo, Ataxia, Nystagmus, and Tinnitus (Refer page 451 of Textbook)

Injury to the vestibular branch of the vestibulocochlear (VIII) nerve may cause **vertigo** (ver-TI-gō) (a subjective feeling that one's own body or the environment is rotating), **ataxia** (a-TAK-sē-a) (muscular incoordination), and **nystagmus** (nis-TAG-mus) (involuntary rapid movement of the eyeball). Injury to the cochlear branch may cause **tinnitus** (ringing in the ears) or deafness. The vestibulocochlear nerve may be injured as a result of conditions such as trauma, lesions, or middle ear infections.

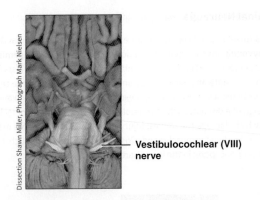

Dissection Shawn Miller, Photograph Mark Nielsen

— **Vestibulocochlear (VIII) nerve**

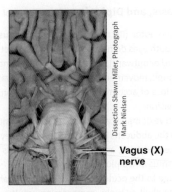

Dissection Shawn Miller, Photograph Mark Nielsen

— **Vagus (X) nerve**

Dysphagia, Aptyalia, and Ageusia (Refer page 451 of Textbook)

Injury to the glossopharyngeal (IX) nerve causes **dysphagia** (dis-FĀ-gē-a), or difficulty in swallowing; **aptyalia** (ap-tē-Ā-lē-a), or reduced secretion of saliva; loss of sensation in the throat; and **ageusia** (a-GOO-sē-a), or loss of taste sensation. The glossopharyngeal nerve may be injured as a result of conditions such as trauma or lesions.

The **pharyngeal** (*gag*) **reflex** is a rapid and intense contraction of the pharyngeal muscles. Except for normal swallowing, the pharyngeal reflex is designed to prevent choking by not allowing objects to enter the throat. The reflex is initiated by contact of an object with the roof of the mouth, back of the tongue, area around the tonsils, and back of the throat. Stimulation of receptors in these areas sends sensory information to the brain via the glossopharyngeal (IX) and vagus (X) nerves. Returning motor information via the same nerves results in contraction of the pharyngeal muscles. People with a hyperactive pharyngeal reflex have difficulty swallowing pills and are very sensitive to various medical and dental procedures.

Paralysis of the Sternocleidomastoid and Trapezius Muscles
(Refer page 453 of Textbook)

If the accessory (XI) nerve is damaged due to conditions such as trauma, lesions, or stroke, the result is **paralysis of the sternocleidomastoid and trapezius muscles** so that the person is unable to raise the shoulders and has difficulty in turning the head.

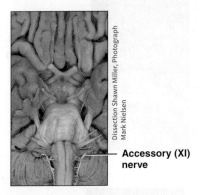

Dissection Shawn Miller, Photograph Mark Nielsen

— **Accessory (XI) nerve**

Dissection Shawn Miller, Photograph Mark Nielsen

— **Glossopharyngeal (IX) nerve**

Dysarthria and Dysphagia (Refer page 455 of Textbook)

Injury to the hypoglossal (XII) nerve results in difficulty in chewing; **dysarthria** (dis-AR-thrē-a), or difficulty in speaking; and **dysphagia** (dis-FĀ-gē-a), or difficulty in swallowing. The tongue, when protruded, curls toward the affected side, and the affected side atrophies. The hypoglossal nerve may be injured as a result of conditions such as trauma, lesions, stroke, amyotrophic lateral sclerosis (Lou Gehrig's disease), or infections in the brainstem.

Vagal Neuropathy, Dysphagia, and Tachycardia (Refer page 452 of Textbook)

Injury to the vagus (X) nerve due to conditions such as trauma or lesions causes **vagal neuropathy**, or interruptions of sensations from many organs in the thoracic and abdominal cavities; **dysphagia** (dis-FĀ-gē-a), or difficulty in swallowing; and **tachycardia** (tak'-i-KAR-dē-a), or increased heart rate.

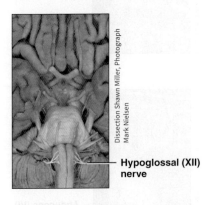

Dissection Shawn Miller, Photograph Mark Nielsen

— **Hypoglossal (XII) nerve**

DISORDERS: HOMEOSTATIC IMBALANCES

Cerebrovascular Accident

The most common brain disorder is a **cerebrovascular accident (CVA)**, also called a *stroke or brain attack*. CVAs affect 500,000 people each year in the United States and represent the third leading cause of death, behind heart attacks and cancer. A CVA is characterized by abrupt onset of persisting neurological symptoms, such as paralysis or loss of sensation, that arise from destruction of brain tissue. Common causes of CVAs are intracerebral hemorrhage (bleeding from a blood vessel in the pia mater or brain), emboli (blood clots), and atherosclerosis of the cerebral arteries (formation of cholesterol-containing plaques that block blood flow).

Among the risk factors implicated in CVAs are high blood pressure, high blood cholesterol, heart disease, narrowed carotid arteries, transient ischemic attacks (TIAs; discussed next), diabetes, smoking, obesity, and excessive alcohol intake.

A clot-dissolving drug called tissue *plasminogen activator* (*tPA*) is used to open up blocked blood vessels in the brain. The drug is most effective when administered *within three hours* of the onset of the CVA, however, and is helpful only for CVAs due to a blood clot (*ischemic CVAs*). Use of tPA can decrease the permanent disability associated with these types of CVAs by 50%. However, tPA should not be administered to individuals with strokes caused by hemorrhaging (*hemorrhagic CVAs*) since it can cause further injury or even death. The distinction between the types of CVA is made on the basis of a CT scan.

New studies show that "cold therapy" might be successful in limiting the amount of residual damage from a CVA. States of hypothermia, such as those experienced by cold-water drowning victims, seem to trigger a survival response in which the body requires less oxygen; application of this principle to stroke victims has showed promise. Some commercial companies now provide "CVA survival kits" including cooling blankets that can be kept in the home.

Transient Ischemic Attacks

A **transient ischemic attack** (**TIA**) (is-KĒ-mik) is an episode of temporary cerebral dysfunction caused by impaired blood flow to part of the brain. Symptoms include dizziness, weakness, numbness, or paralysis in a limb or on one side of the body; drooping of one side of the face; headache; slurred speech or difficulty understanding speech; and/or a partial loss of vision or double vision. Sometimes nausea or vomiting also occurs. The onset of symptoms is sudden and reaches maximum intensity almost immediately. A TIA usually persists for 5 to 10 minutes and only rarely lasts as long as 24 hours. It leaves no permanent neurological deficits. The causes of TIA include blood clots, atherosclerosis, and certain blood disorders. About one-third of patients who experience a TIA will have a CVA eventually. Therapy for TIAs includes drugs such as aspirin, which blocks the aggregation of blood platelets, and anticoagulants; cerebral artery bypass grafting; and carotid endarterectomy (removal of the cholesterol-containing plaques and inner lining of an artery).

Alzheimer's Disease

Severe memory deficits occur in individuals who have **Alzheimer's disease** (**AD**). AD is the most common form of *senile dementia*, the age-related loss of intellectual capabilities (including impairment of memory, judgment, abstract thinking, and changes in personality). The cause of most AD cases is still unknown, but evidence suggests that it is due to a combination of genetic factors, environmental or lifestyle factors, and the aging process. Individuals with AD initially have trouble remembering recent events. They then become confused and forgetful, often repeating questions or getting lost while traveling to familiar places. Disorientation grows; memories of past events disappear; and episodes of paranoia, hallucination, or violent changes in mood

may occur. As their minds continue to deteriorate, people with AD lose their ability to read, write, talk, eat, or walk. The disease culminates in dementia. A person with AD usually dies of some complication that afflicts bedridden patients, such as pneumonia. At autopsy brains of AD victims show four distinct structural abnormalities:

1. ***Loss of neurons that liberate acetylcholine.*** A major center of neurons that liberate acetylcholine is the nucleus basalis, which is below the globus pallidus. Axons of these neurons project widely throughout the cerebral cortex and limbic system. Their destruction is a hallmark of Alzheimer's disease.

2. ***Deterioration of the hippocampus.*** Recall that the hippocampus plays a major role in memory formation.

3. ***Beta-amyloid plaques.*** These are clusters of abnormal proteins deposited outside neurons.

4. ***Neurofibrillary tangles.*** These are abnormal bundles of filaments inside neurons in affected brain regions.

Drugs that inhibit acetylcholinesterase (AChE), the enzyme that inactivates acetylcholine, improve alertness and behavior in some AD patients.

Brain Tumors

A **brain tumor** is an abnormal growth of tissue in the brain that may be malignant or benign. Unlike most other tumors in the body, malignant and benign brain tumors may be equally serious, compressing adjacent tissues and causing a buildup of pressure in the skull. The most common malignant tumors are secondary tumors that metastasize from other cancers in the body, such as those in the lungs, breasts, skin (malignant melanoma), blood (leukemia), and lymphatic organs (lymphoma). Most primary brain tumors (those that originate within the brain) are gliomas, which develop in neuroglia. The symptoms of a brain tumor depend on its size, location, and rate of growth. Among the symptoms are headache, poor balance and coordination, dizziness, double vision, slurred speech, nausea and vomiting, fever, abnormal pulse and breathing rates, personality changes, numbness and weakness of the limbs, and seizures. Treatment options for brain tumors vary with their size, location, and type and may include surgery, radiation therapy, and/or chemotherapy. Unfortunately, chemotherapeutic agents do not readily cross the blood–brain barrier.

Attention Deficit Hyperactivity Disorder

Attention deficit hyperactivity disorder (ADHD) is a learning disorder characterized by poor or short attention span, a consistent level of hyperactivity, and a level of impulsiveness inappropriate for the child's age. ADHD is believed to affect about 5% of children and is diagnosed 10 times more often in boys than in girls. The condition typically begins in childhood and continues into adolescence and adulthood. Symptoms of ADHD develop in early childhood, often before age 4, and include difficulty in organizing and finishing tasks, lack of attention to details, short attention span and inability to concentrate, difficulty following instructions, talking excessively and frequently interrupting others, frequent running or excessive climbing, inability to play quietly alone, and difficulty waiting or taking turns.

The causes of ADHD are not fully understood, but it does have a strong genetic component. Some evidence also suggests that ADHD is related to problems with neurotransmitters. In addition, recent imaging studies have demonstrated that people with ADHD have less nervous tissue in specific regions of the brain such as the frontal and temporal lobes, caudate nucleus, and cerebellum. Treatment may involve remedial education, behavioral modification techniques, restructuring routines, and drugs that calm the child and help focus attention.

MEDICAL TERMINOLOGY

Agnosia (ag-NŌ-zē-a; *a-* = without; *-gnosia* = knowledge) Inability to recognize the significance of sensory stimuli such as sounds, sights, smells, tastes, and touch.

Apraxia (a-PRAK-sē-a; *-praxia* = coordinated) Inability to carry out purposeful movements in the absence of paralysis.

Consciousness (KON-shus-nes) A state of wakefulness in which an individual is fully alert, aware, and oriented, partly as a result of feedback between the cerebral cortex and reticular activating system.

Delirium (dē-LIR-ē-um = off the track) A transient disorder of abnormal cognition and disordered attention accompanied by disturbances of the sleep–wake cycle and psychomotor behavior (hyperactivity or hypoactivity of movements and speech). Also called **acute confusional state (ACS)**.

Dementia (de-MEN-shē-a; *de-* = away from; *-mentia* = mind) Permanent or progressive general loss of intellectual abilities, including impairment of memory, judgment, and abstract thinking and changes in personality.

Encephalitis (en′-sef-a-LĪ-tis) An acute inflammation of the brain caused by either a direct attack by any of several viruses or an allergic reaction to any of the many viruses that are normally harmless to the central nervous system. If the virus affects the spinal cord as well, the condition is called **encephalomyelitis**.

Encephalopathy (en-sef′-a-LOP-a-thē; *encephalo-* = brain; *-pathos* = disease) Any disorder of the brain.

Lethargy (LETH-ar-jē) A condition of functional sluggishness.

Microcephaly (mī-krō-SEF-a-lē; *micro-* = small; *-cephal* = head) A congenital condition that involves the development of a small brain and skull and frequently results in mental retardation.

Prosopagnosia (pros′-ō-pag-NŌ-sē-a; *a-* = without; *-gnosia* = knowledge) Inability to recognize faces, usually caused by damage to the facial recognition area in the inferior temporal lobe of both cerebral hemispheres.

Reye's syndrome (RĪZ) Occurs after a viral infection, particularly chickenpox or influenza, most often in children or teens who have taken aspirin; characterized by vomiting and brain dysfunction (disorientation, lethargy, and personality changes) that may progress to coma and death.

Stupor (STOO-por) Unresponsiveness from which a patient can be aroused only briefly and only by vigorous and repeated stimulation.

SELF-QUIZ QUESTIONS

Fill in the blanks in the following statements.

1. The cerebral hemispheres are connected internally by a broad band of white matter known as the _____.

2. List the five lobes of the cerebrum: _____, _____, _____, _____, _____.

3. The _____ separates the cerebrum into right and left halves.

Indicate whether the following statements are true or false.

4. The brain stem consists of the medulla oblongata, pons, and diencephalon.

5. You are the greatest student of anatomy and physiology, and you are well-prepared for your exam on the brain. As you confidently answer the questions, your brain is exhibiting beta waves.

Choose the one best answer to the following questions.

6. Which of the following is *not* a function of the thalamus?
   (a) relaying information from the cerebellum and basal nuclei to primary motor areas of the cerebral cortex
   (b) helping maintain consciousness
   (c) playing a role in emotions and memory
   (d) regulating body temperature
   (e) relaying sensory impulses to the cerebral cortex

7. Which of the following statements is *false*?
   (a) The blood supply to the brain is provided mainly by the internal carotid and vertebral arteries.
   (b) Neurons in the brain rely almost exclusively on aerobic respiration to produce ATP.
   (c) An interruption of blood flow to the brain for even 20 seconds may impair brain function.
   (d) Glucose supply to the brain must be continuous.
   (e) Low levels of glucose in the blood to the brain may result in unconsciousness.

8. In which of the following ways does cerebrospinal fluid contribute to homeostasis? (1) mechanical protection, (2) chemical protection, (3) electrical protection, (4) circulation, (5) immunity.
   (a) 1, 2, and 3
   (b) 2, 3, and 4
   (c) 3, 4, and 5
   (d) 1, 2, and 4
   (e) 2, 4, and 5

9. Which of the following are functions of the hypothalamus? (1) control of the ANS, (2) production of hormones, (3) regulation of emotional and behavioral patterns, (4) regulation of eating and drinking, (5) control of body temperature, (6) regulation of circadian rhythms.
   (a) 1, 2, 4, and 6
   (b) 2, 3, 5, and 6
   (c) 1, 3, 5, and 6
   (d) 1, 4, 5, and 6
   (e) 1, 2, 3, 4, 5, and 6

10. Which of the following statements is *false*?
    (a) Association tracts transmit nerve impulses between gyri in the same hemisphere.
    (b) Commissural tracts transmit impulses from the gyri in one cerebral hemisphere to the corresponding gyri in the other hemisphere.
    (c) Projection tracts form descending and ascending tracts that transmit impulses from the cerebrum and other parts of the brain to the spinal cord, or from the spinal cord to the brain.
    (d) The internal capsule is an example of a commissural tract.
    (e) The corpus callosum is an example of a commissural tract.

11. Which of the following statements is *true*?
    (a) The right and left hemispheres of the cerebrum are completely symmetrical.
    (b) The left hemisphere controls the left side of the body.
    (c) The right hemisphere is more important for spoken and written language.
    (d) The left hemisphere is more important for musical and artistic awareness.
    (e) Hemispheric lateralization is more pronounced in males than in females.

12. Match the following (some answers will be used more than once):

_____ (a) oculomotor
_____ (b) trigeminal
_____ (c) abducens
_____ (d) vestibulocochlear
_____ (e) accessory
_____ (f) vagus
_____ (g) facial
_____ (h) glossopharyngeal
_____ (i) olfactory
_____ (j) trochlear
_____ (k) optic
_____ (l) hypoglossal
_____ (m) functions in sense of smell
_____ (n) functions in hearing and equilibrium
_____ (o) functions in chewing
_____ (p) functions in facial expression and secretion of saliva and tears
_____ (q) functions in movement of tongue during speech and swallowing
_____ (r) functions in secretion of digestive fluids
_____ (s) functions in secretion of saliva, taste, regulation of blood pressure, and muscle sense
_____ (t) sensory only
_____ (u) functions in eye movement by controlling extrinsic eye muscles
_____ (v) functions in swallowing and head movements

(1) cranial nerve I
(2) cranial nerve II
(3) cranial nerve III
(4) cranial nerve IV
(5) cranial nerve V
(6) cranial nerve VI
(7) cranial nerve VII
(8) cranial nerve VIII
(9) cranial nerve IX
(10) cranial nerve X
(11) cranial nerve XI
(12) cranial nerve XII

13. Match the following (some answers may be used more than once):

_____ (a) emotional brain; involved in olfaction and memory
_____ (b) bridge connecting parts of the brain with each other
_____ (c) sensory relay area
_____ (d) alerts the cerebral cortex to incoming sensory signals
_____ (e) regulates posture and balance
_____ (f) lacks a blood-brain barrier; can monitor chemical changes in the blood
_____ (g) site of decussation of pyramids
_____ (h) site of pneumotaxic and apneustic areas
_____ (i) secretes melatonin
_____ (j) contains sensory, motor, and association areas
_____ (k) responsible for maintaining consciousness and awakening from sleep
_____ (l) controls ANS
_____ (m) contains reflex centers for movements of the eyes, head, and neck in response to visual and other stimuli, and reflex center for movements of the head and trunk in response to auditory stimuli
_____ (n) plays an essential role in awareness and in the acquisition of knowledge; cognition
_____ (o) several groups of nuclei that control large autonomic movements of skeletal muscles and help regulate muscle tone required for specific body movements
_____ (p) produces hormones that regulate endocrine gland function
_____ (q) contains the vital cardiovascular center and medullary rhythmicity center

(1) medulla oblongata
(2) pons
(3) midbrain
(4) cerebellum
(5) pineal gland
(6) thalamus
(7) hypothalamus
(8) cerebrum
(9) limbic system
(10) reticular formation
(11) circumventricular organs
(12) reticular activating system
(13) basal nuclei

14. Match the following:

_____ (a) protrusions in the medulla formed by the large corticospinal tracts

_____ (b) dura mater extension that separates the two cerebral hemispheres

_____ (c) fingerlike extensions of arachnoid mater where CSF is reabsorbed

_____ (d) dura mater extension that separates the two cerebellar hemispheres

_____ (e) located in the hypothalamus; relay stations for reflexes related to smell

_____ (f) folds in the cerebral cortex

_____ (g) shallow grooves in the cerebral cortex

_____ (h) bundles of white matter that relay information between the cerebellum and other parts of the brain

_____ (i) a thick band of sensory and motor tracts that connect the cerebral cortex with the brain stem and spinal cord

_____ (j) dura mater extension that separates the cerebrum from the cerebellum

_____ (k) thin membranous partition between the lateral ventricles

(1) gyri
(2) internal capsule
(3) mammillary bodies
(4) tentorium cerebelli
(5) pyramids
(6) falx cerebelli
(7) septum pellucidum
(8) cerebellar peduncles
(9) falx cerebri
(10) sulci
(11) arachnoid villi

15. Match the following:

_____ (a) allows planning and production of speech

_____ (b) receives impulses for sound

_____ (c) controls voluntary contraction of muscles

_____ (d) allows recognition and evaluation of visual experiences

_____ (e) integration and interpretation of somatic sensations; comparison of past to present sensations

_____ (f) receives impulses for touch, proprioception, pain, and temperature

_____ (g) receives impulses for taste

_____ (h) interpretation of sounds as speech, music, or noise

_____ (i) receives impulses from many sensory and association areas as well as the thalamus and brain stem; allows formation of thoughts so appropriate action can occur

_____ (j) translates words into thoughts

_____ (k) receives impulses for smell

_____ (l) allows awareness of shape, color, and movement

_____ (m) coordinates muscle movement for complex, learned sequential motor activities

_____ (n) involved in scanning eye movements

_____ (o) allows discrimination among different odors

(1) primary visual area
(2) primary auditory area
(3) primary gustatory area
(4) primary olfactory area
(5) primary somatosensory area
(6) primary motor area
(7) somatosensory association area
(8) visual association area
(9) frontal eye field
(10) Broca's area
(11) auditory association area
(12) premotor area
(13) Wernicke's area
(14) common integrative area
(15) orbitofrontal cortex

15 | THE AUTONOMIC NERVOUS SYSTEM

15.1 Comparison of Somatic and Autonomic Nervous Systems

1. The somatic nervous system operates under conscious control; the ANS usually operates without conscious control.
2. Sensory input for the somatic nervous system is mainly from the somatic senses and special senses; sensory input for the ANS is from interoceptors, in addition to somatic senses and special senses.
3. The axons of somatic motor neurons extend from the CNS and synapse directly with an effector. Autonomic motor pathways consist of two motor neurons in series. The axon of the first motor neuron extends from the CNS and synapses in an autonomic ganglion with the second motor neuron; the second neuron synapses with an effector.
4. The output (motor) portion of the ANS has two major divisions: sympathetic and parasympathetic. Most body organs receive dual innervation; usually one ANS division causes excitation and the other causes inhibition. The enteric division consists of nerves and ganglia within the wall of the GI tract.
5. Somatic nervous system effectors are skeletal muscles; ANS effectors include cardiac muscle, smooth muscle, and glands.
6. **Table 15.1** compares the somatic and autonomic nervous systems.

15.2 Anatomy of Autonomic Motor Pathways

1. A preganglionic neuron is the first of the two motor neurons in any autonomic motor pathway; the axon of the preganglionic neuron extends to an autonomic ganglion, where it synapses with a postganglionic neuron, the second neuron in the autonomic motor pathway. Preganglionic neurons are myelinated; postganglionic neurons are unmyelinated.
2. The cell bodies of sympathetic preganglionic neurons are in the lateral gray horns of the 12 thoracic and the first two or three lumbar segments of the spinal cord; the cell bodies of parasympathetic preganglionic neurons are in four cranial nerve nuclei (III, VII, IX, and X) in the brainstem and lateral gray matter of the second through fourth sacral segments of the spinal cord.
3. There are two major groups of autonomic ganglia: sympathetic ganglia and parasympathetic ganglia. Sympathetic ganglia include sympathetic trunk ganglia (on both sides of vertebral column) and prevertebral ganglia (anterior to vertebral column). Parasympathetic ganglia are known as terminal ganglia (near or inside visceral effectors).
4. Sympathetic preganglionic neurons synapse with postganglionic neurons in ganglia of the sympathetic trunk or in prevertebral ganglia; parasympathetic preganglionic neurons synapse with postganglionic neurons in terminal ganglia.

15.3 ANS Neurotransmitters and Receptors

1. Cholinergic neurons release acetylcholine. In the ANS, cholinergic neurons include all sympathetic and parasympathetic preganglionic neurons, sympathetic postganglionic neurons that innervate most sweat glands, and all parasympathetic postganglionic neurons.
2. Acetylcholine binds to cholinergic receptors. The two types of cholinergic receptors, both of which bind acetylcholine, are nicotinic receptors and muscarinic receptors. Nicotinic receptors are present in the plasma membranes of dendrites and cell bodies of both sympathetic and parasympathetic postganglionic neurons, in the plasma membranes of chromaffin cells of the adrenal medullae, and in the motor end plate at the neuromuscular junction. Muscarinic receptors are present in the plasma membranes of all effectors innervated by parasympathetic postganglionic neurons and in most sweat glands innervated by cholinergic sympathetic postganglionic neurons.
3. In the ANS, adrenergic neurons release norepinephrine. Most sympathetic postganglionic neurons are adrenergic.
4. Both epinephrine and norepinephrine bind to adrenergic receptors, which are found on visceral effectors innervated by most sympathetic postganglionic neurons. The two main types of adrenergic receptors are alpha receptors and beta receptors.
5. **Table 15.2** summarizes the types of cholinergic and adrenergic receptors.
6. An agonist is a substance that binds to and activates a receptor, mimicking the effect of a natural neurotransmitter or hormone. An antagonist is a substance that binds to and blocks a receptor, thereby preventing a natural neurotransmitter or hormone from exerting its effect.

15.4 Physiology of the ANS

1. The sympathetic division favors body functions that can support vigorous physical activity and rapid production of ATP (fight-or-flight response); the parasympathetic division regulates activities that conserve and restore body energy.
2. The effects of sympathetic stimulation are longer lasting and more widespread than the effects of parasympathetic stimulation.
3. **Table 15.3** compares structural and functional features of the sympathetic and parasympathetic divisions.
4. **Table 15.4** lists sympathetic and parasympathetic responses.

15.5 Integration and Control of Autonomic Functions

1. An autonomic (visceral) reflex adjusts the activities of smooth muscle, cardiac muscle, and glands.
2. An autonomic (visceral) reflex arc consists of a receptor, a sensory neuron, an integrating center, two autonomic motor neurons, and a visceral effector.

3. The hypothalamus is the major control and integration center of the ANS. It is connected to both the sympathetic and the parasympathetic divisions.

CLINICAL CONNECTIONS

Horner's Syndrome (Refer page 469 of Textbook)

In Horner's syndrome, the sympathetic innervation to one side of the face is lost due to an inherited mutation, an injury, or a disease that affects sympathetic outflow through the superior cervical ganglion. Symptoms occur on the affected side and include ptosis (drooping of the upper eyelid), miosis (constricted pupil), and anhidrosis (lack of sweating).

DISORDERS: HOMEOSTATIC IMBALANCES

Autonomic Dysreflexia

Autonomic dysreflexia (dis'-rē-FLEKS-sē-a) is an exaggerated response of the sympathetic division of the ANS that occurs in about 85% of individuals with spinal cord injury at or above the level of T6. The condition is seen after recovery from spinal shock (see Disorders: Homeostatic Imbalances in Chapter 13) and occurs due to interruption of the control of ANS neurons by higher centers. When certain sensory impulses, such as those resulting from stretching of a full urinary bladder, are unable to ascend the spinal cord, mass stimulation of the sympathetic nerves inferior to the level of injury occurs. Other triggers include stimulation of pain receptors and the visceral contractions resulting from sexual stimulation, labor/delivery, and bowel stimulation. Among the effects of increased sympathetic activity is severe vasoconstriction, which elevates blood pressure. In response, the cardiovascular center in the medulla oblongata (1) increases parasympathetic output via the vagus (X) nerve, which decreases heart rate, and (2) decreases sympathetic output, which causes dilation of blood vessels superior to the level of the injury.

The condition is characterized by a pounding headache; hypertension; flushed, warm skin with profuse sweating above the injury level; pale, cold, dry skin below the injury level; and anxiety. This emergency condition requires immediate intervention. The first approach is to quickly identify and remove the problematic stimulus. If this does not relieve the symptoms, an antihypertensive drug such as clonidine or nitroglycerin can be administered. Untreated autonomic dysreflexia can cause seizures, stroke, or heart attack.

Raynaud Phenomenon

In **Raynaud phenomenon** (rā-NŌ) the digits (fingers and toes) become ischemic (lack blood) after exposure to cold or with emotional stress. The condition is due to excessive sympathetic stimulation of smooth muscle in the arterioles of the digits and a heightened response to stimuli that cause vasoconstriction. When arterioles in the digits vasoconstrict in response to sympathetic stimulation, blood flow is greatly diminished. As a result, the digits may blanch (look white due to blockage of blood flow) or become cyanotic (look blue due to deoxygenated blood in capillaries). In extreme cases, the digits may become necrotic from lack of oxygen and nutrients. With rewarming after cold exposure, the arterioles may dilate, causing the fingers and toes to look red. Many patients with Raynaud phenomenon have low blood pressure. Some have increased numbers of alpha adrenergic receptors. Raynaud is most common in young women and occurs more often in cold climates. Patients with Raynaud phenomenon should avoid exposure to cold, wear warm clothing, and keep the hands and feet warm. Drugs used to treat Raynaud include nifedipine, a calcium channel blocker that relaxes vascular smooth muscle, and prazosin, which relaxes smooth muscle by blocking alpha receptors. Smoking and the use of alcohol or illicit drugs can exacerbate the symptoms of this condition.

MEDICAL TERMINOLOGY

Autonomic nerve neuropathy (noo-ROP-a-thē) A *neuropathy* (disorder of a cranial or spinal nerve) that affects one or more autonomic nerves, with multiple effects on the autonomic nervous system, including constipation, urinary incontinence, impotence, and fainting and low blood pressure when standing (*orthostatic hypotension*) due to decreased sympathetic control of the cardiovascular system. Often caused by long-term diabetes mellitus (*diabetic neuropathy*).

Biofeedback A technique in which an individual is provided with information regarding an autonomic response such as heart rate, blood pressure, or skin temperature. Various electronic monitoring devices provide visual or auditory signals about the autonomic responses. By concentrating on positive thoughts, individuals learn to alter autonomic responses. For example, biofeedback has been used to decrease heart rate and blood pressure and increase skin temperature in order to decrease the severity of migraine headaches.

Dysautonomia (dis-aw-tō-NŌ-mē-a; *dys-* = difficult; *-autonomia* = selfgoverning) An inherited disorder in which the autonomic nervous system functions abnormally, resulting in reduced tear gland secretions, poor vasomotor control, motor incoordination, skin blotching, absence of pain sensation, difficulty in swallowing, hyporeflexia, excessive vomiting, and emotional instability.

Hyperhidrosis (hī'-per-hī-DRŌ-sis; *hyper-* = above or too much; *-hidrosis* = sweat) Excessive or profuse sweating due to intense stimulation of sweat glands.

Mass reflex In cases of severe spinal cord injury above the level of the sixth thoracic vertebra, stimulation of the skin or overfilling of a visceral organ (such as the urinary bladder or colon) below the level of the injury results in intense activation of autonomic and somatic output from the spinal cord as reflex activity returns. The exaggerated response occurs because there is no inhibitory input from the brain. The mass reflex consists of flexor spasms of the lower limbs, evacuation of the urinary bladder and colon, and profuse sweating below the level of the lesion.

Megacolon (*mega-* = big) An abnormally large colon. In congenital mega-colon, parasympathetic nerves to the distal segment of the colon do not develop properly. Loss of motor function in the segment causes massive dilation of the normal proximal colon. The condition results in extreme constipation, abdominal distension, and occasionally, vomiting. Surgical removal of the affected segment of the colon corrects the disorder.

Reflex sympathetic dystrophy (RSD) A syndrome that includes sponta-neous pain, painful hypersensitivity to stimuli such as light touch, and excessive coldness and sweating in the involved body part. The disorder frequently involves the forearms, hands, knees, and feet. It appears that activation of the sympathetic division of the autonomic nervous sys-tem due to traumatized nociceptors as a result of trauma or surgery on bones or joints is involved. Treatment consists of anesthetics and physical therapy. Recent clinical studies also suggest that the drug baclofen can be used to reduce pain and restore normal function to the affected body part. Also called **complex regional pain syndrome type 1.**

Vagotomy (vā-GOT-ō-mē; *-tome* = incision) Cutting the vagus (X) nerve. It is frequently done to decrease the production of hydrochloric acid in per-sons with ulcers.

SELF-QUIZ QUESTIONS

Fill in the blanks in the following statements.

1. Cholinergic neurons release _____ and adrenergic neurons release _____ .

2. Because of the location of the preganglionic cell bodies, the sympathetic division of the ANS is also called the _____ division; the parasympathetic division is also called the _____ division.

Indicate whether the following statements are true or false.

3. The vagus nerves transmit 80% of the outflow of the parasympathetic preganglionic axons.

4. Organs that receive both sympathetic and parasympathetic motor im-pulses are said to have dual innervation.

Choose the one best answer to the following questions.

5. Which of the following statements is *false*?
   (a) A single sympathetic preganglionic fiber may synapse with 20 or more postganglionic fibers, which partly explains why sympathetic responses are widespread throughout the body.
   (b) Parasympathetic effects tend to be localized because parasympa-thetic neurons usually synapse in the terminal ganglia with only four or five postsynaptic neurons (all of which supply a single effector).
   (c) Some sympathetic preganglionic neurons extend to and terminate in the adrenal medullae.
   (d) The parasympathetic preganglionic neurons synapse with the post-ganglionic axons in the prevertebral ganglia.
   (e) Parasympathetic preganglionic neurons emerge from the CNS as part of a cranial nerve or anterior root of a spinal nerve.

6. Which autonomic plexus supplies the large intestine? (1) renal, (2) infe-rior mesenteric, (3) hypogastric, (4) superior mesenteric, (5) celiac.
   (a) 2, 3, and 4 (b) 1, 2, 3, 4, and 5 (c) 3 and 4
   (d) 4 and 5 (e) 2 and 4

7. Which of the following statements are *true*? (1) The somatic nervous sys-tem and the ANS both include sensory and motor neurons. (2) Somatic motor neurons release the neurotransmitter norepinephrine. (3) The effect of an autonomic motor neuron is either excitation or inhibition, but that of a somatic motor neuron is always excitation. (4) Autonomic sensory neurons are mostly associated with interoceptors. (5) Autonomic motor pathways consist of two motor neurons in series. (6) Somatic mo-tor pathways consist of two motor neurons in series.
   (a) 1, 2, 3, 4, and 5 (b) 1, 3, 4, and 5 (c) 2, 3, 5, and 6
   (d) 1, 3, 5, and 6 (e) 2, 4, 5, and 6

8. Which of the following statements is *false*?
   (a) The first neuron in an autonomic pathway is the preganglionic neuron.
   (b) The axons of preganglionic neurons are located in spinal or cranial nerves.
   (c) The postganglionic neuron's cell body is within the CNS.
   (d) Postganglionic neurons relay impulses from autonomic ganglia to visceral effectors.
   (e) All somatic motor neurons release acetylcholine.

9. Which of the following are *true*? (1) Monoamine oxidase enzymatically breaks down norepinephrine. (2) Activation of α_2 and β_2 receptors gen-erally produces excitation in the effectors. (3) A beta blocker works by preventing activation of β receptors by epinephrine and norepinephrine. (4) An agonist is a substance that binds to a receptor and prevents the natural neurotransmitter from exerting its effect. (5) Activation of nico-tinic receptors always causes excitation of the postsynaptic cell.
   (a) 2 and 3 (b) 1, 2, and 3 (c) 2, 4, and 5
   (d) 1, 2, 3, 4, and 5 (e) 1, 3, and 5

10. Which of the following are cholinergic neurons? (1) all sympathetic preganglionic neurons, (2) all parasympathetic preganglionic neurons, (3) all parasympathetic postganglionic neurons, (4) all sympathetic post-ganglionic neurons, (5) some sympathetic postganglionic neurons.
    (a) 1, 2, 3, and 5 (b) 1, 2, 3, and 4 (c) 2, 3, and 5
    (d) 2 and 5 (e) 1, 3, and 5

11. Which of the following statements are *true*? (1) Most sympathetic post-ganglionic axons are adrenergic. (2) Cholinergic receptors are classified as nicotinic and muscarinic. (3) Adrenergic receptors are classified as alpha and beta. (4) Muscarinic receptors are present on all effectors innervated by parasympathetic postganglionic axons. (5) In general, norepinephrine stimulates alpha receptors more vigorously than beta receptors; epinephrine is a potent stimulator of both alpha and beta receptors.
    (a) 1, 2, 3, 4, and 5 (b) 2, 3, 4 and 5 (c) 1, 3, 4, and 5
    (d) 3, 4, and 5 (e) 1, 2, 3, and 4

12. Which of the following are reasons why the effects of sympathetic stimu-lation are longer lasting and more widespread than those of parasym-pathetic stimulation? (1) There is greater divergence of sympathetic postganglionic fibers. (2) There is less divergence of sympathetic post-ganglionic fibers. (3) Acetylcholinesterase quickly inactivates ACh, but norepinephrine lingers in the synaptic cleft for a longer time. (4) Nor-epinephrine and epinephrine secreted into the blood by the adrenal

medullae intensify the actions of the sympathetic division. (5) ACh remains in the synaptic cleft until norepinephrine is produced.

(a) 1 and 3   (b) 1, 3, and 5   (c) 1, 3, and 4

(d) 2, 3, and 4   (e) 2, 3, and 5

13. Place the following components of an autonomic reflex arc in the correct order from beginning to end.

(a) postganglionic neuron   (b) sensory neuron

(c) effector   (d) autonomic ganglion

(e) receptor   (f) preganglionic neuron

(g) integrating center

14. Match the following:

_____ (a) also known as intramural ganglia

_____ (b) includes the celiac, superior mesenteric, and inferior mesenteric ganglia

_____ (c) also called vertebral chain or paravertebral ganglia

_____ (d) lie in a vertical row on either side of the vertebral column

_____ (e) postganglionic fibers, in general, innervate organs below the diaphragm

_____ (f) ganglia located at the end of an autonomic motor pathway close to or actually within the wall of a visceral organ

_____ (g) includes ciliary, pterygopalatine, submandibular, and otic ganglia

_____ (h) extend from base of the skull to the coccyx

_____ (i) myelinated preganglionic fibers that connect the anterior rami of spinal nerves with the ganglia of the sympathetic trunk

_____ (j) also known as collateral ganglia

_____ (k) unmyelinated postganglionic axons that connect the ganglia of the sympathetic trunk to spinal nerves

(1) sympathetic trunk ganglia

(2) prevertebral ganglia

(3) terminal ganglia

(4) white rami communicantes

(5) gray rami communicantes

15. Match the following:

_____ (a) stimulates urination and defecation

_____ (b) prepares the body for emergency situations

_____ (c) fight-or-flight response

_____ (d) promotes digestion and absorption of food

_____ (e) concerned primarily with processes involving the expenditure of energy

_____ (f) controlled by the posterior and lateral portions of the hypothalamus

_____ (g) controlled by the anterior and medial portions of the hypothalamus

_____ (h) causes a decrease in heart rate

(1) increased activity of the sympathetic division of the ANS

(2) increased activity of the parasympathetic division of the ANS

16 | SENSORY, MOTOR, AND INTEGRATIVE SYSTEMS

CHAPTER REVIEW

16.1 Sensation

1. Sensation is the conscious or subconscious awareness of changes in the external or internal environment. Perception is the conscious awareness and interpretation of sensations and is primarily a function of the cerebral cortex.
2. The nature of a sensation and the type of reaction generated vary according to the destination of sensory impulses in the CNS.
3. Each different type of sensation is a sensory modality; usually, a given sensory neuron serves only one modality.
4. General senses include somatic senses (touch, pressure, vibration, warmth, cold, pain, itch, tickle, and proprioception) and visceral senses; special senses include the modalities of smell, taste, vision, hearing, and equilibrium.
5. For a sensation to arise, four events typically occur: stimulation, transduction, generation of impulses, and integration.
6. Simple receptors, consisting of free nerve endings and encapsulated nerve endings, are associated with the general senses; complex receptors are associated with the special senses.
7. Sensory receptors respond to stimuli by producing receptor potentials.
8. **Table 16.1** summarizes the classification of sensory receptors.
9. Adaptation is a decrease in sensitivity during a long-lasting stimulus. Receptors are either rapidly adapting or slowly adapting.

16.2 Somatic Sensations

1. Somatic sensations include tactile sensations (touch, pressure, vibration, itch, and tickle), thermal sensations (warmth and cold), pain, and proprioception.
2. Receptors for tactile, thermal, and pain sensations are located in the skin, subcutaneous layer, and mucous membranes of the mouth, vagina, and anus.
3. Receptors for touch are (a) corpuscles of touch or Meissner corpuscles and hair root plexuses, which are rapidly adapting, and (b) slowly adapting type I cutaneous mechanoreceptors or tactile discs. Type II cutaneous mechanoreceptors or Ruffini corpuscles, which are slowly adapting, are sensitive to stretching.
4. Receptors for pressure include type I and type II cutaneous mechanoreceptors.
5. Receptors for vibration are corpuscles of touch and lamellated corpuscles.
6. Itch receptors, tickle receptors, and thermoreceptors are free nerve endings. Cold receptors are located in the stratum basale of the epidermis; warm receptors are located in the dermis.

7. Pain receptors (nociceptors) are free nerve endings that are located in nearly every body tissue.
8. Nerve impulses for fast pain propagate along medium-diameter, myelinated A fibers; those for slow pain conduct along small-diameter, unmyelinated C fibers.
9. Receptors for proprioceptive sensations (position and movement of body parts) are located in muscles, tendons, joints, and the inner ear. Proprioceptors include muscle spindles, tendon organs, joint kinesthetic receptors, and hair cells of the inner ear.
10. **Table 16.2** summarizes the somatic sensory receptors and the sensations they convey.

16.3 Somatic Sensory Pathways

1. Somatic sensory pathways from receptors to the cerebral cortex involve first-order, second-order, and third-order neurons.
2. Axon collaterals (branches) of somatic sensory neurons simultaneously carry signals into the cerebellum and the reticular formation of the brainstem.
3. Nerve impulses for touch, pressure, vibration, and conscious proprioception in the limbs, trunk, neck, and posterior head ascend to the cerebral cortex along the posterior column–medial lemniscus pathway.
4. Nerve impulses for pain, temperature, itch, and tickle from the limbs, trunk, neck, and posterior head ascend to the cerebral cortex along the anterolateral (spinothalamic) pathway.
5. Nerve impulses for most somatic sensations (tactile, thermal, pain, and proprioceptive) from the face, nasal cavity, oral cavity, and teeth ascend to the cerebral cortex along the trigeminothalamic pathway.
6. Specific regions of the primary somatosensory area (postcentral gyrus) of the cerebral cortex receive somatic sensory input from different parts of the body.
7. The neural pathways to the cerebellum are the anterior and posterior spinocerebellar tracts, which transmit impulses for subconscious proprioception from the trunk and lower limbs.
8. **Table 16.3** summarizes the major somatic sensory pathways.

16.4 Control of Body Movement

1. All excitatory and inhibitory signals that control movement converge on motor neurons, also known as lower motor neurons (LMNs) or the final common pathway.
2. Neurons in four neural circuits participate in control of movement by providing input to lower motor neurons: local circuit

neurons, upper motor neurons, basal nuclei neurons, and cerebellar neurons.

3. The primary motor area (precentral gyrus) of the cortex is a major control region for executing voluntary movements.

4. The axons of upper motor neurons (UMNs) extend from the brain to lower motor neurons via direct and indirect motor pathways.

5. The direct (pyramidal) pathways include the corticospinal pathways and the corticobulbar pathway. The corticospinal pathways convey nerve impulses from the motor cortex to skeletal muscles in the limbs and trunk. The corticobulbar pathway conveys nerve impulses from the motor cortex to skeletal muscles in the head.

6. Indirect (extrapyramidal) pathways extend from several motor centers of the brainstem into the spinal cord. Indirect pathways include the rubrospinal, tectospinal, vestibulospinal, and medial and lateral reticulospinal tracts.

7. **Table 16.4** summarizes the major somatic motor pathways.

8. Neurons of the basal nuclei assist movement by providing input to the upper motor neurons. They help initiate and suppress movements.

9. Vestibular nuclei in the medulla and pons play an important role in regulating posture; the reticular formation helps control posture and muscle tone; the superior colliculus allows the body to respond to sudden visual stimuli and permits rapid movements of the eyes; and the red nucleus permits fine, precise, voluntary movements of the distal parts of the upper limbs.

10. The cerebellum is active in learning and performing rapid, coordinated, highly skilled movements. It also contributes to maintaining balance and posture.

16.5 Integrative Functions of the Cerebrum

1. Sleep and wakefulness are integrative functions that are controlled by the suprachiasmatic nucleus and the reticular activating system (RAS).

2. Non-rapid eye movement (NREM) sleep consists of four stages.

3. Most dreaming occurs during rapid eye movement (REM) sleep.

4. Coma is a state of unconsciousness in which an individual has little or no response to stimuli.

5. Learning is the ability to acquire new information or skills through instruction or experience.

6. Memory is the process by which information acquired through learning is stored and retrieved.

7. Language is a system of vocal sounds and symbols that conveys information.

CLINICAL CONNECTIONS

Phantom Limb Sensation (Refer page 484 of Textbook)

Patients who have had a limb amputated may still experience sensations such as itching, pressure, tingling, or pain as if the limb were still there. This phenomenon is called **phantom limb sensation**. Although the limb has been removed, severed endings of sensory axons are still present in the remaining stump. If these severed endings are activated, the cerebral cortex interprets the sensation as coming from the sensory receptors in the nonexisting (phantom) limb. Another explanation for phantom limb sensation is that the area of the cerebral cortex that previously received sensory input from the missing limb undergoes extensive functional reorganization that allows it to respond to stimuli from another body part. The remodeling of this cortical area is thought to give rise to false sensory perceptions from the missing limb. Phantom limb pain can be very distressing to an amputee. Many report that the pain is severe or extremely intense, and that it often does not respond to traditional pain medication therapy. In such cases, alternative treatments may include electrical nerve stimulation, acupuncture, and biofeedback.

Acupuncture (Refer page 485 of Textbook)

Acupuncture is a type of therapy that originated in China over 2000 years ago. It is based on the idea that vital energy called *qi* (pronounced chee) flows through the body along pathways called *meridians*. Practitioners of acupuncture believe that illness results when the flow of qi along one or more meridians is blocked or out of balance. Acupuncture is performed by inserting fine needles into the skin at specific locations in order to unblock and rebalance the flow of qi. A main purpose for using acupuncture is to provide pain relief. According to one theory, acupuncture relives pain by activating sensory neurons that ultimately trigger the release of neurotransmitters that function as analgesics such as endorphins, enkephalins, and dynorphins (see Section 12.5 Neurotransmitters). In contrast, many Western practitioners view the acupuncture points as places to stimulate nerves, muscles, and connective tissue. Studies have shown that acupuncture is a safe procedure as long as it is administered by a trained professional who uses a sterile needle for each application site. Therefore, many members of the medical community consider acupuncture to be a viable alternative to traditional methods for relieving pain.

Syphilis (Refer page 490 of Textbook)

Syphilis is a sexually transmitted disease caused by the bacterium *Treponema pallidum*. Because it is a bacterial infection, it can be treated with antibiotics. However, if the infection is not treated, the third stage of syphilis typically causes debilitating neurological symptoms. A common outcome is progressive degeneration of the posterior portions of the spinal cord, including the posterior columns, posterior spinocerebellar tracts, and posterior roots. Somatic sensations are lost, and the person's gait becomes uncoordinated and jerky because proprioceptive impulses fail to reach the cerebellum.

Paralysis (Refer page 493 of Textbook)

Damage or disease of *lower* motor neurons produces **flaccid paralysis** (FLAS-id or FLAK-sid) of muscles on the same side of the body. There is neither voluntary nor reflex action of the innervated muscle fibers, muscle tone is decreased or lost, and the muscle remains limp or flaccid. Injury or disease of *upper* motor neurons in the cerebral cortex removes inhibitory influences that some of these neurons have on lower motor neurons, which causes **spastic paralysis** of muscles on the opposite side of the body. In this condition muscle tone is increased, reflexes are exaggerated, and pathological reflexes such as the Babinski sign appear (see Clinical Connection: Reflexes and Diagnosis in Chapter 13).

Amyotrophic Lateral Sclerosis (Refer page 495 of Textbook)

Amyotrophic lateral sclerosis (ALS) (ā'-mī-ō-TRŌF-ik; *a-* = without; *-myo-* = muscle; *-trophic* = nourishment) is a progressive degenerative disease that attacks motor areas of the cerebral cortex, axons of upper motor neurons in the lateral white columns (corticospinal and rubrospinal tracts), and lower motor neuron cell bodies. It causes progressive muscle weakness and atrophy. ALS often begins in sections of the spinal cord that serve the hands and arms but rapidly spreads to involve the whole body and face, without affecting intellect or sensations. Death typically occurs in 2 to 5 years. ALS is commonly known as *Lou Gehrig's disease*, after the New York Yankees baseball player who died from it at age 37 in 1941.

Inherited mutations account for about 15% of all cases of ALS (familial ALS). Noninherited (sporadic) cases of ALS appear to have several implicating factors. According to one theory, there is a buildup in the synaptic cleft of the neurotransmitter glutamate released by motor neurons due to a mutation of the protein that normally deactivates and recycles the neurotransmitter. The excess glutamate causes motor neurons to malfunction and eventually die. The drug riluzole, which is used to treat ALS, reduces damage to motor neurons by decreasing the release of glutamate. Other factors may include damage to motor neurons by free radicals, autoimmune responses, viral infections, deficiency of nerve growth factor, apoptosis (programmed cell death), environmental toxins, and trauma.

In addition to riluzole, ALS is treated with drugs that relieve symptoms such as fatigue, muscle pain and spasticity, excessive saliva, and difficulty sleeping. The only other treatment is supportive care provided by physical, occupational, and speech therapists; nutritionists; social workers; and home care and hospice nurses.

Disorders of the Basal Nuclei (Refer page 498 of Textbook)

Disorders of the basal nuclei can affect body movements, cognition, and behavior. Uncontrollable shaking (tremor) and muscle rigidity (stiffness) are hallmark signs of **Parkinson's disease (PD)** (see Disorders: Homeostatic Imbalances at the end of this chapter). In this disorder, dopamine-releasing neurons that extend from the substantia nigra to the putamen and caudate nucleus degenerate.

Huntington disease (HD) is an inherited disorder in which the caudate nucleus and putamen degenerate, with loss of neurons that normally release GABA or acetylcholine. A key sign of HD is **chorea** (KŌ-rē-a = a dance), in which rapid, jerky movements occur involuntarily and without purpose. Progressive mental deterioration also occurs. Symptoms of HD often do not appear until age 30 or 40. Death occurs 10 to 20 years after symptoms first appear.

Tourette syndrome is a disorder that is characterized by involuntary body movements (motor tics) and the use of inappropriate or unnecessary sounds or words (vocal tics). Although the cause is unknown, research suggests that this disorder involves a dysfunction of the cognitive neural circuits between the basal nuclei and the prefrontal cortex.

Some psychiatric disorders, such as schizophrenia and obsessive–compulsive disorder, are thought to involve dysfunction of the behavioral neural circuits between the basal nuclei and the limbic system. In **schizophrenia**, excess dopamine activity in the brain causes a person to experience delusions, distortions of reality, paranoia, and hallucinations. People who have **obsessive–compulsive disorder (OCD)** experience repetitive thoughts (obsessions) that cause repetitive behaviors (compulsions) that they feel obligated to perform. For example, a person with OCD might have repetitive thoughts about someone breaking into the house; these thoughts might drive that person to check the doors of the house over and over again (for minutes or hours at a time) to make sure that they are locked.

Sleep Disorders (Refer page 500 of Textbook)

Sleep disorders affect over 70 million Americans each year. Common sleep disorders include insomnia, sleep apnea, and narcolepsy. A person with **insomnia** (in-SOM-nē-a) has difficulty in falling asleep or staying asleep. Possible causes of insomnia include stress, excessive caffeine intake, disruption of circadian rhythms (for example, working the night shift instead of the day shift at your job), and depression. **Sleep apnea** (AP-nē-a) is a disorder in which a person repeatedly stops breathing for 10 or more seconds while sleeping. Most often, it occurs because a loss of muscle tone in pharyngeal muscles allows the airway to collapse. **Narcolepsy** (NAR-kō-lep-sē) is a condition in which REM sleep cannot be inhibited during waking periods. As a result, involuntary periods of sleep that last about 15 minutes occur throughout the day. Recent studies have revealed that people with narcolepsy have a deficiency of the neuropeptide *orexin*, which is also known as **hypocretin**. Orexin is released from certain neurons of the hypothalamus and has a role in promoting wakefulness.

Amnesia (Refer page 501 of Textbook)

Amnesia (am-NĒ-zē-a = forgetfulness) refers to the lack or loss of memory. It is a total or partial inability to remember past experiences. In *anterograde amnesia*, there is memory loss for events that occur *after* the trauma or disease that caused the condition. In other words, it is an inability to form new memories. In *retrograde amnesia*, there is a memory loss for events that occurred *before* the trauma or disease that caused the condition. In other words, it is an inability to recall past events.

Aphasia (Refer page 501 of Textbook)

Much of what we know about language areas comes from studies of patients with language or speech disturbances that have resulted from brain damage. Injury to language areas of the cerebral cortex results in **aphasia** (a-FĀ-zē-a), an inability to use or comprehend words. Damage to Broca's area results in **expressive aphasia**, an inability to properly articulate or form words. People with this type of aphasia know what they wish to say but have difficulty speaking. Damage to Wernicke's area results in **receptive aphasia**, characterized by faulty understanding of spoken or written words. A person experiencing this type of aphasia may fluently produce strings of words that have no meaning ("word salad"). For example, someone with receptive aphasia might say, "I car river dinner light rang pencil jog." The underlying deficit may be **word deafness** (an inability to understand spoken words), **word blindness** (an inability to understand written words), or both.

DISORDERS: HOMEOSTATIC IMBALANCES

Parkinson's Disease

Parkinson's disease (PD) is a progressive disorder of the CNS that typically affects its victims around age 60. Neurons that extend from the substantia nigra to the putamen and caudate nucleus, where they release the neurotransmitter dopamine (DA), degenerate in PD. The caudate nucleus of the basal nuclei contains neurons that liberate the neurotransmitter acetylcholine (ACh). Although the level of ACh does not change as the level of DA declines, the imbalance of neurotransmitter activity—too little DA and too much ACh—is thought to cause most of the symptoms. The cause of PD is unknown, but toxic environmental chemicals, such as pesticides, herbicides, and carbon monoxide, are suspected contributing agents. Only 5% of PD patients have a family history of the disease.

In PD patients, involuntary skeletal muscle contractions often interfere with voluntary movement. For instance, the muscles of the upper limb may alternately contract and relax, causing the hand to shake. This shaking, called **tremor**, is the most common symptom of PD. Also, muscle tone may increase greatly, causing rigidity of the involved body part. Rigidity of the facial muscles gives the face a masklike appearance. The expression is characterized by a wide-eyed, unblinking stare and a slightly open mouth with uncontrolled drooling.

Motor performance is also impaired by **bradykinesia** (brady- = slow), slowness of movements. Activities such as shaving, cutting food, and buttoning a shirt take longer and become increasingly more difficult as the disease progresses. Muscular movements also exhibit **hypokinesia** (hypo- = under), decreasing range of motion. For example, words are written smaller, letters are poorly formed, and eventually handwriting becomes illegible. Often, walking is impaired; steps become shorter and shuffling, and arm swing diminishes. Even speech may be affected.

Treatment of PD is directed toward increasing levels of DA and decreasing levels of ACh. Although people with PD do not manufacture enough dopamine, taking it orally is useless because DA cannot cross the blood–brain barrier. Even though symptoms are partially relieved by a drug developed in the 1960s called levodopa (L-dopa), a precursor of DA, the drug does not slow the progression of the disease. As more and more affected brain cells die, the drug becomes useless. Another drug, called selegiline (Deprenyl®), is used to inhibit monoamine oxidase, an enzyme that degrades dopamine. This drug slows progression of PD and may be used together with levodopa. Anticholinergic drugs such as benztropine and trihexyphenidyl can also be used to block the effects of ACh at some of the synapses between basal nuclei neurons. This helps to restore the balance between ACh and DA. Anticholinergic drugs effectively reduce symptomatic tremor, rigidity, and drooling.

For more than a decade, surgeons have sought to reverse the effects of Parkinson's disease by transplanting dopamine-rich fetal nervous tissue into the basal nuclei (usually the putamen) of patients with severe PD. Only a few postsurgical patients have shown any degree of improvement, such as less rigidity and improved quickness of motion. Another surgical technique that has produced improvement for some patients is *pallidotomy*, in which a part of the globus pallidus that generates tremors and produces muscle rigidity is destroyed. In addition, some patients are being treated with a surgical procedure called *deep-brain stimulation (DBS)*, which involves the implantation of electrodes into the subthalamic nucleus. The electrical currents released by the implanted electrodes reduce many of the symptoms of PD.

MEDICAL TERMINOLOGY

Cerebral palsy (CP) A motor disorder that results in the loss of muscle control and coordination; caused by damage of the motor areas of the brain during fetal life, birth, or infancy. Radiation during fetal life, temporary lack of oxygen during birth, and hydrocephalus during infancy may also cause cerebral palsy.

Pain threshold The smallest intensity of a painful stimulus at which a person perceives pain. All individuals have the same pain threshold.

Pain tolerance The greatest intensity of painful stimulation that a person is able to tolerate. Individuals vary in their tolerance to pain.

Synesthesia (sin-es-THĒ-zē-a; *syn-* = together; *-aisthesis* = sensation) A condition in which sensations of two or more modalities accompany one another. In some cases, a stimulus for one sensation is perceived as a stimulus for another; for example, a sound produces a sensation of color. In other cases, a stimulus from one part of the body is experienced as coming from a different part.

SELF-QUIZ QUESTIONS

Fill in the blanks in the following statements.

1. _____ is the conscious or subconscious awareness of external or internal stimuli; _____ is the conscious awareness and interpretation of sensory input.

2. The term used to describe the crossing over of axons from one side of the brain or spinal cord to the other side is _____ .

Indicate whether the following statements are true or false.

3. Touch, pressure, and pain are all classified as tactile sensations.

4. Awakening from sleep involves increased activity in the reticular activating system.

Choose the one best answer to the following questions.

5. A nurse touches the lower back of a patient, but the patient does not feel the sensation. Which of the following could explain the lack of sensation? (1) The stimulus was not in the receptive field. (2) The generator potential has not reached threshold. (3) There is damage to the somatosensory region of the cerebral cortex. (4) The nurse was stimulating a proprioceptor. (5) A slowly adapting receptor has been stimulated.
   (a) 1, 3 and 5
   (b) 3, 4, and 5
   (c) 1, 2, and 3
   (d) 2, 3, and 4
   (e) 1 only

6. Which of the following statements is *false*? (1) Upper motor neurons transmit impulses from the CNS to skeletal muscle fibers. (2) Lower motor neurons have their cell bodies in the brain stem and spinal cord. (3) Local circuit neurons receive input from somatic sensory receptors and help coordinate rhythmic activity in specific muscle groups. (4) The activity of upper motor neurons is influenced by both the basal ganglia and cerebellum. (5) The cerebellum helps to monitor differences between intended movements and actual movements for coordination, posture, and balance.

7. Which of the following statements are *true*? (1) Slow pain is a result of impulse propagation along myelinated A nerve fibers. (2) Visceral pain occurs when nociceptors in the skin are stimulated. (3) Referred pain is pain felt in an area far from the stimulated organ. (4) Nociceptors exhibit very little adaptation. (5) Nociceptors are located in every body tissue.
   (a) 1, 3, 4, and 5
   (b) 2, 3, and 5
   (c) 1 and 5
   (d) 3 and 4
   (e) 3, 4, and 5

8. You cannot "hear" with your eyes because
   (a) hearing is a somatic sense and vision is a special sense.
   (b) the sensory neurons for sight carry information only for the modality of vision.
   (c) the impulses for hearing are transmitted to the somatosensory area of the cerebral cortex.
   (d) hearing receptors are selective and vision receptors are not.
   (e) hearing receptors produce a generator potential and vision receptors produce a receptor potential.

9. Which of the following statements is *false*?
   (a) First-order sensory neurons carry signals from the somatic receptors into either the brain stem or spinal cord.
   (b) Second-order neurons carry signals from the spinal cord and brain stem to the thalamus.
   (c) Third-order neurons project to the primary somatosensory area of the cortex, where conscious perception of the sensation results.
   (d) The somatic sensory pathways to the cerebellum are the posterior column–medial lemniscus pathway and the anterolateral pathway.
   (e) Axons of second-order neurons decussate (cross) in the spinal cord or brain stem before ascending to the thalamus.

10. Which of the following is not an aspect of cerebellar function?
   (a) monitoring movement intention
   (b) monitoring actual movement
   (c) comparing intent with actual performance
   (d) sending out corrective signals
   (e) directing sensory input to effectors

11. During REM sleep (1) neuronal activity in the pons and midbrain is high, (2) most somatic motor neurons are inhibited, (3) most dreaming occurs, (4) sleepwalking can occur, (5) there is an increase in heart rate and blood pressure.
   (a) 1, 2, 4, and 5
   (b) 2, 3, and 5
   (c) 1, 2, 3, 4, and 5
   (d) 2, 3, and 4
   (e) 1, 2, and 3

12. Which of the following statements is *incorrect*?
- (a) The graded potentials produced by receptors that serve the senses of touch, pressure, stretching, vibration, pain, proprioception, and smell are generator potentials.
- (b) The graded potentials produced by receptors that serve the special senses of vision, hearing, equilibrium, and taste are receptor potentials.
- (c) When a generator potential is large enough to reach threshold, it generates one or more nerve impulses in its first-order sensory neuron.
- (d) A receptor potential generates nerve impulses in a second-order neuron.
- (e) The amplitude of both generator and receptor potentials varies with the intensity of the stimulus.

13. Match the following (some answers may be used more than once):

_____ (a) receptors located in muscles, tendons, joints, and the inner ear	(1) exteroceptors
_____ (b) receptors located in blood vessels, visceral organs, muscles, and the nervous system	(2) interoceptors
_____ (c) receptors that detect temperature changes	(3) proprioceptors
_____ (d) receptors that detect light that strikes the retina of the eye	(4) mechanoreceptors
_____ (e) receptors located at or near the external surface of the body	(5) thermoreceptors
_____ (f) bare dendrites associated with pain, thermal, tickle, itch, and some touch sensations	(6) nociceptors
_____ (g) receptors that provide information about body position, muscle tension, and position and activity of joints	(7) photoreceptors
_____ (h) receptors that sense osmotic pressures of body fluids	(8) chemoreceptors
_____ (i) receptors that detect chemicals in the mouth, nose, and body fluids	(9) free nerve endings
_____ (j) receptors that detect mechanical pressure or stretching	(10) encapsulated nerve endings
_____ (k) receptors that respond to stimuli resulting from physical or chemical damage to tissues	(11) osmoreceptors
_____ (l) dendrites enclosed in a connective tissue capsule	

14. Match the following:

_____ (a) specialized groupings of muscle fibers interspersed among regular skeletal muscle fibers and oriented parallel to them; monitor changes in the length of a skeletal muscle	(1) Meissner corpuscles
_____ (b) inform the CNS about changes in muscle tension	(2) Merkel discs
_____ (c) widely distributed free nerve ending receptors for pain	(3) Ruffini corpuscles
_____ (d) encapsulated receptors for touch located in the dermal papillae; found in hairless skin, eyelids, tip of the tongue, and lips	(4) Pacinian corpuscles
_____ (e) lamellated corpuscles that detect pressure	(5) cold receptors
_____ (f) type II cutaneous mechanoreceptors; most sensitive to stretching that occurs as digits or limbs are moved	(6) warm receptors
	(7) nociceptors
	(8) tendon organs
	(9) joint kinesthetic receptors
	(10) muscle spindles

_____ (g) located in the stratum basale and activated by low temperatures

_____ (h) located in the dermis and activated by high temperatures

_____ (i) found within and around the articular capsules of synovial joints; respond to pressure and acceleration and deceleration of joints

_____ (j) type I cutaneous mechanoreceptors that function in touch

15. Match the following:

_____ (a) located in the precentral gyrus, this is the major control region of the cerebral cortex for initiation of voluntary movements	(1) posterior column
_____ (b) direct pathways conveying impulses from the cerebral cortex to the spinal cord that result in precise, voluntary movements	(2) anterolateral (spinothalamic) pathway
_____ (c) contains motor neurons that control skilled movements of the hands and feet	(3) spinocerebellar tracts
_____ (d) tracts include rubrospinal, tectospinal, vestibulospinal, lateral reticulospinal, and medial reticulospinal	(4) lateral corticospinal tract
_____ (e) contain neurons that help initiate and terminate movements; can suppress unwanted movements; influence muscle tone	(5) anterior corticospinal tract
_____ (f) carries impulses for pain, temperature, tickle, and itch	(6) corticobulbar tracts
_____ (g) the major routes relaying proprioceptive input to the cerebellum; critical for posture, balance, and coordination of skilled movements	(7) extrapyramidal pathways
_____ (h) composed of axons of first-order neurons; include the gracile fasciculus and cuneate fasciculus	(8) pyramidal pathways
_____ (i) contains motor neurons that coordinate movements of the axial skeleton	(9) primary motor area
_____ (j) contains axons that convey impulses for precise, voluntary movements of the eyes, tongue, and neck, plus chewing, facial expression, and speech	(10) basal nuclei
_____ (k) conveys sensations of touch, conscious proprioception, pressure, and vibration to the cerebral cortex	(11) posterior column–medial lemniscus pathway
_____ (l) carries impulses for most somatic sensations from the face, nasal cavity, oral cavity, and teeth	(12) trigeminothalamic pathway

17 THE SPECIAL SENSES

17.1 Olfaction: Sense of Smell

1. The receptors for olfaction, which are bipolar neurons, are in the nasal epithelium along with olfactory glands, which produce mucus that dissolves odorants.
2. In olfactory reception, a receptor potential develops and triggers one or more nerve impulses.
3. The threshold of smell is low, and adaptation to odors occurs quickly.
4. Axons of olfactory receptor cells form the olfactory (I) nerves, which convey nerve impulses to the olfactory bulbs, olfactory tracts, limbic system, and cerebral cortex (temporal and frontal lobes).

17.2 Gustation: Sensation of Taste

1. The receptors for gustation, the gustatory receptor cells, are located in taste buds.
2. Dissolved chemicals, called tastants, stimulate gustatory receptor cells by flowing through ion channels in the plasma membrane or by binding to receptors attached to G proteins in the membrane.
3. Receptor potentials developed in gustatory receptor cells cause the release of neurotransmitter, which can generate nerve impulses in first-order sensory neurons.
4. The threshold varies with the taste involved, and adaptation to taste occurs quickly.
5. Gustatory receptor cells trigger nerve impulses in the facial (VII), glossopharyngeal (IX), and vagus (X) nerves. Taste signals then pass to the medulla oblongata, thalamus, and cerebral cortex (parietal lobe).

17.3 Vision: An Overview

1. More than half of the sensory receptors in the human body are located in the eyes.
2. The eyes are responsible for the detection of visible light, the part of the electromagnetic spectrum with wavelengths ranging from about 400 to 700 nm.

17.4 Accessory Structures of the Eye

1. Accessory structures of the eyes include the eyebrows, eyelids, eyelashes, lacrimal apparatus, and extrinsic eye muscles.
2. The lacrimal apparatus consists of structures that produce and drain tears.

17.5 Anatomy of the Eyeball

1. The eye is constructed of three layers: (a) fibrous tunic (sclera and cornea), (b) vascular tunic (choroid, ciliary body, and iris), and (c) retina.
4. The retina consists of a pigmented layer and a neural layer that includes a photoreceptor layer, bipolar cell layer, ganglion cell layer, horizontal cells, and amacrine cells.
5. The anterior cavity contains aqueous humor; the vitreous chamber contains the vitreous body.

17.6 Physiology of Vision

1. Image formation on the retina involves refraction of light rays by the cornea and lens, which focus an inverted image on the fovea centralis of the retina.
2. For viewing close objects, the lens increases its curvature (accommodation) and the pupil constricts to prevent light rays from entering the eye through the periphery of the lens.
3. The near point of vision is the minimum distance from the eye at which an object can be clearly focused with maximum accommodation.
4. In convergence, the eyeballs move medially so they are both directed toward an object being viewed.
5. The first step in vision is the absorption of light by photopigments in rods and cones and isomerization of cis-retinal. Receptor potentials in rods and cones decrease the release of inhibitory neurotransmitter, which induces graded potentials in bipolar cells and horizontal cells.
6. Horizontal cells transmit inhibitory signals between photoreceptors and bipolar cells; bipolar or amacrine cells transmit excitatory signals to ganglion cells, which depolarize and initiate nerve impulses.
7. Impulses from ganglion cells are conveyed into the optic (II) nerve, through the optic chiasm and optic tract, to the thalamus. From the thalamus, impulses for vision propagate to the cerebral cortex (occipital lobe). Axon collaterals of retinal ganglion cells extend to the midbrain and hypothalamus.

17.7 Hearing

1. The external (outer) ear consists of the auricle, external auditory canal, and tympanic membrane (eardrum).
2. The middle ear consists of the auditory tube, ossicles, oval window, and round window.

3. The internal (inner) ear consists of the bony labyrinth and membranous labyrinth. The internal ear contains the spiral organ (organ of Corti), the organ of hearing.

4. Sound waves enter the external auditory canal, strike the tympanic membrane, pass through the ossicles, strike the oval window, set up waves in the perilymph, strike the vestibular membrane and scala tympani, increase pressure in the endolymph, vibrate the basilar membrane, and stimulate hair bundles on the spiral organ (organ of Corti).

5. Hair cells convert mechanical vibrations into a receptor potential, which releases neurotransmitter that can initiate nerve impulses in first-order sensory neurons.

6. Sensory axons in the cochlear branch of the vestibulocochlear (VIII) nerve terminate in the medulla oblongata. Auditory signals then pass to the inferior colliculus, thalamus, and temporal lobes of the cerebral cortex.

17.8 Equilibrium

1. The maculae of the utricle and saccule detect linear acceleration or deceleration and head tilt.

2. The cristae in the semicircular ducts detect rotational acceleration or deceleration.

3. Most vestibular branch axons of the vestibulocochlear nerve enter the brainstem and terminate in the medulla and pons; other axons enter the cerebellum.

17.9 Development of the Eyes and Ears

1. The eyes begin their development about 22 days after fertilization from ectoderm of the lateral walls of the prosencephalon (forebrain).

2. The ears begin their development about 22 days after fertilization from a thickening of ectoderm on either side of the rhombencephalon (hindbrain). The sequence of development of the ear is internal ear, middle ear, and external ear.

17.10 Aging and the Special Senses

1. Most people do not experience problems with the senses of smell and taste until about age 50.

2. Among the age-related changes to the eyes are presbyopia, cataracts, difficulty adjusting to light, macular disease, glaucoma, dry eyes, and decreased sharpness of vision.

3. With age there is a progressive loss of hearing, and tinnitus occurs more frequently.

CLINICAL CONNECTIONS

Hyposmia (Refer page 507 of Textbook)

Women often have a keener sense of smell than men do, especially at the time of ovulation. Smoking seriously impairs the sense of smell in the short term and may cause long-term damage to olfactory receptors. With aging the sense of smell deteriorates. **Hyposmia** (hī-POZ-mē-a; *osmi* = smell, odor), a reduced ability to smell, affects half of those over age 65 and 75% of those over age 80. Hyposmia also can be caused by neurological changes, such as a head injury, Alzheimer's disease, or Parkinson's disease; certain drugs, such as antihistamines, analgesics, or steroids; and the damaging effects of smoking.

Taste Aversion (Refer page 510 of Textbook)

Probably because of taste projections to the hypothalamus and limbic system, there is a strong link between taste and pleasant or unpleasant emotions. Sweet foods evoke reactions of pleasure while bitter ones cause expressions of disgust, even in newborn babies. This phenomenon is the basis for **taste aversion**, in which people and animals quickly learn to avoid a food if it upsets the digestive system. The advantage of avoiding foods that cause such illness is longer survival. However, the drugs and radiation treatments used to combat cancer often cause nausea and gastrointestinal upset regardless of what foods are consumed. Thus, cancer patients may lose their appetite because they develop taste aversions for most foods.

Age-Related Macular Disease (Refer page 515 of Textbook)

Age-related macular disease (AMD), also known as *macular degeneration*, is a degenerative disorder of the retina in persons 50 years of age and older. In AMD, abnormalities occur in the region of the macula lutea, which is ordinarily the area of most acute vision. Victims of advanced AMD retain their peripheral vision but lose the ability to see straight ahead. For instance, they cannot see facial features to identify a person in front of them. AMD is the leading cause of blindness in those over age 75, afflicting 13 million Americans, and is 2.5 times more common in pack-a-day smokers than in nonsmokers. Initially, a person may experience blurring and distortion at the center of the visual field. In "dry" AMD, central vision gradually diminishes because the pigmented layer atrophies and degenerates. There is no effective treatment. In about 10% of cases, dry AMD progresses to "wet" AMD, in which new blood vessels form in the choroid and leak plasma or blood under the retina. Vision loss can be slowed by using laser surgery to destroy the leaking blood vessels.

Detached Retina (Refer page 517 of Textbook)

A **detached retina** may occur due to trauma, such as a blow to the head, in various eye disorders, or as a result of age-related degeneration. The detachment occurs between the neural portion of the retina and the pigmented epithelium. Fluid accumulates between these layers, forcing the thin, pliable retina to billow outward. The result is distorted vision and blindness in the corresponding field of vision. The retina may be reattached by laser surgery or cryosurgery (localized application of extreme cold), and reattachment must be accomplished quickly to avoid permanent damage to the retina.

Presbyopia (Refer page 519 of Textbook)

With aging, the lens loses elasticity and thus its ability to curve to focus on objects that are close. Therefore, older people cannot read print at the same close range as can younger people. This condition is called **presbyopia** (prez-bē-Ō-pē-a; *presby-* = old; *-opia* = pertaining to the eye or vision). By age 40 the near point of vision may have increased to 20 cm (8 in.), and at age 60 it may be as much as 80 cm (31 in.). Presbyopia usually begins in the mid-40s. At about that age, people who have not previously worn glasses begin to need them for reading. Those who already wear glasses typically start to need bifocals, lenses that can focus for both distant and close vision.

LASIK (Refer page 520 of Textbook)

An increasingly popular alternative to wearing glasses or contact lenses is refractive surgery to correct the curvature of the cornea for conditions such as farsightedness, nearsightedness, and astigmatism. The most common type of refractive surgery is **LASIK** (laser-assisted in-situ keratomileusis). After anesthetic drops are placed in the eye, a circular flap of tissue from the center of the cornea is cut. The flap is folded out of the way, and the underlying layer of cornea is reshaped with a laser, one microscopic layer at a time. A computer assists the physician in removing very precise layers of the cornea. After the sculpting is complete, the corneal flap is repositioned over the treated area. A patch is placed over the eye overnight, and the flap quickly reattaches to the rest of the cornea.

Color Blindness and Night Blindness (Refer page 521 of Textbook)

Most forms of **color blindness**, an inherited inability to distinguish between certain colors, result from the absence or deficiency of one of the three types of cones. The most common type is *red–green color blindness*, in which red cones or green cones are missing. As a result, the person cannot distinguish between red and green. Prolonged vitamin A deficiency and the resulting below-normal amount of rhodopsin may cause **night blindness** or *nyctalopia* (nik′-ta-LŌ-pē-a), an inability to see well at low light levels.

Loud Sounds and Hair Cell Damage (Refer page 532 of Textbook)

Exposure to loud music and the engine roar of jet planes, revved-up motorcycles, lawn mowers, and vacuum cleaners damages hair cells of the cochlea. Because prolonged noise exposure causes hearing loss, employers in the United States must require workers to use hearing protectors when occupational noise levels exceed 90 dB. Rock concerts and even inexpensive headphones can easily produce sounds over 110 dB. Continued exposure to high-intensity sounds is one cause of **deafness**, a significant or total hearing loss. The louder the sounds, the more rapid the hearing loss. Deafness usually begins with loss of sensitivity for high-pitched sounds. If you are listening to music through ear buds and bystanders can hear it, the decibel level is in the damaging range. Most people fail to notice their progressive hearing loss until destruction is extensive and they begin having difficulty understanding speech. Wearing earplugs with a noise-reduction rating of 30 dB while engaging in noisy activities can protect the sensitivity of your ears.

Cochlear Implants (Refer page 534 of Textbook)

A **cochlear implant** is a device that translates sounds into electrical signals that can be interpreted by the brain. Such a device is useful for people with deafness that is caused by damage to hair cells in the cochlea. The external parts of a cochlear implant consist of (1) a *microphone* worn around the ear that picks up sound waves, (2) a *sound processor*, which may be placed in a shirt pocket, that converts sound waves into electrical signals, and (3) a *transmitter*, worn behind the ear, which receives signals from the sound processor and passes them to an internal receiver. The internal parts of a cochlear implant are the (1) *internal receiver*, which relays signals to (2) *electrodes* implanted in the cochlea, where they trigger nerve impulses in sensory neurons in the cochlear branch of the vestibulocochlear (VIII) nerve. These artificially induced nerve impulses propagate over their normal pathways to the brain. The perceived sounds are crude compared to normal hearing, but they provide a sense of rhythm and loudness; information about certain noises, such as those made by telephones and automobiles; and the pitch and cadence of speech. Some patients hear well enough with a cochlear implant to use the telephone.

Motion Sickness (Refer page 538 of Textbook)

Motion sickness is a condition that results when there is a conflict among the senses with regard to motion. For example, the vestibular apparatus senses angular and vertical motion, while the eyes and proprioceptors in muscles and joints determine the position of the body in space. If you are in the cabin of a moving ship, your vestibular apparatus informs the brain that there is movement from waves. But your eyes don't see any movement. This leads to the conflict among the senses. Motion sickness can also be experienced in other situations that involve movement, for example, in a car or airplane or on a train or amusement park ride.

Symptoms of motion sickness include paleness, restlessness, excess salivation, nausea, dizziness, cold sweats, headache, and malaise that may progress to vomiting. Once the motion is stopped, the symptoms disappear. If it is not possible to stop the motion, you might try sitting in the front seat of a car, the forward car of a train, the upper deck on a boat, or the wing seats in a plane. Looking at the horizon and not reading also help. Medications for motion sickness are usually taken in advance of travel and include scopolamine in time-release patches or tablets, dimenhydrinate (Dramamine®), and meclizine (Bonine®).

DISORDERS: HOMEOSTATIC IMBALANCES

Cataracts

A common cause of blindness is a loss of transparency of the lens known as a **cataract** (KAT-a-rakt = waterfall). The lens becomes cloudy (less transparent) due to changes in the structure of the lens proteins. Cataracts often occur with aging but may also be caused by injury, excessive exposure to ultraviolet rays, certain medications (such as long-term use of steroids), or complications of other diseases (for example, diabetes). People who smoke also have increased risk of developing cataracts. Fortunately, sight can usually be restored by surgical removal of the old lens and implantation of a new artificial one.

Glaucoma

Glaucoma (glaw-KŌ-ma) is the most common cause of blindness in the United States, afflicting about 2% of the population over age 40. In many cases, glaucoma is due to an abnormally high intraocular pressure as a result of a buildup of aqueous humor within the anterior cavity. The fluid compresses the lens into the vitreous body and puts pressure on the neurons of the retina. Persistent pressure results in a progression from mild visual impairment to irreversible destruction of neurons of the retina, damage to the optic nerve, and blindness. Glaucoma is

painless, and the other eye compensates largely, so a person may experience considerable retinal damage and loss of vision before the condition is diagnosed. Because glaucoma occurs more often with advancing age, regular measurement of intraocular pressure is an increasingly important part of an eye exam as people grow older. Risk factors include race (African Americans are more susceptible), increasing age, family history, and past eye injuries and disorders.

Some individuals have another form of glaucoma called **normal-tension** (*low-tension*) **glaucoma**. In this condition, there is damage to the optic nerve with a corresponding loss of vision, even though intraocular pressure is normal. Although the cause is unknown, it appears to be related to a fragile optic nerve, vasospasm of blood vessels around the optic nerve, and ischemia due to narrowed or obstructed blood vessels around the optic nerve. The incidence of normal-tension glaucoma is higher among Japanese and Koreans and among females.

Deafness

Deafness is significant or total hearing loss. **Sensorineural deafness** (sen′-so-rē-NOO-ral) is caused by either impairment of hair cells in the cochlea or damage of the cochlear branch of the vestibulocochlear (VIII) nerve. This type of deafness may be caused by atherosclerosis, which reduces blood supply to the ears; by repeated exposure to loud noise, which destroys hair cells of the spiral organ; by certain drugs such as aspirin and streptomycin; and/or by genetic factors. **Conduction deafness** is caused by impairment of the external and middle ear mechanisms for transmitting sounds to the cochlea. Causes of conduction deafness include otosclerosis, the deposition of new bone around the oval window; impacted cerumen; injury to the eardrum; and aging, which often results in thickening of the eardrum and stiffening of the joints of the auditory ossicles. A hearing test called *Weber's test* is used to distinguish between sensorineural and conduction deafness. In the test, the stem of a vibrating fork is held to the forehead. In people with normal hearing, the sound is heard equally in both ears. If the sound is heard best in the affected ear, the deafness is probably of the conduction type; if the sound is heard best in the normal ear, it is probably of the sensorineural type.

Ménière's Disease

Ménière's disease (men'-ē-ĀRZ) results from an increased amount of endolymph that enlarges the membranous labyrinth. Among the symptoms are fluctuating hearing loss (caused by distortion of the basilar membrane of the cochlea) and roaring tinnitus (ringing). Spinning or whirling vertigo (dizziness) is also characteristic of Ménière's disease. Almost total destruction of hearing may occur over a period of years.

Otitis Media

Otitis media (ō-TĪ-tis MĒ-dē-a) is an acute infection of the middle ear caused mainly by bacteria and associated with infections of the nose and throat. Symptoms include pain, malaise, fever, and a reddening and outward bulging of the eardrum, which may rupture unless prompt treatment is received. (This may involve draining pus from the middle ear.) Bacteria passing into the auditory tube from the nasopharynx are the primary cause of middle ear infections. Children are more susceptible than adults to middle ear infections because their auditory tubes are almost horizontal, which decreases drainage. If otitis media occurs frequently, a surgical procedure called **tympanotomy** (tim'-pa-NOT-ō-mē; *tympano-* = drum; *-tome* = incision) is often employed. This consists of the insertion of a small tube into the eardrum to provide a pathway for the drainage of fluid from the middle ear.

MEDICAL TERMINOLOGY

Ageusia (a-GOO-sē-a; *a-* = without; *-geusis* = taste) Loss of the sense of taste.

Amblyopia (am'-blē-Ō-pē-a; *ambly-* = dull or dim) Term used to describe the loss of vision in an otherwise normal eye that, because of muscle imbalance, cannot focus in synchrony with the other eye. Sometimes called "wandering eyeball" or a "lazy eye."

Anosmia (an-OZ-mē-a; *a-* = without; *-osmi* = smell, odor) Total lack of the sense of smell.

Barotrauma (bar'-ō-TRAW-ma; *baros-* = weight) Damage or pain, mainly affecting the middle ear, as a result of pressure changes. It occurs when pressure on the outer side of the tympanic membrane is higher than on the inner side, for example, when flying in an airplane or diving. Swallowing or holding your nose and exhaling with your mouth closed usually opens the auditory tubes, allowing air into the middle ear to equalize the pressure.

Blepharitis (blef-a-RĪ-tis; *blephar-* = eyelid; *-itis* = inflammation of) An inflammation of the eyelid.

Conjunctivitis (pinkeye) An inflammation of the conjunctiva; when caused by bacteria such as pneumococci, staphylococci, or *Haemophilus influenzae*, it is very contagious and more common in children. Conjunctivitis may also be caused by irritants, such as dust, smoke, or pollutants in the air, in which case it is not contagious.

Corneal abrasion (KOR-nē-al a-BRĀ -zhun) A scratch on the surface of the cornea, for example, from a speck of dirt or damaged contact lenses. Symptoms include pain, redness, watering, blurry vision, sensitivity to bright light, and frequent blinking.

Corneal transplant A procedure in which a defective cornea is removed and a donor cornea of similar diameter is sewn in. It is the most common and most successful transplant operation. Since the cornea is avascular, antibodies in the blood that might cause rejection do not enter the transplanted tissue, and rejection rarely occurs. The shortage of donor corneas has been partially overcome by the development of artificial corneas made of plastic.

Diabetic retinopathy (ret-i-NOP-a-thē; *retino-* = retina; *-pathos* = suffering) Degenerative disease of the retina due to diabetes mellitus, in which blood vessels in the retina are damaged or new ones grow and interfere with vision.

Exotropia (ek'-sō-TRŌ-pē-a; *ex-* = out; *-tropia* = turning) Turning outward of the eyes.

Keratitis (ker'-a-TĪ-tis; *kerat-* = cornea) An inflammation or infection of the cornea.

Miosis (mī-Ō-sis) Constriction of the pupil.

Mydriasis (mi-DRĪ-a-sis) Dilation of the pupil.

Nystagmus (nis-TAG-mus; *nystagm-* = nodding or drowsy) A rapid involuntary movement of the eyeballs, possibly caused by a disease of the central nervous system. It is associated with conditions that cause vertigo.

Otalgia (ō-TAL-jē-a; *oto-* = ear; *-algia* = pain) Earache.

Photophobia (fō'-tō-FŌ-bē-a; *photo-* = light; *-phobia* = fear) Abnormal visual intolerance to light.

Ptosis (TŌ-sis = fall) Falling or drooping of the eyelid (or slippage of any organ below its normal position).

Retinoblastoma (ret-i-nō-blas-TŌ-ma; *-oma* = tumor) A tumor arising from immature retinal cells; it accounts for 2% of childhood cancers.

Scotoma (skō-TŌ-ma = darkness) An area of reduced or lost vision in the visual field.

Strabismus (stra-BIZ-mus; *strabismos* = squinting) Misalignment of the eyeballs so that the eyes do not move in unison when viewing an object; the affected eye turns either medially or laterally with respect to the normal eye and the result is double vision (diplopia). It may be caused by physical trauma, vascular injuries, or tumors of the extrinsic eye muscle or the oculomotor (III), trochlear (IV), or abducens (VI) cranial nerves.

Tinnitus (ti-NĪ-tus) A ringing, roaring, or clicking in the ears.

Tonometer (tō-NOM-ē-ter; *tono-* = tension or pressure; *-metron* = measure) An instrument for measuring pressure, especially intraocular pressure.

Trachoma (tra-KŌ-ma) A serious form of conjunctivitis and the greatest single cause of blindness in the world. It is caused by the bacterium *Chlamydia trachomatis*. The disease produces an excessive growth of subconjunctival tissue and invasion of blood vessels into the cornea, which progresses until the entire cornea is opaque.

Vertigo (VER-ti-gō = dizziness) A sensation of spinning or movement in which the world seems to revolve or the person seems to revolve in space, often associated with nausea and, in some cases, vomiting. It may be caused by arthritis of the neck or an infection of the vestibular apparatus.

SELF-QUIZ QUESTIONS

Fill in the blanks in the following statements.

1. The five primary taste sensations are _____, _____, _____, _____, and _____.

2. _____ equilibrium refers to the maintenance of the position of the body relative to the force of gravity; _____ equilibrium refers to the maintenance of body position in response to rotational acceleration or deceleration.

Indicate whether the following statements are true or false.

3. Of all of the special senses, only smell and taste sensations project both to higher cortical areas and to the limbic system.

4. The ability to change the curvature of the lens for near vision is convergence.

Choose the one best answer to the following questions.

5. Which of the following are *true*? (1) The sites of olfactory transduction are the olfactory hairs. (2) The olfactory bulbs transmit impulses to the temporal lobe of the cerebral cortex. (3) The axons of olfactory receptors pass through the olfactory foramina in the cribriform plate of the ethmoid bone. (4) The olfactory nerves are bundles of axons that terminate in the olfactory tracts. (5) Within the olfactory bulbs, the first-order neurons synapse with the second-order neurons.
   (a) 1, 2, and 4   (b) 2, 3, 4, and 5   (c) 1, 2, 3, 4, and 5
   (d) 1, 3, and 5   (e) 1, 2, 3, and 5

6. Which of the following statements is *incorrect*?
   (a) Olfactory receptors respond to the chemical stimulation of an odorant molecule by producing a receptor potential.
   (b) Basal stem cells continually produce new olfactory receptors.
   (c) Adaptation to odors is rapid and occurs in both olfactory receptors and the CNS.
   (d) Production of nasal mucus by olfactory glands serves to moisten the olfactory epithelium and dissolve odorants.
   (e) The orbitofrontal area is an important region for odor identification and discrimination.

7. Which of the following statements is *incorrect*?
   (a) Taste is a chemical sense.
   (b) The receptors for taste sensations are found in taste buds located on the tongue, the soft palate, the pharynx, and the epiglottis.
   (c) Gustatory hairs are the sites of taste transduction.
   (d) The threshold for bitter substances is the highest.
   (e) Complete adaptation to taste can occur in 1 to 5 minutes.

8. When viewing an object close to your eyes, which of the following are required for proper image formation on the retina? (1) increased curvature of the lens, (2) contraction of the ciliary muscle, (3) divergence of the eyeballs, (4) refraction of light at the anterior and posterior surfaces of the cornea, (5) constriction of the pupil by contraction of the extrinsic eye muscles.
   (a) 1, 2, 3, 4, and 5   (b) 1, 2, and 4   (c) 1, 2, 3, and 4
   (d) 2, 4, and 5   (e) 2, 3, and 4

9. Which of the following are *mismatched*?
   (a) fungiform papillae: scattered over the entire tongue's surface.
   (b) filiform papillae: contain taste buds in early childhood.
   (c) vallate papillae: each houses 100–300 taste buds.
   (d) foliate papillae: located in trenches on the lateral margins of the tongue.
   (e) fungiform papillae: each houses about five taste buds.

10. Place in order the structures involved in the visual pathway.
    (a) optic tract   (b) ganglion cells   (c) cornea
    (d) lens   (e) bipolar cells   (f) optic nerve
    (g) visual cortex   (h) vitreous body   (i) optic chiasm
    (j) aqueous humor   (k) pupil   (l) photoreceptors
    (m) thalamus

11. Which of the following statements is *incorrect*?
    (a) Retinal is the light-absorbing portion of all visual photopigments.
    (b) The only photopigment in rods is rhodopsin, but three different cone photopigments are present in the retina.
    (c) Retinal is a derivative of vitamin C.
    (d) Color vision results from different colors of light selectively activating different cone photopigments.
    (e) Bleaching and regeneration of the photopigments account for much but not all of the sensitivity changes during light and dark adaptation.

12. Which of the following is the *correct* sequence for the auditory pathway?
    (a) external auditory canal, tympanic membrane, auditory ossicles, oval window, cochlea and spiral organ
    (b) tympanic membrane, external auditory canal, auditory ossicles, cochlea and spiral organ, round window
    (c) auditory ossicles, tympanic membrane, cochlea and spiral organ, round window, oval window, external auditory canal
    (d) auricle, tympanic membrane, round window, cochlea and spiral organ, oval window
    (e) external auditory canal, tympanic membrane, auditory ossicles, internal auditory canal, spiral organ, oval window

13. Match the following:
    _____ (a) upper and lower eyelids; shade the eyes during sleep, spread lubricating secretions over the eyeballs
    _____ (b) produces and drains tears
    _____ (c) arch transversely above the eyeballs and help protect the eyeballs from foreign objects, perspiration, and the direct rays of the sun
    _____ (d) move the eyeball medially, laterally, superiorly, or inferiorly
    _____ (e) a thick fold of connective tissue that gives form and support to the eyelids
    _____ (f) modified sebaceous glands; secretion helps keep eyelids from adhering to one another
    _____ (g) project from the border of each eyelid; help protect the eyeballs from foreign objects, perspiration, and direct rays of the sun
    _____ (h) a thin, protective mucous membrane that lines the inner aspect of the eyelids and passes from the eyelids onto the surface of the eyeball, where it covers the sclera

    (1) palpebrae
    (2) tarsal or Meibomian glands
    (3) conjunctiva
    (4) eyelashes
    (5) lacrimal apparatus
    (6) extrinsic eye muscles
    (7) eyebrows
    (8) tarsal plate

14. Match the following:

_____ (a) colored portion of the eyeball; regulates the amount of light entering the posterior part of the eyeball

_____ (b) innermost layer of the eyeball; beginning of the visual pathway; contains rods and cones

_____ (c) biconvex transparent structure that fine tunes focusing of light rays for clear vision

_____ (d) circular band of smooth muscle that alters the shape of the lens for near or far vision

_____ (e) watery fluid in the anterior cavity that helps nourish the lens and cornea; helps maintain shape of the eyeball

_____ (f) the hole in the center of the iris

_____ (g) white of the eye; gives shape to the eyeball, makes it more rigid, protects its inner parts

_____ (h) avascular superficial layer of the eyeball; includes cornea and sclera

_____ (i) middle, vascularized layer of the eyeball; includes choroid, ciliary body, and iris

(1) sclera
(2) ciliary muscle
(3) iris
(4) pupil
(5) retina
(6) fibrous tunic
(7) aqueous humor
(8) lens
(9) vascular tunic

15. Match the following:

_____ (a) partition between external auditory canal and middle ear

_____ (b) oval central portion of the bony labyrinth; contains utricle and saccule

_____ (c) receptor for static equilibrium; also contributes to some aspects of dynamic equilibrium; consists of hair cells and supporting cells

_____ (d) contains hair cells which are the receptors for hearing

_____ (e) ear bones: malleus, incus, stapes

_____ (f) the pressure equalization tube that connects the middle ear to the nasopharynx

_____ (g) contains the spiral organ

_____ (h) fluid found within the membranous labyrinth; pressure waves in this fluid cause vibration of the basilar membrane

_____ (i) receptor organs for equilibrium; the saccule, utricle, and semicircular canals

_____ (j) swollen enlargement in semicircular canals; contains structures involved in dynamic equilibrium

_____ (k) opening between the middle ear and internal ear; is enclosed by a membrane called the secondary tympanic membrane

_____ (l) the flap of elastic cartilage covered by skin that captures sound waves; the pinna

_____ (m) fluid found inside bony labyrinth; bulging of the oval window causes pressure waves in this fluid

_____ (n) opening between the middle and inner ear; receives base of stapes

(1) auricle
(2) tympanic membrane
(3) auditory ossicles
(4) vestibular apparatus
(5) ampulla
(6) cochlea
(7) perilymph
(8) oval window
(9) round window
(10) auditory or eustachian tube
(11) vestibule
(12) endolymph
(13) spiral organ
(14) macula

18 THE ENDOCRINE SYSTEM

Introduction

1. Hormones regulate the activity of smooth muscle, cardiac muscle, and some glands; alter metabolism; spur growth and development; influence reproductive processes; and participate in circadian (daily) rhythms.

18.1 Comparison of Control by the Nervous and Endocrine Systems

1. The nervous system controls homeostasis through nerve impulses and neurotransmitters, which act locally and quickly. The endocrine system uses hormones, which act more slowly in distant parts of the body. (See **Table 18.1**.)
2. The nervous system controls neurons, muscle cells, and glandular cells; the endocrine system regulates virtually all body cells.

18.2 Endocrine Glands

1. Exocrine glands (sudoriferous, sebaceous, mucous, and digestive) secrete their products through ducts into body cavities or onto body surfaces. Endocrine glands secrete hormones into interstitial fluid. Then, the hormones diffuse into the blood.
2. The endocrine system consists of endocrine glands (pituitary, thyroid, parathyroid, adrenal, and pineal glands) and other hormone-secreting tissues (hypothalamus, thymus, pancreas, ovaries, testes, kidneys, stomach, liver, small intestine, skin, heart, adipose tissue, and placenta).

18.3 Hormone Activity

1. Hormones affect only specific target cells that have receptors to recognize (bind) a given hormone. The number of hormone receptors may decrease (down-regulation) or increase (up-regulation).
2. Circulating hormones enter the bloodstream; local hormones (paracrines and autocrines) act locally on neighboring cells.
3. Chemically, hormones are either lipid-soluble (steroids, thyroid hormones, and nitric oxide) or water-soluble (amines; peptides, proteins, and glycoproteins; and eicosanoids). (See **Table 18.2**.)
4. Water-soluble hormone molecules circulate in the watery blood plasma in a "free" form (not attached to plasma proteins); most lipid-soluble hormones are bound to transport proteins synthesized by the liver.

18.4 Mechanisms of Hormone Action

1. Lipid-soluble steroid hormones and thyroid hormones affect cell function by altering gene expression.

2. Water-soluble hormones alter cell function by activating plasma membrane receptors, which elicit production of a second messenger that activates various enzymes inside the cell.
3. Hormonal interactions can have three types of effects: permissive, synergistic, or antagonistic.

18.5 Homeostatic Control of Hormone Secretion

1. Hormone secretion is controlled by signals from the nervous system, chemical changes in blood, and other hormones.
2. Negative feedback systems regulate the secretion of many hormones.

18.6 Hypothalamus and Pituitary Gland

1. The hypothalamus is the major integrating link between the nervous and endocrine systems. The hypothalamus and pituitary gland regulate virtually all aspects of growth, development, metabolism, and homeostasis. The pituitary gland is located in the hypophyseal fossa and is divided into two main portions: the anterior pituitary (glandular portion) and the posterior pituitary (nervous portion).
2. Secretion of anterior pituitary hormones is stimulated by releasing hormones and suppressed by inhibiting hormones from the hypothalamus.
3. The blood supply to the anterior pituitary is from the superior hypophyseal arteries. Hypothalamic releasing and inhibiting hormones enter the primary plexus and flow to the secondary plexus in the anterior pituitary by the hypophyseal portal veins.
4. The anterior pituitary consists of somatotrophs that produce growth hormone (GH), lactotrophs that produce prolactin (PRL), corticotrophs that secrete adrenocorticotropic hormone (ACTH) and melanocyte-stimulating hormone (MSH), thyrotrophs that secrete thyroid-stimulating hormone (TSH), and gonadotrophs that synthesize follicle-stimulating hormone (FSH) and luteinizing hormone (LH). (See **Tables 18.3** and **18.4**.)
5. Growth hormone (GH) stimulates body growth through insulinlike growth factors (IGFs). Secretion of GH is inhibited by GHIH (growth hormone–inhibiting hormone, or somatostatin) and promoted by GHRH (growth hormone–releasing hormone).
6. TSH regulates thyroid gland activities. Its secretion is stimulated by TRH (thyrotropin-releasing hormone) and suppressed by GHIH.
7. FSH and LH regulate the activities of the gonads—ovaries and testes. Their secretion is controlled by GnRH (gonadotropin-releasing hormone).

8. Prolactin (PRL) helps initiate milk secretion. Prolactin-inhibiting hormone (PIH) suppresses secretion of PRL; prolactin-releasing hormone (PRH) stimulates PRL secretion.

9. ACTH regulates the activities of the adrenal cortex and is controlled by CRH (corticotropin-releasing hormone). Dopamine inhibits secretion of MSH.

10. The posterior pituitary contains axon terminals of neurosecretory cells whose cell bodies are in the hypothalamus. Hormones made by the hypothalamus and stored in the posterior pituitary are oxytocin (OT), which stimulates contraction of the uterus and ejection of milk from the breasts, and antidiuretic hormone (ADH), which stimulates water reabsorption by the kidneys and constriction of arterioles. (See **Table 18.5**.) Oxytocin secretion is stimulated by uterine stretching and suckling during nursing; ADH secretion is controlled by osmotic pressure of the blood and blood volume.

18.7 Thyroid Gland

1. The thyroid gland is located inferior to the larynx.

2. It consists of thyroid follicles composed of follicular cells, which secrete the thyroid hormones thyroxine (T_4) and triiodothyronine (T_3), and parafollicular cells, which secrete calcitonin (CT).

3. Thyroid hormones are synthesized from iodine and tyrosine within thyroglobulin (TGB). They are transported in the blood bound to plasma proteins, mostly thyroxine-binding globulin (TBG).

4. Secretion is controlled by TRH from the hypothalamus and thyroid-stimulating hormone (TSH) from the anterior pituitary.

5. Thyroid hormones regulate oxygen use and metabolic rate, cellular metabolism, and growth and development.

6. Calcitonin (CT) can lower the blood level of calcium ions (Ca^{2+}) and promote deposition of Ca^{2+} into bone matrix. Secretion of CT is controlled by the Ca^{2+} level in the blood. (See **Table 18.6**.)

18.8 Parathyroid Glands

1. The parathyroid glands are embedded in the posterior surfaces of the lateral lobes of the thyroid gland. They consist of chief cells and oxyphil cells.

2. Parathyroid hormone (PTH) regulates the homeostasis of calcium, magnesium, and phosphate ions by increasing blood calcium and magnesium levels and decreasing blood phosphate levels. PTH secretion is controlled by the level of calcium in the blood. (See **Table 18.7**.)

18.9 Adrenal Glands

1. The adrenal glands are located superior to the kidneys. They consist of an outer adrenal cortex and inner adrenal medulla.

2. The adrenal cortex is divided into a zona glomerulosa, a zona fasciculata, and a zona reticularis; the adrenal medulla consists of chromaffin cells and large blood vessels.

3. Cortical secretions include mineralocorticoids, glucocorticoids, and androgens.

4. Mineralocorticoids (mainly aldosterone) increase sodium and water reabsorption and decrease potassium reabsorption. Secretion is controlled by the renin–angiotensin–aldosterone (RAA) pathway and by K^+ level in the blood.

5. Glucocorticoids (mainly cortisol) promote protein breakdown, gluconeogenesis, and lipolysis; help resist stress; and serve as anti-inflammatory substances. Their secretion is controlled by ACTH.

6. Androgens secreted by the adrenal cortex stimulate growth of axillary and pubic hair, aid the prepubertal growth spurt, and contribute to libido.

7. The adrenal medulla secretes epinephrine and norepinephrine (NE), which are released during stress and produce effects similar to sympathetic responses. (See **Table 18.8**.)

18.10 Pancreatic Islets

1. The pancreas lies in the curve of the duodenum. It has both endocrine and exocrine functions.

2. The endocrine portion consists of pancreatic islets or islets of Langerhans, made up of four types of cells: alpha, beta, delta, and F cells.

3. Alpha cells secrete glucagon, beta cells secrete insulin, delta cells secrete somatostatin, and F cells secrete pancreatic polypeptide.

4. Glucagon increases blood glucose level; insulin decreases blood glucose level. Secretion of both hormones is controlled by the level of glucose in the blood. (See **Table 18.9**.)

18.11 Ovaries and Testes

1. The ovaries are located in the pelvic cavity and produce estrogens, progesterone, and inhibin. These sex hormones govern the development and maintenance of female secondary sex characteristics, reproductive cycles, pregnancy, lactation, and normal female reproductive functions. (See **Table 18.10**.)

2. The testes lie inside the scrotum and produce testosterone and inhibin. These sex hormones govern the development and maintenance of male secondary sex characteristics and normal male reproductive functions. (See **Table 18.10**.)

18.12 Pineal Gland and Thymus

1. The pineal gland is attached to the roof of the third ventricle of the brain. It consists of secretory cells called pinealocytes, neuroglia, and endings of sympathetic postganglionic axons.

2. The pineal gland secretes melatonin, which contributes to setting the body's biological clock (controlled in the suprachiasmatic nucleus). During sleep, plasma levels of melatonin increase.

3. The thymus secretes several hormones related to immunity.

4. Thymosin, thymic humoral factor (THF), thymic factor (TF), and thymopoietin promote the maturation of T cells.

18.13 Other Endocrine Tissues and Organs, Eicosanoids, and Growth Factors

1. Body tissues other than those normally classified as endocrine glands contain endocrine tissue and secrete hormones; they include the gastrointestinal tract, placenta, kidneys, skin, and heart. (See **Table 18.11**.)

2. Prostaglandins and leukotrienes are eicosanoids that act as local hormones in most body tissues.

3. Growth factors are local hormones that stimulate cell growth and division. (See **Table 18.12**.)

18.14 The Stress Response

1. Productive stress is termed eustress, and harmful stress is termed distress.
2. If stress is extreme, it triggers the stress response (general adaptation syndrome), which occurs in three stages: the fight-or-flight response, resistance reaction, and exhaustion.
3. The stimuli that produce the stress response are called stressors. Stressors include surgery, poisons, infections, fever, and strong emotional responses.
4. The fight-or-flight response is initiated by nerve impulses from the hypothalamus to the sympathetic division of the autonomic nervous system and the adrenal medulla. This response rapidly increases circulation, promotes ATP production, and decreases nonessential activities.
5. The resistance reaction is initiated by releasing hormones secreted by the hypothalamus, most importantly CRH, TRH, and GHRH. Resistance reactions are longer lasting and accelerate breakdown reactions to provide ATP for counteracting stress.
6. Exhaustion results from depletion of body resources during the resistance stage.
7. Stress may trigger certain diseases by inhibiting the immune system. An important link between stress and immunity is interleukin-l, produced by macrophages; it stimulates secretion of ACTH.

18.15 Development of the Endocrine System

1. The development of the endocrine system is not as localized as in other systems because endocrine organs develop in widely separated parts of the embryo.
2. The pituitary gland, adrenal medulla, and pineal gland develop from ectoderm; the adrenal cortex develops from mesoderm; and the thyroid gland, parathyroid glands, pancreas, and thymus develop from endoderm.

18.16 Aging and the Endocrine System

1. Although some endocrine glands shrink as we get older, their performance may or may not be compromised.
2. Production of growth hormone, thyroid hormones, cortisol, aldosterone, and estrogens decreases with advancing age.
3. With aging, the blood levels of TSH, LH, FSH, and PTH rise.
4. The pancreas releases insulin more slowly with age, and receptor sensitivity to glucose declines.
5. After puberty, thymus size begins to decrease, and thymic tissue is replaced by adipose and areolar connective tissue.

CLINICAL CONNECTIONS

Blocking Hormone Receptors (Refer page 545 of Textbook)

Synthetic hormones that **block the receptors** for some naturally occurring hormones are available as drugs. For example, RU486 (mifepristone), which is used to induce abortion, binds to the receptors for progesterone (a female sex hormone) and prevents progesterone from exerting its normal effect, in this case preparing the lining of the uterus for implantation. When RU486 is given to a pregnant woman, the uterine conditions needed for nurturing an embryo are not maintained, embryonic development stops, and the embryo is sloughed off along with the uterine lining. This example illustrates an important endocrine principle: If a hormone is prevented from interacting with its receptors, the hormone cannot perform its normal functions.

Administering Hormones (Refer page 548 of Textbook)

Both steroid hormones and thyroid hormones are effective when taken by mouth. They are not split apart during digestion and easily cross the intestinal lining because they are lipid-soluble. By contrast, peptide and protein hormones, such as insulin, are not effective oral medications because digestive enzymes destroy them by breaking their peptide bonds. This is why people who need insulin must take it by injection.

Diabetogenic Effect of GH (Refer page 556 of Textbook)

One symptom of excess growth hormone (GH) is hyperglycemia. Persistent hyperglycemia in turn stimulates the pancreas to secrete insulin continually. Such excessive stimulation, if it lasts for weeks or months, may cause "beta-cell burnout," a greatly decreased capacity of pancreatic beta cells to synthesize and secrete insulin. Thus, excess secretion of growth hormone may have a **diabetogenic effect** (dī'-a-bet'-o-JEN-ik); that is, it causes diabetes mellitus (lack of insulin activity).

Oxytocin and Childbirth (Refer page 559 of Textbook)

Years before oxytocin was discovered, it was common practice in midwifery to let a first-born twin nurse at the mother's breast to speed the birth of the second child. Now we know why this practice is helpful—it stimulates the release of oxytocin. Even after a single birth, nursing promotes expulsion of the placenta (afterbirth) and helps the uterus regain its smaller size. Synthetic oxytocin (Pitocin) often is given to induce labor or to increase uterine tone and control hemorrhage just after giving birth.

Congenital Adrenal Hyperplasia (Refer page 570 of Textbook)

Congenital adrenal hyperplasia (CAH) (hī-per-PLĀ-zē-a) is a genetic disorder in which one or more enzymes needed for synthesis of cortisol are absent. Because the cortisol level is low, secretion of ACTH by the anterior pituitary is high due to lack of negative feedback inhibition. ACTH in turn stimulates growth and secretory activity of the adrenal cortex. As a result, both adrenal glands are enlarged. However, certain steps leading to synthesis of cortisol are blocked. Thus, precursor molecules accumulate, and some of these are weak androgens that can undergo conversion to testosterone. The result is **virilism** (VIR-i-lizm), or masculinization. In a female, virile characteristics include growth of a beard, development of a much deeper voice and a masculine distribution of body hair, growth of the clitoris so it may resemble a penis, atrophy of the breasts, and increased muscularity that produces a masculine physique. In prepubertal males, the syndrome causes the same characteristics as in females, plus rapid development of the male sexual organs and emergence of male sexual desires. In adult males, the virilizing effects of CAH are usually completely obscured by the normal virilizing effects of the testosterone secreted by the testes. As a result, CAH is often difficult to diagnose in adult males. Treatment involves cortisol therapy, which inhibits ACTH secretion and thus reduces production of adrenal androgens.

Seasonal Affective Disorder and Jet Lag (Refer page 575 of Textbook)

Seasonal affective disorder (SAD) is a type of depression that afflicts some people during the winter months, when day length is short. It is thought to be due, in part, to overproduction of melatonin. Full-spectrum bright-light therapy—repeated doses of several hours of exposure to artificial light as bright as sunlight—provides relief for some people. Three to six hours of exposure to bright light also appears to speed recovery from jet lag, the fatigue suffered by travelers who quickly cross several time zones.

Nonsteroidal Anti-inflammatory Drugs (Refer page 576 of Textbook)

In 1971, scientists solved the long-standing puzzle of how aspirin works. Aspirin and related **nonsteroidal anti-inflammatory drugs (NSAIDs)**, such as ibuprofen (Motrin®), inhibit cyclooxygenase (COX), a key enzyme involved in prostaglandin synthesis. NSAIDs are used to treat a wide variety of inflammatory disorders, from rheumatoid arthritis to tennis elbow. The success of NSAIDs in reducing fever, pain, and inflammation shows how prostaglandins contribute to these woes.

Post-Traumatic Stress Disorder (Refer page 579 of Textbook)

Post-traumatic stress disorder (PTSD) is an anxiety disorder that may develop in an individual who has experienced, witnessed, or learned about a physically or psychologically distressing event. The immediate cause of PTSD appears to be the specific stressors associated with the events. Among the stressors are terrorism, hostage taking, imprisonment, military duty, serious accidents, torture, sexual or physical abuse, violent crimes, school shootings, massacres, and natural disasters. In the United States, PTSD affects 10% of females and 5% of males. Symptoms of PTSD include reexperiencing the event through nightmares or flashbacks; avoidance of any activity, person, place, or event associated with the stressors; loss of interest and lack of motivation; poor concentration; irritability; and insomnia. Treatment may include the use of antidepressants, mood stabilizers, and antianxiety and antipsychotic agents.

DISORDERS: HOMEOSTATIC IMBALANCES

Disorders of the endocrine system often involve either **hyposecretion** (*hypo-* = too little or under), inadequate release of a hormone, or **hypersecretion** (*hyper-* = too much or above), excessive release of a hormone. In other cases, the problem is faulty hormone receptors, an inadequate number of receptors, or defects in second-messenger systems. Because hormones are distributed in the blood to target tissues throughout the body, problems associated with endocrine dysfunction may also be widespread.

Pituitary Gland Disorders

PITUITARY DWARFISM, GIANTISM, AND ACROMEGALY Several disorders of the anterior pituitary involve growth hormone (GH). Hyposecretion of GH during the growth years slows bone growth, and the epiphyseal plates close before normal height is reached. This condition is called **pituitary dwarfism** (see Clinical Connection: Hormonal Abnormalities That Affect Height in Chapter 6). Other organs of the body also fail to grow, and the body proportions are childlike. Treatment requires administration of GH during childhood, before the epiphyseal plates close.

Hypersecretion of GH during childhood causes **giantism,** an abnormal increase in the length of long bones. The person grows to be very tall, but body proportions are about normal. **Figure SG18.1a** shows identical twins; one brother developed giantism due to a pituitary tumor. Hypersecretion of GH during adulthood is called **acromegaly** (ak'-rō-MEG-a-lē). Although GH cannot produce further lengthening of the long bones because the epiphyseal plates are already closed, the bones of the hands, feet, cheeks, and jaws thicken and other tissues enlarge. In addition, the eyelids, lips, tongue, and nose enlarge, and the skin thickens and develops furrows, especially on the forehead and soles (**Figure SG18.1b**).

DIABETES INSIPIDUS The most common abnormality associated with dysfunction of the posterior pituitary is **diabetes insipidus (DI)** (dī-a-BĒ-tēz in-SIP-i-dus; *diabetes* = overflow; *insipidus* = tasteless). This disorder is due to defects in antidiuretic hormone (ADH) receptors or an inability to secrete ADH. *Neurogenic diabetes insipidus* results from hyposecretion of ADH, usually caused by a brain tumor, head trauma, or brain surgery that damages the posterior pituitary or the hypothalamus. In *nephrogenic diabetes insipidus*, the kidneys do not respond to ADH. The ADH receptors may be nonfunctional, or the kidneys may be damaged. A common symptom of both forms of DI is excretion of large volumes of urine, with resulting dehydration and thirst. Bed-wetting is common in afflicted children. Because so much water is lost in the urine, a person with DI may die of dehydration if deprived of water for only a day or so.

Treatment of neurogenic diabetes insipidus involves hormone replacement, usually for life. Either subcutaneous injection or nasal spray application of ADH analogs is effective. Treatment of nephrogenic diabetes insipidus is more complex and depends on the nature of the kidney dysfunction. Restriction of salt in the diet and, paradoxically, the use of certain diuretic drugs, are helpful.

Thyroid Gland Disorders

Thyroid gland disorders affect all major body systems and are among the most common endocrine disorders. **Congenital hypothyroidism**, hyposecretion of thyroid hormones that is present at birth, has devastating consequences if not treated promptly. Previously termed *cretinism*, this condition causes severe mental retardation and stunted bone growth. At birth, the baby typically is normal because lipid-soluble maternal thyroid hormones crossed the placenta during pregnancy and allowed normal development. Most states require testing of all newborns to ensure adequate thyroid function. If congenital hypothyroidism exists, oral thyroid hormone treatment must be started soon after birth and continued for life.

Hypothyroidism during the adult years produces **myxedema** (mix-e-DĒ-ma), which occurs about five times more often in females than in males. A hallmark of this disorder is edema (accumulation of interstitial fluid) that causes the facial tissues to swell and look puffy. A person with myxedema has a slow heart rate, low body temperature, sensitivity to cold, dry hair and skin, muscular weakness, general lethargy, and a tendency to gain weight easily. Because the brain has already reached maturity, mental retardation does not occur, but the person may be less alert. Oral thyroid hormones reduce the symptoms.

The most common form of hyperthyroidism is **Graves disease**, which also occurs seven to ten times more often in females than in males, usually before age 40. Graves disease is an autoimmune disorder in which the person produces antibodies that mimic the action of thyroid-stimulating hormone (TSH). The antibodies continually stimulate the thyroid gland to grow and produce thyroid hormones. A primary sign is an enlarged thyroid, which may be two to three times its normal size. Graves patients often have a peculiar edema behind the eyes, called **exophthalmos** (ek'-sof-THAL-mos), which causes the eyes to protrude (**Figure SG18.1d**). Treatment may include surgical removal of part or all of the thyroid gland (thyroidectomy), the use of radioactive iodine (^{131}I) to selectively destroy thyroid tissue, and the use of antithyroid drugs to block synthesis of thyroid hormones.

A **goiter** (GOY-ter; *guttur* = throat) is simply an enlarged thyroid gland. It may be associated with hyperthyroidism, hypothyroidism, or **euthyroidism**

Figure SG18.1 Various endocrine disorders.

Disorders of the endocrine system often involve hyposecretion or hypersecretion of hormones.

(a) A 22-year-old man with pituitary giantism shown beside his identical twin

From New England Journal of Medicine, Massachusettes Medical Society, February 18, 1999, vol.340, No. 7, page 524.

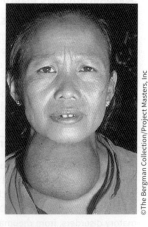

(b) Acromegaly (excess GH during adulthood)

©The Bergman Collection/Project Masters, Inc

(c) Goiter (enlargement of thyroid gland)

©The Bergman Collection/Project Masters, Inc

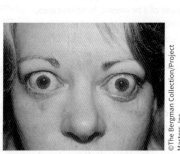

(d) Exophthalmos (excess thyroid hormones, as in Graves disease)

©The Bergman Collection/Project Masters, Inc

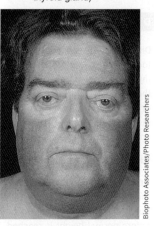

(e) Cushing's syndrome (excess glucocorticoids)

Biophoto Associates/Photo Researchers

Q Which endocrine disorder is due to antibodies that mimic the action of TSH?

(ū-THĪ-royd-izm; *eu* = good), which means normal secretion of thyroid hormone. In some places in the world, dietary iodine intake is inadequate; the resultant low level of thyroid hormone in the blood stimulates secretion of TSH, which causes thyroid gland enlargement (**Figure SG18.1c**).

Parathyroid Gland Disorders

Hypoparathyroidism (hī-pō-par′-a-THĪ-royd-izm)—too little parathyroid hormone—leads to a deficiency of blood Ca^{2+}, which causes neurons and muscle fibers to depolarize and produce action potentials spontaneously. This leads to twitches, spasms, and **tetany** (maintained contraction) of skeletal muscle. The leading cause of hypoparathyroidism is accidental damage to the parathyroid glands or to their blood supply during thyroidectomy surgery.

Hyperparathyroidism, an elevated level of parathyroid hormone, most often is due to a tumor of one of the parathyroid glands. An elevated level of PTH causes excessive resorption of bone matrix, raising the blood levels of calcium and phosphate ions and causing bones to become soft and easily fractured. High blood calcium level promotes formation of kidney stones. Fatigue, personality changes, and lethargy are also seen in patients with hyperparathyroidism.

Adrenal Gland Disorders

CUSHING'S SYNDROME Hypersecretion of cortisol by the adrenal cortex produces **Cushing's syndrome** (**Figure SG18.1e**). Causes include a tumor of the adrenal gland that secretes cortisol, or a tumor elsewhere that secretes adrenocorticotropic hormone (ACTH), which in turn stimulates excessive secretion of cortisol. The condition is characterized by breakdown of muscle proteins and redistribution of body fat, resulting in spindly arms and legs accompanied by a rounded "moon face," "buffalo hump" on the back, and pendulous (hanging) abdomen. Facial skin is flushed, and the skin covering the abdomen develops stretch marks. The person also bruises easily, and wound healing is poor. The elevated level of cortisol causes hyperglycemia, osteoporosis, weakness, hypertension, increased susceptibility to infection, decreased resistance to stress, and mood swings. People who need long-term glucocorticoid therapy—for instance, to prevent rejection of a transplanted organ—may develop a cushingoid appearance.

ADDISON'S DISEASE Hyposecretion of glucocorticoids and aldosterone causes **Addison's disease** (*chronic adrenocortical insufficiency*). The majority of cases are autoimmune disorders in which antibodies cause adrenal cortex destruction or block binding of ACTH to its receptors. Pathogens, such as the bacterium that causes tuberculosis, also may trigger adrenal cortex destruction. Symptoms, which typically do not appear until 90% of the adrenal cortex has been destroyed, include mental lethargy, anorexia, nausea and vomiting, weight loss, hypoglycemia, and muscular weakness. Loss of aldosterone leads to elevated potassium and decreased sodium in the blood, low blood pressure, dehydration, decreased cardiac output, arrhythmias, and even cardiac arrest. The skin may have a "bronzed" appearance that often is mistaken for a suntan. Such was true in the case of President John F. Kennedy, whose Addison's disease was known to only a few while he was alive. Treatment consists of replacing glucocorticoids and mineralocorticoids and increasing sodium in the diet.

PHEOCHROMOCYTOMAS Usually benign tumors of the chromaffin cells of the adrenal medulla, called **pheochromocytomas** (fē-ō-krō′-mō-si-TŌ-mas; *pheo-* = dusky; *-chromo-* = color; *-cyto-* = cell), cause hypersecretion of epinephrine and norepinephrine. The result is a prolonged version of the fight-or-flight response: rapid heart rate, high blood pressure, high levels of glucose in blood and urine, an elevated basal metabolic rate (BMR), flushed face, nervousness, sweating, and decreased gastrointestinal motility. Treatment is surgical removal of the tumor.

Pancreatic Islet Disorders

The most common endocrine disorder is **diabetes mellitus** (MEL-i-tus; *melli-* = honey sweetened), caused by an inability to produce or use insulin. Diabetes mellitus is the fourth leading cause of death by disease in the United States, primarily because of its damage to the cardiovascular system. Because insulin is unavailable to aid transport of glucose into body cells, blood glucose level is high and glucose "spills" into the urine (glucosuria). Hallmarks of diabetes mellitus are the three "polys": *polyuria*, excessive urine production due to an inability of the kidneys to reabsorb water; *polydipsia*, excessive thirst; and *polyphagia*, excessive eating.

Both genetic and environmental factors contribute to onset of the two types of diabetes mellitus—type 1 and type 2—but the exact mechanisms are still unknown. **Type 1 diabetes**, previously known as *insulin-dependent diabetes mellitus (IDDM)*, occurs because the person's immune system destroys the pancreatic beta cells. As a result, the pancreas produces little or no insulin. Type 1 diabetes usually develops in people younger than age 20 and it persists throughout life. By the time symptoms of type 1 diabetes arise, 80–90% of the islet beta cells have been destroyed. Type 1 diabetes is most common in northern Europe, especially in Finland, where nearly 1% of the population develops type 1 diabetes by 15 years of age. In the United States, type 1 diabetes is 1.5–2.0 times more common in whites than in African American or Asian populations.

The cellular metabolism of an untreated type 1 diabetic is similar to that of a starving person. Because insulin is not present to aid the entry of glucose into body cells, most cells use fatty acids to produce ATP. Stores of triglycerides in adipose tissue are catabolized to yield fatty acids and glycerol. The by-products of fatty acid breakdown—organic acids called ketones or ketone bodies—accumulate. Buildup of ketones causes blood pH to fall, a condition known as **ketoacidosis** (kē′-tō-as-i-DŌ-sis). Unless treated quickly, ketoacidosis can cause death.

The breakdown of stored triglycerides also causes weight loss. As lipids are transported by the blood from storage depots to cells, lipid particles are deposited on the walls of blood vessels, leading to atherosclerosis and a multitude of cardiovascular problems, including cerebrovascular insufficiency, ischemic heart disease, peripheral vascular disease, and gangrene. A major complication of diabetes is loss of vision due either to cataracts (excessive glucose attaches to lens proteins, causing cloudiness) or to damage to blood vessels of the retina. Severe kidney problems also may result from damage to renal blood vessels.

Type 1 diabetes is treated through self-monitoring of blood glucose level (up to 7 times daily), regular meals containing 45–50% carbohydrates and less than 30% fats, exercise, and periodic insulin injections (up to 3 times a day). Several implantable pumps are available to provide insulin without the need for repeated injections. Because they lack a reliable glucose sensor, however, the person must self-monitor blood glucose level to determine insulin doses. It is also possible to successfully transplant a pancreas, but immunosuppressive drugs must then be taken for life. Another promising approach under investigation is transplantation of isolated islets in semipermeable hollow tubes. The tubes allow glucose and insulin to enter and leave but prevent entry of immune system cells that might attack the islet cells.

Type 2 diabetes, formerly known as *non-insulin-dependent diabetes mellitus (NIDDM)*, is much more common than type 1, representing more than 90% of all cases. Type 2 diabetes most often occurs in obese people who are over age 35. However, the number of obese children and teenagers with type 2 diabetes is increasing. Clinical symptoms are mild, and the high glucose levels in the blood often can be controlled by diet, exercise, and weight loss. Sometimes, drugs such as *glyburide* (DiaBeta®) and metformin (Fortamet®) are used to stimulate secretion of insulin by pancreatic beta cells. Although some type 2 diabetics need insulin, many have a sufficient amount (or even a surplus) of insulin in the blood. For these people, diabetes arises not from a shortage of insulin but because target cells become less sensitive to it due to down-regulation of insulin receptors.

Hyperinsulinism most often results when a diabetic injects too much insulin. The main symptom is **hypoglycemia**, decreased blood glucose level, which occurs because the excess insulin stimulates too much uptake of glucose by body cells. The resulting hypoglycemia stimulates the secretion of epinephrine, glucagon, and growth hormone. As a consequence, anxiety, sweating, tremor, increased heart rate, hunger, and weakness occur. When blood glucose falls, brain cells are deprived of the steady supply of glucose they need to function effectively. Severe hypoglycemia leads to mental disorientation, convulsions, unconsciousness, and shock. Shock due to an insulin overdose is termed **insulin shock**. Death can occur quickly unless blood glucose is restored to normal levels. From a clinical standpoint, a diabetic suffering from either a hyperglycemia or a hypoglycemia crisis can have very similar symptoms—mental changes, coma, seizures, and so on. It is important to quickly and correctly identify the cause of the underlying symptoms and treat them appropriately.

MEDICAL TERMINOLOGY

Gynecomastia (gī′-ne-kō-MAS-tē-a; *gyneco-* = woman; *-mast-* = breast) Excessive development of mammary glands in a male. Sometimes a tumor of the adrenal gland may secrete sufficient amounts of estrogen to cause the condition.

Hirsutism (HER-soo-tizm; *hirsut-* = shaggy) Presence of excessive body and facial hair in a male pattern, especially in women; may be due to excess androgen production due to tumors or drugs.

Thyroid crisis (storm) A severe state of hyperthyroidism that can be life-threatening. It is characterized by high body temperature, rapid heart rate, high blood pressure, gastrointestinal symptoms (abdominal pain, vomiting, diarrhea), agitation, tremors, confusion, seizures, and possibly coma.

Virilizing adenoma (*aden-* = gland; *-oma* = tumor) Tumor of the adrenal gland that liberates excessive androgens, causing virilism (masculinization) in females. Occasionally, adrenal tumor cells liberate estrogens to the extent that a male patient develops gynecomastia. Such a tumor is called a **feminizing adenoma**.

SELF-QUIZ QUESTIONS

Fill in the blanks in the following statements.

1. The three stages of the stress response or general adaptation syndrome, in order of occurrence, are _____, _____, and _____.

2. The _____ is the major integrating link between the nervous and endocrine systems, acts as an endocrine gland itself, and helps control the stress response.

3. Down-regulation makes a target cell _____ sensitive to a hormone while up-regulation makes a target cell _____ sensitive to a hormone.

Indicate whether the following statements are true or false.

4. If the effect of two or more hormones acting together is greater than the sum of each acting alone, then the two hormones are said to have a permissive effect.

5. In the direct gene activation method of hormone action, the hormone enters the target cell and binds to an intracellular receptor. The activated receptor–hormone complex then alters gene expression to produce the protein that causes the physiological responses that are characteristic of the hormone.

Choose the one best answer to the following questions.

6. Which of the following comparisons are *true*? (1) Nerve impulses produce their effects quickly; hormonal responses generally are slower. (2) Nervous system effects are brief; endocrine system effects are longer lasting. (3) The nervous system controls homeostasis through mediator molecules called neurotransmitters; the endocrine system works through mediator molecules called hormones. (4) The nervous system can stimulate or inhibit the release of hormones; some hormones are released by neurons as neurotransmitters. (5) Neurotransmitters transmit impulses directly; hormones must bind to receptors on or in target cells in order to exert their effects.
   (a) 1, 2, 3, 4, and 5
   (b) 1, 2, 3, and 4
   (c) 2, 3, 4, and 5
   (d) 2, 4, and 5
   (e) 1, 4, and 5

7. Insulin and thyroxine arrive at an organ at the same time. Thyroxine causes an effect on the organ but insulin does not. Why?
   (a) Insulin is a lipid-soluble hormone and thyroxine is not.
   (b) The target cells in the organ have up-regulated for thyroxine.
   (c) Thyroxine is a local hormone and insulin is a circulating hormone.
   (d) Thyroxine inhibits the action of insulin.
   (e) The organ's cells have receptors for thyroxine but not for insulin.

8. Which of the following is not a category of water-soluble hormones?
   (a) peptides
   (b) amines
   (c) eicosanoids
   (d) steroids
   (e) proteins

9. Place in correct order the action of a water-soluble hormone on its target cell. (1) Adenylate cyclase is activated, catalyzing the conversion of ATP to cAMP. (2) Enzymes catalyze reactions that produce a physiological response attributed to the hormone. (3) The hormone binds to a membrane receptor. (4) Activated protein kinases phosphorylate cellular proteins. (5) The hormone–receptor complex activates G proteins. (6) Cyclic AMP activates protein kinases.
   (a) 3, 5, 1, 6, 4, 2
   (b) 3, 1, 5, 6, 4, 2
   (c) 5, 1, 4, 2, 3, 6
   (d) 3, 4, 5, 1, 6, 2
   (e) 6, 3, 5, 1, 4, 2

10. Hormones (1) generally utilize negative feedback mechanisms to regulate their secretion; (2) will only affect target cells far removed from the hormone-producing secretory cells; (3) must bind to transport proteins in order to circulate in the blood; (4) may be released in low concentrations but can produce large effects in the target cells because of amplification; (5) can regulate the responsiveness of the target tissue by controlling the number of receptor sites for the hormone.
   (a) 1, 2, and 3
   (b) 1, 2, 4, and 5
   (c) 2, 3, and 4
   (d) 2, 3, 4, and 5
   (e) 1, 4, and 5

11. The pituitary gland (1) is located in the cribriform plate of the ethmoid bone, (2) is linked to the hypothalamus by the infundibulum, (3) has a posterior portion that contains axon terminals from hypothalamic neurosecretory cells, (4) produces releasing and inhibiting hormones, (5) has a vascular connection with the hypothalamus known as the hypophyseal portal system.
   (a) 1, 2, and 4
   (b) 2, 3, 4, and 5
   (c) 2, 3, and 5
   (d) 1, 2, 3, 4, and 5
   (e) 2, 4, and 5

12. The class of adrenal gland hormones that provides resistance to stress, produces anti-inflammatory effects, and promotes normal metabolism to ensure adequate quantities of ATP is
   (a) glucocorticoids.
   (b) mineralocorticoids.
   (c) androgens.
   (d) catecholamines.
   (e) gonadocorticoids.

13. Match the following:
   _____ (a) increases blood Ca^{2+} level
   _____ (b) increases blood glucose level
   _____ (c) decreases blood Ca^{2+} level
   _____ (d) decreases blood glucose level
   _____ (e) initiates and maintains milk secretion by the mammary glands
   _____ (f) regulates the body's biological clock
   _____ (g) stimulates sex hormone production; triggers ovulation
   _____ (h) augments fight-or-flight responses
   _____ (i) regulates metabolism and resistance to stress
   _____ (j) helps control water and electrolyte homeostasis
   _____ (k) suppresses release of FSH
   _____ (l) stimulates growth of axillary and pubic hair
   _____ (m) promotes T cell maturation
   _____ (n) regulates oxygen use, basal metabolic rate, cellular metabolism, and growth and development
   _____ (o) stimulates protein synthesis, inhibits protein breakdown, stimulates lipolysis, and retards use of glucose for ATP production

   (1) insulin
   (2) glucagon
   (3) inhibin
   (4) follicle-stimulating hormone
   (5) luteinizing hormone
   (6) thyroxine and triiodothyronine
   (7) calcitonin
   (8) parathormone
   (9) melanocyte-stimulating hormone
   (10) oxytocin
   (11) antidiuretic hormone
   (12) prolactin
   (13) human growth hormone
   (14) hypothalamic regulating hormones
   (15) aldosterone
   (16) thyroid-stimulating hormone
   (17) androgens
   (18) epinephrine and norepinephrine

_____ (p) inhibits water loss through the kidneys

_____ (q) stimulates egg and sperm formation

_____ (r) enhances uterine contractions during labor; stimulates milk ejection

_____ (s) stimulates and inhibits secretion of anterior pituitary hormones

_____ (t) increases skin pigmentation when present in excess

_____ (u) stimulates synthesis and release of T_3 and T_4

_____ (v) local hormones involved in inflammation, smooth muscle contraction, blood flow, and inflammation

(19) prostaglandins
(20) melatonin
(21) thymosin
(22) cortisol

14. Match the following hormone-secreting cells to the hormone(s) they release:

_____ (a) ACTH and MSH

_____ (b) TSH

_____ (c) glucagon

_____ (d) PTH

_____ (e) glucocorticoids

_____ (f) calcitonin

_____ (g) insulin

_____ (h) androgens

_____ (i) progesterone

_____ (j) FSH and LH

_____ (k) epinephrine and norepinephrine

_____ (l) hGH

_____ (m) testosterone

_____ (n) mineralocorticoids

_____ (o) thyroxine and triiodothyronine

_____ (p) PRL

(1) beta cells of pancreatic islet
(2) alpha cells of pancreatic islet
(3) follicular cells of thyroid gland
(4) parafollicular cells of thyroid gland
(5) testes
(6) ovary
(7) somatotrophs
(8) thyrotrophs
(9) gonadotrophs
(10) corticotrophs
(11) lactotrophs
(12) chief cells
(13) chromaffin cells
(14) zona glomerulosa cells
(15) zona fasciculata cells
(16) zona reticularis cells

15. Match the endocrine disorder to the problem that produced the disorder:

_____ (a) hyposecretion of insulin or down-regulation of insulin receptors

_____ (b) hypersecretion of hGH before closure of epiphyseal plates

_____ (c) hyposecretion of thyroid hormone that is present at birth

_____ (d) hypersecretion of glucocorticoids

_____ (e) hyposecretion of hGH before closure of epiphyseal plates

_____ (f) hypersecretion of epinephrine and norepinephrine

_____ (g) hypersecretion of hGH after closure of epiphyseal plates

_____ (h) hyposecretion of glucocorticoids and aldosterone

_____ (i) hyposecretion of ADH

_____ (j) hypersecretion of melatonin

_____ (k) hyposecretion of thyroid hormone in adults

_____ (l) hyperthyroidism, an autoimmune disease

(1) giantism
(2) acromegaly
(3) pituitary dwarfism
(4) diabetes insipidus
(5) myxedema
(6) Graves' disease
(7) Cushing's syndrome
(8) seasonal affective disorder
(9) Addison's disease
(10) pheochromocy-tomas
(11) congenital hypothyroidism
(12) diabetes mellitus

19 THE CARDIOVASCULAR SYSTEM: THE BLOOD

Introduction

1. The cardiovascular system consists of the blood, heart, and blood vessels.
2. Blood is a liquid connective tissue that consists of cells and cell fragments surrounded by a liquid extracellular matrix (blood plasma).

19.1 Functions and Properties of Blood

1. Blood transports oxygen, carbon dioxide, nutrients, wastes, and hormones.
2. It helps regulate pH, body temperature, and water content of cells.
3. It provides protection through clotting and by combating toxins and microbes through certain phagocytic white blood cells or specialized blood plasma proteins.
4. Physical characteristics of blood include a viscosity greater than that of water; a temperature of 38°C (100.4°F); and a pH of 7.35–7.45.
5. Blood constitutes about 8% of body weight, and its volume is 4–6 liters in adults.
6. Blood is about 55% blood plasma and 45% formed elements.
7. The hematocrit is the percentage of total blood volume occupied by red blood cells.
8. Blood plasma consists of 91.5% water and 8.5% solutes. Principal solutes include proteins (albumins, globulins, fibrinogen), nutrients, vitamins, hormones, respiratory gases, electrolytes, and waste products.
9. The formed elements in blood include red blood cells (erythrocytes), white blood cells (leukocytes), and platelets.

19.2 Formation of Blood Cells

1. Hemopoiesis is the formation of blood cells from hemopoietic stem cells in red bone marrow.
2. Myeloid stem cells form RBCs, platelets, granulocytes, and monocytes. Lymphoid stem cells give rise to lymphocytes.
3. Several hemopoietic growth factors stimulate differentiation and proliferation of the various blood cells.

19.3 Red Blood Cells

1. Mature RBCs are biconcave discs that lack nuclei and contain hemoglobin.
2. The function of the hemoglobin in red blood cells is to transport oxygen and some carbon dioxide.

3. RBCs live about 120 days. A healthy male has about 5.4 million RBCs/μL of blood; a healthy female has about 4.8 million/μL.
4. After phagocytosis of aged RBCs by macrophages, hemoglobin is recycled.
5. RBC formation, called erythropoiesis, occurs in adult red bone marrow of certain bones. It is stimulated by hypoxia, which stimulates the release of erythropoietin by the kidneys.
6. A reticulocyte count is a diagnostic test that indicates the rate of erythropoiesis.

19.4 White Blood Cells

1. WBCs are nucleated cells. The two principal types are granulocytes (neutrophils, eosinophils, and basophils) and agranulocytes (lymphocytes and monocytes).
2. The general function of WBCs is to combat inflammation and infection. Neutrophils and macrophages (which develop from monocytes) do so through phagocytosis.
3. Eosinophils combat the effects of histamine in allergic reactions, phagocytize antigen–antibody complexes, and combat parasitic worms. Basophils liberate heparin, histamine, and serotonin in allergic reactions that intensify the inflammatory response.
4. B lymphocytes, in response to the presence of foreign substances called antigens, differentiate into plasma cells that produce antibodies. Antibodies attach to the antigens and render them harmless. This antigen–antibody response combats infection and provides immunity. T lymphocytes destroy foreign invaders directly. Natural killer cells attack infectious microbes and tumor cells.
5. Except for lymphocytes, which may live for years, WBCs usually live for only a few hours or a few days. Normal blood contains 5000–10,000 WBCs/μL.

19.5 Platelets

1. Platelets are disc-shaped cell fragments that splinter from megakaryocytes. Normal blood contains 150,000–400,000 platelets/μL.
2. Platelets help stop blood loss from damaged blood vessels by forming a platelet plug.

19.6 Stem Cell Transplants from Bone Marrow and Cord Blood

1. Bone marrow transplants involve removal of red bone marrow as a source of stem cells from the iliac crest.

2. In a cord-blood transplant, stem cells from the placenta are removed from the umbilical cord.

3. Cord-blood transplants have several advantages over bone marrow transplants.

19.7 Hemostasis

1. Hemostasis refers to the stoppage of bleeding.

2. It involves vascular spasm, platelet plug formation, and blood clotting (coagulation).

3. In vascular spasm, the smooth muscle of a blood vessel wall contracts, which slows blood loss.

4. Platelet plug formation involves the aggregation of platelets to stop bleeding.

5. A clot is a network of insoluble protein fibers (fibrin) in which formed elements of blood are trapped.

6. The chemicals involved in clotting are known as clotting (coagulation) factors.

7. Blood clotting involves a cascade of reactions that may be divided into three stages: formation of prothrombinase, conversion of prothrombin into thrombin, and conversion of soluble fibrinogen into insoluble fibrin.

8. Clotting is initiated by the interplay of the extrinsic and intrinsic pathways of blood clotting.

9. Normal coagulation requires vitamin K and is followed by clot retraction (tightening of the clot) and ultimately fibrinolysis (dissolution of the clot).

10. Clotting in an unbroken blood vessel is called thrombosis. A thrombus that moves from its site of origin is called an embolus.

19.8 Blood Groups and Blood Types

1. ABO and Rh blood groups are genetically determined and based on antigen–antibody responses.

2. In the ABO blood group, the presence or absence of A and B antigens on the surface of RBCs determines blood type.

3. In the Rh system, individuals whose RBCs have Rh antigens are classified as Rh^+; those who lack the antigen are Rh^-.

4. Hemolytic disease of the newborn (HDN) can occur when an Rh^- mother is pregnant with an Rh^+ fetus.

5. Before blood is transfused, a recipient's blood is typed and then either cross-matched to potential donor blood or screened for the presence of antibodies.

CLINICAL CONNECTIONS

Withdrawing Blood (Refer page 585 of Textbook)

Blood samples for laboratory testing may be obtained in several ways. The most common procedure is **venipuncture** (vēn′-i-PUNK-chur), withdrawal of blood from a vein using a needle and collecting tube, which contains various additives. A tourniquet is wrapped around the arm above the venipuncture site, which causes blood to accumulate in the vein. This increased blood volume makes the vein stand out. Opening and closing the fist further causes it to stand out, making the venipuncture more successful. A common site for venipuncture is the median cubital vein anterior to the elbow (see **Figure 21.26c**). Another method of withdrawing blood is through a **finger** or **heel stick.** Diabetic patients who monitor their daily blood sugar typically perform a finger stick, and it is often used for drawing blood from infants and children. In an **arterial stick**, blood is withdrawn from an artery; this test is used to determine the level of oxygen in oxygenated blood.

Bone Marrow Examination (Refer page 588 of Textbook)

Sometimes a sample of red bone marrow must be obtained in order to diagnose certain blood disorders, such as leukemia and severe anemias. **Bone marrow examination** may involve *bone marrow aspiration* (withdrawal of a small amount of red bone marrow with a fine needle and syringe) or a *bone marrow biopsy* (removal of a core of red bone marrow with a larger needle).

Both types of samples are usually taken from the iliac crest of the hip bone, although samples are sometimes aspirated from the sternum. In young children, bone marrow samples are taken from a vertebra or tibia (shin bone). The tissue or cell sample is then sent to a pathology lab for analysis. Specifically, laboratory technicians look for signs of neoplastic (cancer) cells or other diseased cells to assist in diagnosis.

Medical Uses of Hemopoietic Growth Factors (Refer page 589 of Textbook)

Hemopoietic growth factors made available through recombinant DNA technology hold tremendous potential for medical uses when a person's natural ability to form new blood cells is diminished or defective. The artificial form of erythropoietin (epoetin alfa) is very effective in treating the diminished red blood cell production that accompanies end-stage kidney disease. Granulocyte–macrophage colony-stimulating factor and granulocyte CSF are given to stimulate white blood cell formation in cancer patients who are undergoing chemotherapy, which kills red bone marrow cells as well as cancer cells because both cell types are undergoing mitosis. (Recall that white blood cells help protect against disease.) Thrombopoietin shows great promise for preventing the depletion of platelets, which are needed to help blood clot, during chemotherapy. CSFs and thrombopoietin also improve the outcome of patients who receive bone marrow transplants. Hemopoietic growth factors are also used to treat thrombocytopenia in neonates, other clotting disorders, and various types of anemia.

Iron Overload and Tissue Damage (Refer page 591 of Textbook)

Because free iron ions (Fe^{2+} and Fe^{3+}) bind to and damage molecules in cells or in the blood, transferrin and ferritin act as protective "protein escorts" during transport and storage of iron ions. As a result, plasma contains virtually no free iron. Furthermore, only small amounts are available inside body cells for use in synthesis of iron-containing molecules such as the cytochrome pigments needed for ATP production in mitochondria (see **Figure 25.9**). In cases of **iron overload**, the amount of iron present in the body builds up. Because we have no method for eliminating excess iron, any condition that increases dietary iron absorption can cause iron overload. At some point, the proteins transferrin and ferritin become saturated with iron ions, and free iron level rises. Common consequences of iron overload are diseases of the liver, heart, pancreatic islets, and gonads. Iron overload also allows certain iron-dependent microbes to flourish. Such microbes normally are not pathogenic, but they multiply rapidly and can cause lethal effects in a short time when free iron is present.

Reticulocyte Count (Refer page 592 of Textbook)

The rate of erythropoiesis is measured by a **reticulocyte count**. Normally, a little less than 1% of the oldest RBCs are replaced by newcomer reticulocytes on any given day. It then takes 1 to 2 days for the reticulocytes to lose the last vestiges of endoplasmic reticulum and become mature RBCs. Thus, reticulocytes account for about 0.5–1.5% of all RBCs in a normal blood sample. A low "retic" count in a person who is anemic might indicate a shortage of erythropoietin or an inability of the red bone marrow to respond to EPO, perhaps because of a nutritional deficiency or leukemia. A high "retic" count might indicate a good red bone marrow response to previous blood loss or to iron therapy in someone who had been iron-deficient. It could also point to illegal use of epoetin alfa by an athlete.

Blood Doping (Refer page 592 of Textbook)

Delivery of oxygen to muscles is a limiting factor in muscular feats from weightlifting to running a marathon. As a result, increasing the oxygen-carrying capacity of the blood enhances athletic performance, especially in endurance events. Because RBCs transport oxygen, athletes have tried several means of increasing their RBC count, known as **blood doping** or *artificially induced polycythemia* (an abnormally high number of RBCs), to gain a competitive edge. Athletes have enhanced their RBC production by injecting epoetin alfa (Procrit® or Epogen®), a drug that is used to treat anemia by stimulating the production of RBCs by red bone marrow. Practices that increase the number of RBCs are dangerous because they raise the viscosity of the blood, which increases the resistance to blood flow and makes the blood more difficult for the heart to pump. Increased viscosity also contributes to high blood pressure and increased risk of stroke. During the 1980s, at least 15 competitive cyclists died from heart attacks or strokes linked to suspected use of epoetin alfa. Although the International Olympics Committee bans the use of epoetin alfa, enforcement is difficult because the drug is identical to naturally occurring erythropoietin (EPO).

So-called **natural blood doping** is seemingly the key to the success of marathon runners from Kenya. The average altitude throughout Kenya's highlands is about 6000 feet (1829 meters) above sea level; other areas of Kenya are even higher. Altitude training greatly improves fitness, endurance, and performance. At these higher altitudes, the body increases the production of red blood cells, which means that exercise greatly oxygenates the blood. When these runners compete in Boston, for example, at an altitude just above sea level, their bodies contain more erythrocytes than do the bodies of competitors who trained in Boston. A number of training camps have been established in Kenya and now attract endurance athletes from all over the world.

Complete Blood Count (Refer page 595 of Textbook)

A **complete blood count (CBC)** is a very valuable test that screens for anemia and various infections. Usually included are counts of RBCs, WBCs, and platelets per microliter of whole blood; hematocrit; and differential white blood cell count. The amount of hemoglobin in grams per milliliter of blood also is determined. Normal hemoglobin ranges are as follows: infants, 14–20 g/100 mL of blood; adult females, 12–16 g/100 mL of blood; and adult males, 13.5–18 g/100 mL of blood.

Aspirin and Thrombolytic Agents (Refer page 601 of Textbook)

In patients with heart and blood vessel disease, the events of hemostasis may occur even without external injury to a blood vessel. At low doses, **aspirin**

inhibits vasoconstriction and platelet aggregation by blocking synthesis of thromboxane A2. It also reduces the chance of thrombus formation. Due to these effects, aspirin reduces the risk of transient ischemic attacks (TIA), strokes, myocardial infarction, and blockage of peripheral arteries.

Thrombolytic agents (throm′-bō-LIT-ik) are chemical substances that are injected into the body to dissolve blood clots that have already formed to restore circulation. They either directly or indirectly activate plasminogen. The first thrombolytic agent, approved in 1982 for dissolving clots in the coronary arteries of the heart, was **streptokinase**, which is produced by streptococcal bacteria. A genetically engineered version of human **tissue plasminogen activator (tPA)** is now used to treat victims of both heart attacks and brain attacks (strokes) that are caused by blood clots.

Hemolytic Disease of the Newborn (Refer page 603 of Textbook)

The most common problem with Rh incompatibility, **hemolytic disease of the newborn (HDN)**, may arise during pregnancy (**Figure SG19.1**). Normally, no direct contact occurs between maternal and fetal blood while a woman is pregnant. However, if a small amount of Rh⁺ blood leaks from the fetus through the placenta into the bloodstream of an Rh⁻ mother, the mother will start to make anti-Rh antibodies. Because the greatest possibility of fetal blood leakage into the maternal circulation occurs at delivery, the firstborn baby usually is not affected. If the mother becomes pregnant again, however,

Figure SG19.1 Development of hemolytic disease of the newborn (HDN). (a) At birth, a small quantity of fetal blood usually leaks across the placenta into the maternal bloodstream. A problem can arise when the mother is Rh⁻ and the baby is Rh⁺, having inherited an allele for the Rh antigens from the father. (b) On exposure to Rh antigen, the mother's immune system responds by making anti-Rh antibodies. (c) During a subsequent pregnancy, the maternal antibodies cross the placenta into the fetal blood. If the second fetus is Rh⁺, the ensuing antigen–antibody reaction causes agglutination and hemolysis of fetal RBCs. The result is HDN.

HDN occurs when maternal anti-Rh antibodies cross the placenta and cause hemolysis of fetal RBCs.

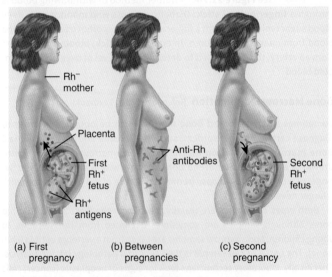

(a) First pregnancy

(b) Between pregnancies

(c) Second pregnancy

Q Why is the firstborn baby unlikely to have HDN?

her anti-Rh antibodies can cross the placenta and enter the bloodstream of the fetus. If the fetus is Rh⁻, there is no problem, because Rh⁻ blood does not have the Rh antigen. If the fetus is Rh⁺, however, agglutination and hemolysis brought on by fetal–maternal incompatibility may occur in the fetal blood.

An injection of anti-Rh antibodies called anti-Rh gamma globulin (RhoGAM®) can be given to prevent HDN. Rh⁻ women should receive RhoGAM® before delivery, and soon after every delivery, miscarriage, or abortion. These antibodies bind to and inactivate the fetal Rh antigens before the mother's immune system can respond to the foreign antigens by producing her own anti-Rh antibodies.

Anticoagulants (Refer page 603 of Textbook)

Patients who are at increased risk of forming blood clots may receive anti-coagulants. Examples are heparin or warfarin. Heparin is often administered during hemodialysis and open-heart surgery. **Warfarin** (*Coumadin®*) acts as an antagonist to vitamin K and thus blocks synthesis of four clotting factors. Warfarin is slower acting than heparin. To prevent clotting in donated blood, blood banks and laboratories often add substances that remove Ca^{2+}; examples are EDTA (ethylenediaminetetraacetic acid) and CPD (citrate phosphate dextrose).

DISORDERS: HOMEOSTATIC IMBALANCES

Anemia

Anemia (a-NĒ-mē-a) is a condition in which the oxygen-carrying capacity of blood is reduced. All of the many types of anemia are characterized by reduced numbers of RBCs or a decreased amount of hemoglobin in the blood. The person feels fatigued and is intolerant of cold, both of which are related to lack of oxygen needed for ATP and heat production. Also, the skin appears pale, due to the low content of red-colored hemoglobin circulating in skin blood vessels. Among the most important causes and types of anemia are the following:

- *Inadequate absorption of iron, excessive loss of iron, increased iron requirement, or insufficient intake of iron* causes **iron-deficiency anemia**, the most common type of anemia. Women are at greater risk for iron-deficiency anemia due to menstrual blood losses and increased iron demands of the growing fetus during pregnancy. Gastrointestinal losses, such as those that occur with malignancy or ulceration, also contribute to this type of anemia.

- *Inadequate intake of vitamin B₁₂ or folic acid* causes **megaloblastic anemia**, in which red bone marrow produces large, abnormal red blood cells (megaloblasts). It may also be caused by drugs that alter gastric secretion or are used to treat cancer.

- *Insufficient hemopoiesis* resulting from an inability of the stomach to produce intrinsic factor, which is needed for absorption of vitamin B₁₂ in the small intestine, causes **pernicious anemia**.

- *Excessive loss of RBCs* through bleeding resulting from large wounds, stomach ulcers, or especially heavy menstruation leads to **hemorrhagic anemia**.

- *RBC plasma membranes rupture prematurely* in **hemolytic anemia**. The released hemoglobin pours into the plasma and may damage the filtering units (glomeruli) in the kidneys. The condition may result from inherited defects such as abnormal red blood cell enzymes, or from outside agents such as parasites, toxins, or antibodies from incompatible transfused blood.

- *Deficient synthesis of hemoglobin* occurs in **thalassemia** (thal'-a-SĒ-mē-a), a group of hereditary hemolytic anemias. The RBCs are small (microcytic), pale (hypochromic), and short-lived. Thalassemia occurs primarily in populations from countries bordering the Mediterranean Sea.

- *Destruction of red bone marrow* results in **aplastic anemia**. It is caused by toxins, gamma radiation, and certain medications that inhibit enzymes needed for hemopoiesis.

Sickle Cell Disease

The RBCs of a person with **sickle cell disease (SCD)** contain Hb-S, an abnormal kind of hemoglobin. When Hb-S gives up oxygen to the interstitial fluid, it forms long, stiff, rodlike structures that bend the erythrocyte into a sickle shape (**Figure SG19.2**). The sickled cells rupture easily. Even though erythropoiesis is stimulated by the loss of the cells, it cannot keep pace with hemolysis. Signs and symptoms of SCD are caused by the sickling of red blood cells. When red blood cells sickle, they break down prematurely (sickled cells die in about 10 to 20 days). This leads to anemia, which can cause shortness of breath, fatigue, paleness, and delayed growth and development in children. The rapid breakdown and loss of blood cells may also cause *jaundice*, yellowing of the eyes and skin. Sickled cells do not move easily through blood vessels, and they tend to stick together and form clumps that cause blockages in blood vessels. This deprives body organs of sufficient oxygen and causes pain (for example, in bones and the abdomen); serious infections; and organ damage, especially in the lungs, brain, spleen, and kidneys. Other symptoms of SCD include fever, rapid heart rate, swelling and inflammation of the hands and/or feet, leg ulcers, eye damage, excessive thirst, frequent urination, and painful and prolonged erections in males. Almost all individuals with SCD have painful episodes that can last from hours to days. Some people have one episode every few years; others have several episodes a year. The episodes may range from mild to those that require hospitalization. Any activity that reduces the amount of oxygen

Figure SG19.2 Red blood cells from a person with sickle cell disease.

The red blood cells of a person with sickle cell disease contain an abnormal type of hemoglobin.

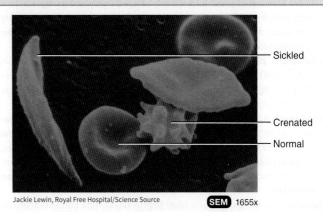

Jackie Lewin, Royal Free Hospital/Science Source   SEM 1655x

— Sickled
— Crenated
— Normal

Q What are some symptoms of sickle cell disease?

in the blood, such as vigorous exercise, may produce **a sickle cell crisis** (worsening of the anemia, pain in the abdomen and long bones of the limbs, fever, and shortness of breath).

Sickle cell disease is inherited. People with two sickle cell genes have severe anemia; those with only one defective gene have the sickle cell trait. Sickle cell genes are found primarily among populations (or their descendants) that live in the malaria belt around the world, including parts of Mediterranean Europe, sub-Saharan Africa, and tropical Asia. The genes responsible for the tendency of the RBCs to sickle also alter the permeability of the plasma membranes of sickled cells, causing potassium ions to leak out. Low levels of potassium kill the malaria parasites that may infect sickled cells. Because of this effect, a person with one normal gene and one sickle cell gene has a higher-than-average resistance to malaria. The possession of a single sickle cell gene thus confers a survival benefit.

Treatment of SCD consists of administration of analgesics to relieve pain, fluid therapy to maintain hydration, oxygen to reduce oxygen deficiency, antibiotics to counter infections, and blood transfusions. People who suffer from SCD have normal fetal hemoglobin (Hb-F), a slightly different form of hemoglobin that predominates at birth and is present in small amounts after birth. In some patients with sickle cell disease, a drug called hydroxyurea promotes transcription of the normal Hb-F gene, elevates the level of Hb-F, and reduces the chance that the RBCs will sickle. Unfortunately, this drug also has toxic effects on the bone marrow; thus, its safety for long-term use is questionable.

Hemophilia

Hemophilia (hē-mō-FIL-ē-a; -philia = loving) is an inherited deficiency of clotting in which bleeding may occur spontaneously or after only minor trauma. It is the oldest known hereditary bleeding disorder; descriptions of the disease are found as early as the second century A.D. Hemophilia usually affects males and is sometimes referred to as "the royal disease" because many descendants of Queen Victoria, beginning with one of her sons, were affected by the disease. Different types of hemophilia are due to deficiencies of different blood clotting factors and exhibit varying degrees of severity, ranging from mild to severe bleeding tendencies. Hemophilia is characterized by spontaneous or traumatic subcutaneous and intramuscular hemorrhaging, nosebleeds, blood in the urine, and hemorrhages in joints that produce pain and tissue damage. Treatment involves transfusions of fresh blood plasma or

concentrates of the deficient clotting factor to relieve the tendency to bleed. Another treatment is the drug desmopressin (DDAVP), which can boost the levels of the clotting factors.

Leukemia

The term **leukemia** (loo-KĒ-mē-a; leuko- = white) refers to a group of red bone marrow cancers in which abnormal white blood cells multiply uncontrollably. The accumulation of the cancerous white blood cells in red bone marrow interferes with the production of red blood cells, white blood cells, and platelets. As a result, the oxygen-carrying capacity of the blood is reduced, an individual is more susceptible to infection, and blood clotting is abnormal. In most leukemias, the cancerous white blood cells spread to the lymph nodes, liver, and spleen, causing them to enlarge. All leukemias produce the usual symptoms of anemia (fatigue, intolerance to cold, and pale skin). In addition, weight loss, fever, night sweats, excessive bleeding, and recurrent infections may occur.

In general, leukemias are classified as **acute** (symptoms develop rapidly) and **chronic** (symptoms may take years to develop). Leukemias are also classified on the basis of the type of white blood cell that becomes malignant. **Lymphoblastic leukemia** (lim-fō-BLAS-tik) involves cells derived from lymphoid stem cells (lymphoblasts) and/or lymphocytes. **Myelogenous leukemia** (mī-e-LOJ-e-nus) involves cells derived from myeloid stem cells (myeloblasts). Combining onset of symptoms and cells involved, there are four types of leukemia:

1. **Acute lymphoblastic leukemia (ALL)** is the most common leukemia in children, but adults can also get it.
2. **Acute myelogenous leukemia (AML)** affects both children and adults.
3. **Chronic lymphoblastic anemia (CLA)** is the most common leukemia in adults, usually those older than 55.
4. **Chronic myelogenous leukemia (CML)** occurs mostly in adults.

The cause of most types of leukemia is unknown. However, certain risk factors have been implicated. These include exposure to radiation or chemotherapy for other cancers, genetics (some genetic disorders such as Down syndrome), environmental factors (smoking and benzene), and microbes such as the human T cell leukemia–lymphoma virus-1 (HTLV-1) and the Epstein–Barr virus.

Treatment options include chemotherapy, radiation, stem cell transplantation, interferon, antibodies, and blood transfusion.

MEDICAL TERMINOLOGY

Acute normovolemic hemodilution (nor-mō-vō-LĒ-mik hē-mō-di-LOO-shun) Removal of blood immediately before surgery and its replacement with a cell-free solution to maintain sufficient blood volume for adequate circulation. At the end of surgery, once bleeding has been controlled, the collected blood is returned to the body.

Autologous preoperative transfusion (aw-TOL-o-gus trans-FŪ-zhun; auto- = self) Donating one's own blood; can be done up to 6 weeks before elective surgery. Also called **predonation**. This procedure eliminates the risk of incompatibility and blood-borne disease.

Blood bank A facility that collects and stores a supply of blood for future use by the donor or others. Because blood banks have additional and diverse functions (immunohematology reference work, continuing medical education, bone and tissue storage, and clinical consultation), they are more appropriately referred to as **centers of transfusion medicine**.

Cyanosis (sī-a-NŌ-sis; cyano- = blue) Slightly bluish/dark-purple skin discoloration, most easily seen in the nail beds and mucous membranes, due to

an increased quantity of *methemoglobin*, hemoglobin not combined with oxygen in systemic blood.

Gamma globulin (GLOB-ū-lin) Solution of immunoglobulins from blood consisting of antibodies that react with specific pathogens, such as viruses. It is prepared by injecting the specific virus into animals, removing blood from the animals after antibodies have accumulated, isolating the antibodies, and injecting them into a human to provide short-term immunity.

Hemochromatosis (hē-mō-krō-ma-TŌ-sis; chroma = color) Disorder of iron metabolism characterized by excessive absorption of ingested iron and excess deposits of iron in tissues (especially the liver, heart, pituitary gland, gonads, and pancreas) that result in bronze discoloration of the skin, cirrhosis, diabetes mellitus, and bone and joint abnormalities.

Hemorrhage (HEM-or-ij; rhegnynai = bursting forth) Loss of a large amount of blood; can be either internal (from blood vessels into tissues) or external (from blood vessels directly to the surface of the body).

Jaundice (*jaund-* = yellow) An abnormal yellowish discoloration of the sclerae of the eyes, skin, and mucous membranes due to excess bilirubin (yellow-orange pigment) in the blood. The three main categories of jaundice are *prehepatic jaundice*, due to excess production of bilirubin; *hepatic jaundice*, abnormal bilirubin processing by the liver caused by congenital liver disease, cirrhosis (scar tissue formation) of the liver, or hepatitis (liver inflammation); and *extrahepatic jaundice*, due to blockage of bile drainage by gallstones or cancer of the bowel or pancreas.

Phlebotomist (fle-BOT-ō-mist; *phlebo-* = vein; *-tom* = cut) A technician who specializes in withdrawing blood.

Septicemia (sep′-ti-SĒ-mē-a; *septic-* = decay; *-emia* = condition of blood) Toxins or disease-causing bacteria in the blood. Also called "blood poisoning."

Thrombocytopenia (throm′-bō-sī-tō-PĒ-nē-a; *-penia* = poverty) Very low platelet count that results in a tendency to bleed from capillaries.

Venesection (vē′-ne-SEK-shun; *ven-* = vein) Opening of a vein for withdrawal of blood. Although **phlebotomy** (fle-BOT-ō-mē) is a synonym for venesection, in clinical practice phlebotomy refers to therapeutic bloodletting, such as the removal of some blood to lower its viscosity in a patient with polycythemia.

Whole blood Blood containing all formed elements, plasma, and plasma solutes in natural concentrations.

SELF-QUIZ QUESTIONS

Fill in the blanks in the following statements.

1. Plasma minus its clotting proteins is termed _____.

2. _____ is the consolidation or tightening of the fibrin clot that helps to bring the edges of a damaged vessel closer together.

Indicate whether the following statements are true or false.

3. Hemoglobin functions in transporting both oxygen and carbon dioxide and in regulating blood pressure.

4. The most numerous white blood cells in a differential white blood cell count of a healthy individual are the neutrophils.

Choose the one best answer to the following questions.

5. Which of the following are *not* required for clot formation? (1) vitamin K, (2) calcium, (3) prostacyclin, (4) plasmin, (5) fibrinogen.
   (a) 1, 2, and 5
   (b) 3, 4, and 5
   (c) 4 and 5
   (d) 1, 2, and 3
   (e) 3 and 4

6. Place the steps involved in hemostasis in the correct order. (1) conversion of fibrinogen into fibrin, (2) conversion of prothrombin into thrombin, (3) adhesion and aggregation of platelets on damaged vessel, (4) prothrombinase formed by extrinsic or intrinsic pathway, (5) reduction of blood loss by initiation of a vascular spasm.
   (a) 5, 3, 4, 2, 1
   (b) 5, 4, 3, 1, 2
   (c) 3, 5, 4, 2, 1
   (d) 5, 3, 2, 1, 4
   (e) 5, 3, 2, 4, 1

7. Which of the following statements explain why red blood cells (RBCs) are highly specialized for oxygen transport? (1) RBCs contain hemoglobin. (2) RBCs lack a nucleus. (3) RBCs have many mitochondria and thus generate ATP aerobically. (4) The biconcave shape of RBCs provides a large surface area for the inward and outward diffusion of gas molecules. (5) RBCs can carry up to four oxygen molecules for each hemoglobin molecule.
   (a) 1, 2, 3, and 5
   (b) 1, 2, 4, and 5
   (c) 2, 3, 4, and 5
   (d) 1, 3, and 5
   (e) 2, 4, and 5

8. Which of the following are *true*? (1) White blood cells leave the bloodstream by emigration. (2) Adhesion molecules help white blood cells stick to the endothelium, which aids emigration. (3) Neutrophils and macrophages are active in phagocytosis. (4) The attraction of phagocytes to microbes and inflamed tissue is termed chemotaxis. (5) Leukopenia is an increase in white blood cell count that occurs during infection.
   (a) 1, 2, 4, and 5
   (b) 2, 3, 4, and 5
   (c) 1, 2, 3, and 4
   (d) 1, 3, and 5
   (e) 1, 2, and 4

9. A person with type A Rh⁻ blood can receive a blood transfusion from which of the following types? (1) A Rh⁺, (2) B Rh⁻, (3) AB Rh⁻, (4) O Rh⁻, (5) A Rh⁻.
   (a) 1 only
   (b) 3 only
   (c) 4 only
   (d) 4 and 5
   (e) 1 and 5

10. A person with type B positive blood receives a transfusion of type AB positive blood. What will happen?
    (a) The recipient's antibodies will react with the donor's red blood cells.
    (b) The donor's antigens will destroy the recipient's antibodies.
    (c) The donor's antibodies will react with and destroy all of the recipient's red blood cells.
    (d) The recipient's blood type will change from Rh⁺ to Rh⁻.
    (e) These blood types are compatible, and the transfusion will be accepted.

11. What happens to the iron (Fe^{3+}) that is released during the breakdown of damaged red blood cells?
    (a) It is used to synthesize proteins.
    (b) It is transported to the liver where it becomes part of bile.
    (c) It is converted into urobilin and excreted in urine.
    (d) It attaches to transferrin and is transported to bone marrow for use in hemoglobin synthesis.
    (e) It is utilized by intestinal bacteria to convert bilirubin into urobilinogen.

12. Which of the following would *not* cause an increase in erythropoietin?
    (a) anemia
    (b) high altitude
    (c) hemorrhage
    (d) donating blood to a blood bank
    (e) polycythemia

13. Match the following:
    ____ (a) the percentage of total blood volume occupied by red blood cells
    ____ (b) the percentage of each type of white blood cell
    ____ (c) measures numbers of RBCs, WBCs, platelets per μ of blood; hematocrit; and differential WBC count
    ____ (d) measures the rate of erythropoiesis
    ____ (e) withdrawal of blood from a vein using a needle and collecting tube
    ____ (f) withdrawal of a small amount of red bone marrow with a fine needle and syringe
    ____ (g) removal of a core of red bone marrow with a large needle

    (1) reticulocyte count
    (2) bone marrow biopsy
    (3) venipuncture
    (4) hematocrit
    (5) bone marrow aspiration
    (6) complete blood count
    (7) differential white blood cell count

14. Match the following:

_____ (a) contain hemoglobin and function in gas transport

_____ (b) cell fragments enclosed by a piece of the cell membrane of megakaryocytes; contain clotting factors

_____ (c) white blood cell showing a kidney-shaped nucleus; capable of phagocytosis

_____ (d) monocytes that roam the tissues and gather at sites of infection or inflammation

_____ (e) occur as B cells, T cells, and natural killer cells

_____ (f) give rise to red blood cells, monocytes, neutrophils, eosinophils, basophils, and platelets

_____ (g) cells that give rise to all the formed elements of blood; derived from mesenchyme

(1) lymphocytes
(2) monocytes
(3) pluripotent stem cells
(4) red blood cells
(5) myeloid stem cells
(6) platelets
(7) wandering macrophages

15. Match the following:

_____ (a) tissue protein that leaks into the blood from cells outside blood vessels and initiates the formation of prothrombinase

_____ (b) an anticoagulant

_____ (c) its formation is initiated by either the extrinsic or intrinsic pathway or both; catalyzes the conversion of prothrombin to thrombin

_____ (d) forms the threads of a clot; produced from fibrinogen

_____ (e) can dissolve a clot by digesting fibrin threads

_____ (f) serves as the catalyst to form fibrin; formed from prothrombin

(1) prothrombinase
(2) thrombin
(3) fibrin
(4) thromboplastin
(5) plasmin
(6) heparin

20 THE CARDIOVASCULAR SYSTEM: THE HEART

CHAPTER REVIEW

20.1 Anatomy of the Heart

1. The heart is located in the mediastinum; about two-thirds of its mass is to the left of the midline. It is shaped like a cone lying on its side. Its apex is the pointed, inferior part; its base is the broad, superior part.

2. The pericardium is the membrane that surrounds and protects the heart; it consists of an outer fibrous layer and an inner serous pericardium, which is composed of a parietal layer and a visceral layer. Between the parietal and visceral layers of the serous pericardium is the pericardial cavity, a potential space filled with a few milliliters of lubricating pericardial fluid that reduces friction between the two membranes.

3. Three layers make up the wall of the heart: epicardium, myocardium, and endocardium. The epicardium consists of mesothelium and connective tissue, the myocardium is composed of cardiac muscle tissue, and the endocardium consists of endothelium and connective tissue.

4. The heart chambers include two superior chambers, the right and left atria, and two inferior chambers, the right and left ventricles. External features of the heart include the auricles, the coronary sulcus between the atria and ventricles, and the anterior and posterior sulci between the ventricles on the anterior and posterior surfaces of the heart, respectively.

5. The right atrium receives blood from the superior vena cava, inferior vena cava, and coronary sinus. It is separated internally from the left atrium by the interatrial septum, which contains the fossa ovalis. Blood exits the right atrium through the tricuspid valve.

6. The right ventricle receives blood from the right atrium. Separated internally from the left ventricle by the interventricular septum, it pumps blood through the pulmonary valve into the pulmonary trunk.

7. Oxygenated blood enters the left atrium from the pulmonary veins and exits through the bicuspid (mitral) valve.

8. The left ventricle pumps oxygenated blood through the aortic valve into the aorta.

9. The thickness of the myocardium of the four chambers varies according to the chamber's function. The left ventricle, with the highest workload, has the thickest wall.

10. The fibrous skeleton of the heart is dense connective tissue surrounding and supporting the heart valves.

20.2 Heart Valves and Circulation of Blood

1. Heart valves prevent backflow of blood within the heart. The atrioventricular (AV) valves, which lie between atria and ventricles, are the tricuspid valve on the right side of the heart and the bicuspid (mitral) valve on the left. The semilunar (SL) valves are the aortic valve, at the entrance to the aorta, and the pulmonary valve, at the entrance to the pulmonary trunk.

2. The left side of the heart is the pump for systemic circulation, the circulation of blood throughout the body except for the air sacs of the lungs. The left ventricle ejects blood into the aorta, and blood then flows into systemic arteries, arterioles, capillaries, venules, and veins, which carry it back to the right atrium.

3. The right side of the heart is the pump for pulmonary circulation, the circulation of blood through the lungs. The right ventricle ejects blood into the pulmonary trunk, and blood then flows into pulmonary arteries, pulmonary capillaries, and pulmonary veins, which carry it back to the left atrium.

4. The coronary circulation provides blood flow to the myocardium. Its main arteries are the left and right coronary arteries; its main veins are the cardiac veins and the coronary sinus.

20.3 Cardiac Muscle Tissue and the Cardiac Conduction System

1. Cardiac muscle fibers usually contain a single centrally located nucleus. Compared with skeletal muscle fibers, cardiac muscle fibers have more and larger mitochondria, slightly smaller sarcoplasmic reticulum, and wider transverse tubules, which are located at Z discs.

2. Cardiac muscle fibers are connected end-to-end via intercalated discs. Desmosomes in the discs provide strength, and gap junctions allow muscle action potentials to conduct from one muscle fiber to its neighbors.

3. Autorhythmic fibers form the conduction system, cardiac muscle fibers that spontaneously depolarize and generate action potentials.

4. Components of the conduction system are the sinoatrial (SA) node (pacemaker), atrioventricular (AV) node, atrioventricular (AV) bundle (bundle of His), bundle branches, and Purkinje fibers.

5. Phases of an action potential in a ventricular contractile fiber include rapid depolarization, a long plateau, and repolarization.

6. Cardiac muscle tissue has a long refractory period, which prevents tetanus.

7. The record of electrical changes during each cardiac cycle is called an electrocardiogram (ECG). A normal ECG consists of a P wave (atrial depolarization), a QRS complex (onset of ventricular depolarization), and a T wave (ventricular repolarization).

8. The P–Q interval represents the conduction time from the beginning of atrial excitation to the beginning of ventricular excitation. The S–T segment represents the time when ventricular contractile fibers are fully depolarized.

20.4 The Cardiac Cycle

1. A cardiac cycle consists of the systole (contraction) and diastole (relaxation) of both atria, plus the systole and diastole of both ventricles. With an average heartbeat of 75 beats/min, a complete cardiac cycle requires 0.8 sec.

2. The phases of the cardiac cycle are (a) atrial systole, (b) ventricular systole, and (c) relaxation period.

3. S1, the first heart sound (lubb), is caused by blood turbulence associated with the closing of the atrioventricular valves. S2, the second sound (dupp), is caused by blood turbulence associated with the closing of semilunar valves.

20.5 Cardiac Output

1. Cardiac output (CO) is the amount of blood ejected per minute by the left ventricle into the aorta (or by the right ventricle into the pulmonary trunk). It is calculated as follows: CO (mL/min) = stroke volume (SV) in mL/beat × heart rate (HR) in beats/min.

2. Stroke volume (SV) is the amount of blood ejected by a ventricle during each systole.

3. Cardiac reserve is the difference between a person's maximum CO and his or her CO at rest.

4. Stroke volume is related to preload (stretch on the heart before it contracts), contractility (forcefulness of contraction), and

afterload (pressure that must be exceeded before ventricular ejection can begin).

5. According to the Frank–Starling law of the heart, a greater preload (end-diastolic volume) stretching cardiac muscle fibers just before they contract increases their force of contraction until the stretching becomes excessive.

6. Nervous control of the cardiovascular system originates in the cardiovascular center in the medulla oblongata.

7. Sympathetic impulses increase heart rate and force of contraction; parasympathetic impulses decrease heart rate.

8. Heart rate is affected by hormones (epinephrine, norepinephrine, thyroid hormones), ions (Na^+, K^+, Ca^{2+}), age, gender, physical fitness, and body temperature.

20.6 Exercise and the Heart

1. Sustained exercise increases oxygen demand on muscles.

2. Among the benefits of aerobic exercise are increased cardiac output, decreased blood pressure, weight control, and increased fibrinolytic activity.

20.7 Help for Failing Hearts

1. A cardiac (heart) transplant is the replacement of a severely damaged heart with a normal one.

2. Cardiac assist devices and procedures include the intra-aortic balloon pump, the ventricular assist device, cardiomyoplasty, and a skeletal muscle assist device.

20.8 Development of the Heart

1. The heart develops from mesoderm.

2. The endocardial tubes develop into the four-chambered heart and great vessels of the heart.

CLINICAL CONNECTIONS

Cardiopulmonary Resuscitation (Refer page 606 of Textbook)

Cardiopulmonary resuscitation (CPR) (kar-dē-ō-PUL-mo-nar'-ē rē-sus-i-TĀ-shun) refers to an emergency procedure for establishing a normal heartbeat and rate of breathing. Standard CPR uses a combination of cardiac compression and artificial ventilation of the lungs via mouth-to-mouth respiration, and for many years this combination was the sole method of CPR. Recently, however, hands-only CPR has become the preferred method.

Because the heart lies between two rigid structures—the sternum and vertebral column—pressure on the chest (compression) can be used to force blood out of the heart and into the circulation. After calling 911, hands-only CPR should be administered. In the procedure, chest compressions should be given hard and fast at a rate of 100 per minute and two inches deep in adults. This should be continued until trained medical professionals arrive or an automated external defibrillator is available. Standard CPR is still recommended for infants and children, as well as anyone who suffers from lack of oxygen, for example, victims of near-drowning, drug overdose, or carbon monoxide poisoning.

It is estimated that hands-only CPR saves about 20% more lives than the standard method. Moreover, hands-only CPR boosts the survival rate from 18% to 34% compared to the traditional method or none at all. It is also easier for an emergency dispatcher to give instructions limited to hands-only CPR

to frightened, nonmedical bystanders. Finally, as public fear of contracting contagious diseases such as HIV, hepatitis, and tuberculosis continues to rise, bystanders are much more likely to perform hands-only CPR rather than treatment involving the standard method.

Pericarditis (Refer page 607 of Textbook)

Inflammation of the pericardium is called **pericarditis** (per'-i-kar-DĪ-tis). The most common type, *acute pericarditis*, begins suddenly and has no known cause in most cases but is sometimes linked to a viral infection. As a result of irritation to the pericardium, there is chest pain that may extend to the left shoulder and down the left arm (often mistaken for a heart attack) and *pericardial friction rub* (a scratchy or creaking sound heard through a stethoscope as the visceral layer of the serous pericardium rubs against the parietal layer of the serous pericardium). Acute pericarditis usually lasts for about 1 week and is treated with drugs that reduce inflammation and pain, such as ibuprofen or aspirin.

Chronic pericarditis begins gradually and is long-lasting. In one form of this condition, there is a buildup of pericardial fluid. If a great deal of fluid accumulates, this is a life-threatening condition because the fluid compresses the heart, a condition called *cardiac tamponade* (tam'-pon-ĀD). As a result of the compression, ventricular filling is decreased, cardiac output is reduced, venous return to the heart is diminished, blood pressure falls, and breathing is difficult. In most cases, the cause of chronic pericarditis involving cardiac tamponade is unknown, but it sometimes results from conditions such as cancer and tuberculosis. Treatment consists of draining the excess fluid through a needle passed into the pericardial cavity.

Myocarditis and Endocarditis (Refer page 608 of Textbook)

Myocarditis (mī-ō-kar-DĪ-tis) is an inflammation of the myocardium that usually occurs as a complication of a viral infection, rheumatic fever, or exposure to radiation or certain chemicals or medications. Myocarditis often has no symptoms. However, if they do occur, they may include fever, fatigue, vague chest pain, irregular or rapid heartbeat, joint pain, and breathlessness. Myocarditis is usually mild and recovery occurs within 2 weeks. Severe cases can lead to cardiac failure and death. Treatment consists of avoiding vigorous exercise, a low-salt diet, electrocardiographic monitoring, and treatment of the cardiac failure. **Endocarditis** (en′-dō-kar-DĪ-tis) refers to an inflammation of the endocardium and typically involves the heart valves. Most cases are caused by bacteria (bacterial endocarditis). Signs and symptoms of endocarditis include fever, heart murmur, irregular or rapid heartbeat, fatigue, loss of appetite, night sweats, and chills. Treatment is with intravenous antibiotics.

Heart Valve Disorders (Refer page 615 of Textbook)

When heart valves operate normally, they open fully and close completely at the proper times. A narrowing of a heart valve opening that restricts blood flow is known as **stenosis** (sten-Ō-sis = a narrowing); failure of a valve to close completely is termed **insufficiency** (in′-su-FISH-en-sē) or *incompetence*. In **mitral stenosis**, scar formation or a congenital defect causes narrowing of the mitral valve. One cause of **mitral insufficiency**, in which there is backflow of blood from the left ventricle into the left atrium, is **mitral valve prolapse (MVP)**. In MVP one or both cusps of the mitral valve protrude into the left atrium during ventricular contraction. Mitral valve prolapse is one of the most common valvular disorders, affecting as much as 30% of the population. It is more prevalent in women than in men, and does not always pose a serious threat. In **aortic stenosis** the aortic valve is narrowed, and in **aortic insufficiency** there is backflow of blood from the aorta into the left ventricle.

Certain infectious diseases can damage or destroy the heart valves. One example is **rheumatic fever**, an acute systemic inflammatory disease that usually occurs after a streptococcal infection of the throat. The bacteria trigger an immune response in which antibodies produced to destroy the bacteria instead attack and inflame the connective tissues in joints, heart valves, and other organs. Even though rheumatic fever may weaken the entire heart wall, most often it damages the mitral and aortic valves.

If daily activities are affected by symptoms and if a heart valve cannot be repaired surgically, then the valve must be replaced. Tissue valves may be provided by human donors or pigs; sometimes, mechanical replacements are used. In any case, valve replacement involves open heart surgery. The aortic valve is the most commonly replaced heart valve.

Myocardial Ischemia and Infarction (Refer page 616 of Textbook)

Partial obstruction of blood flow in the coronary arteries may cause **myocardial ischemia** (is-KĒ-mē-a; *ische-* = to obstruct; *-emia* = in the blood), a condition of reduced blood flow to the myocardium. Usually, ischemia causes **hypoxia** (hī-POKS-ē-a = reduced oxygen supply), which may weaken cells without killing them. **Angina pectoris** (an-JĪ-na, or AN-ji-na, PEK-tō-ris), which literally means "strangled chest," is a severe pain that usually accompanies myocardial ischemia. Typically, sufferers describe it as a tightness or squeezing sensation, as though the chest were in a vise. The pain associated with angina pectoris is often referred to the neck, chin, or down the left arm to the elbow. **Silent myocardial ischemia**, ischemic episodes without pain, is

particularly dangerous because the person has no forewarning of an impending heart attack.

A complete obstruction to blood flow in a coronary artery may result in a **myocardial infarction (MI)** (in-FARK-shun), commonly called a *heart attack*. *Infarction* means the death of an area of tissue because of interrupted blood supply. Because the heart tissue distal to the obstruction dies and is replaced by noncontractile scar tissue, the heart muscle loses some of its strength. Depending on the size and location of the infarcted (dead) area, an infarction may disrupt the conduction system of the heart and cause sudden death by triggering ventricular fibrillation. Treatment for a myocardial infarction may involve injection of a thrombolytic (clot-dissolving) agent such as streptokinase or tPA, plus heparin (an anticoagulant), or performing coronary angioplasty or coronary artery bypass grafting. Fortunately, heart muscle can remain alive in a resting person if it receives as little as 10–15% of its normal blood supply.

Regeneration of Heart Cells (Refer page 619 of Textbook)

As noted earlier in the chapter, the heart of an individual who survives a heart attack often has regions of infarcted (dead) cardiac muscle tissue that typically are replaced with noncontractile fibrous scar tissue over time. Our inability to repair damage from a heart attack has been attributed to a lack of stem cells in cardiac muscle and to the absence of mitosis in mature cardiac muscle fibers. A recent study of heart transplant recipients by American and Italian scientists, however, provides evidence for significant replacement of heart cells. The researchers studied men who had received a heart from a female, and then looked for the presence of a Y chromosome in heart cells. (All female cells except gametes have two X chromosomes and lack the Y chromosome.) Several years after the transplant surgery, between 7% and 16% of the heart cells in the transplanted tissue, including cardiac muscle fibers and endothelial cells in coronary arterioles and capillaries, had been replaced by the recipient's own cells, as evidenced by the presence of a Y chromosome. The study also revealed cells with some of the characteristics of stem cells in both transplanted hearts and control hearts. Evidently, stem cells can migrate from the blood into the heart and differentiate into functional muscle and endothelial cells. The hope is that researchers can learn how to "turn on" such regeneration of heart cells to treat people with heart failure or cardiomyopathy (diseased heart).

Artificial Pacemakers (Refer page 621 of Textbook)

If the SA node becomes damaged or diseased, the slower AV node can pick up the pacemaking task. Its spontaneous pacing rate is 40 to 60 times per minute. If the activity of both nodes is suppressed, the heartbeat may still be maintained by autorhythmic fibers in the ventricles—the AV bundle, a bundle branch, or Purkinje fibers. However, the pacing rate is so slow (20–35 beats per minute) that blood flow to the brain is inadequate. When this condition occurs, normal heart rhythm can be restored and maintained by surgically implanting an **artificial pacemaker**, a device that sends out small electrical currents to stimulate the heart to contract. A pacemaker consists of a battery and impulse generator and is usually implanted beneath the skin just inferior to the clavicle. The pacemaker is connected to one or two flexible leads (wires) that are threaded through the superior vena cava and then passed into the various chambers of the heart. Many of the newer pacemakers, referred to as *activity-adjusted pacemakers*, automatically speed up the heartbeat during exercise.

Heart Murmurs (Refer page 627 of Textbook)

Heart sounds provide valuable information about the mechanical operation of the heart. A **heart murmur** is an abnormal sound consisting of a clicking, rushing, or gurgling noise that either is heard before, between, or after the normal heart sounds, or may mask the normal heart sounds. Heart murmurs in children are extremely common and usually do not represent a health condition. Murmurs are most frequently discovered in children between the ages of 2 and 4. These types of heart murmurs are referred to as *innocent* or *functional heart murmurs*; they often subside or disappear with growth. Although some heart murmurs in adults are innocent, most often an adult murmur indicates a valve disorder. When a heart valve exhibits stenosis, the heart murmur is heard while the valve should be fully open but is not. For example, mitral stenosis (see Clinical Connection: Heart Valve Disorders) produces a murmur during the relaxation period, between S2 and the next S1. An incompetent heart valve, by contrast, causes a murmur to appear when the valve should be fully closed but is not. So, a murmur due to mitral incompetence (see Clinical Connection: Heart Valve Disorders) occurs during ventricular systole, between S1 and S2.

DISORDERS: HOMEOSTATIC IMBALANCES

Coronary Artery Disease

Coronary artery disease (CAD) is a serious medical problem that affects about 7 million people annually. Responsible for nearly three-quarters of a million deaths in the United States each year, it is the leading cause of death for both men and women. CAD results from the effects of the accumulation of atherosclerotic plaques (described shortly) in coronary arteries, which leads to a reduction in blood flow to the myocardium. Some individuals have no signs or symptoms; others experience angina pectoris (chest pain), and still others suffer heart attacks.

Risk Factors for CAD People who possess combinations of certain risk factors are more likely to develop CAD. *Risk factors* (characteristics, symptoms, or signs present in a disease-free person that are statistically associated with a greater chance of developing a disease) include smoking, high blood pressure, diabetes, high cholesterol levels, obesity, "type A" personality, sedentary lifestyle, and a family history of CAD. Most of these can be modified by changing diet and other habits or can be controlled by taking medications. However, other risk factors are unmodifiable (beyond our control), including genetic predisposition (family history of CAD at an early age), age, and gender. For example, adult males are more likely than adult females to develop CAD; after age 70 the risks are roughly equal. Smoking is undoubtedly the number-one risk factor in all CAD-associated diseases, roughly doubling the risk of morbidity and mortality.

Development of Atherosclerotic Plaques Although the following discussion applies to coronary arteries, the process can also occur in arteries outside the heart. Thickening of the walls of arteries and loss of elasticity are the main characteristics of a group of diseases called **arteriosclerosis** (ar-tē-rē-ō-skle-RŌ-sis; *sclero-* = hardening). One form of arteriosclerosis is **atherosclerosis** (ath-er-ō-skle-RŌ-sis), a progressive disease characterized by the formation in the walls of large and medium-sized arteries of lesions called **atherosclerotic plaques** (ath-er-ō-skle-RO-tik) (**Figure SG20.1**).

To understand how atherosclerotic plaques develop, you will need to learn about the role of molecules produced by the liver and small intestine called **lipoproteins**. These spherical particles consist of an inner core of triglycerides and other lipids and an outer shell of proteins, phospholipids, and cholesterol. Like most lipids, cholesterol does not dissolve in water and must be made water-soluble in order to be transported in the blood. This is accomplished by combining it with lipoproteins. Two major lipoproteins are **low-density lipoproteins (LDLs) and high-density lipoproteins (HDLs)**. LDLs transport cholesterol from the liver to body cells for use in cell membrane repair and the production of steroid hormones and bile salts. However, excessive amounts of LDLs promote atherosclerosis, so the cholesterol in these particles is commonly known as "bad cholesterol." HDLs, on the other hand, remove excess cholesterol from body cells and transport it to the liver for elimination. Because HDLs decrease blood cholesterol level, the cholesterol in HDLs is commonly referred to as "good cholesterol." Basically, you want your LDL concentration to be low and your HDL concentration to be high.

Inflammation, a defensive response of the body to tissue damage, plays a key role in the development of atherosclerotic plaques. As a result of tissue damage, blood vessels dilate and increase their permeability, and phagocytes, including macrophages, appear in large numbers. The formation of

Figure SG20.1 Photomicrographs of a transverse section of a normal artery and one partially obstructed by an atherosclerotic plaque.

Inflammation plays a key role in the development of atherosclerotic plaques.

Unobstructed lumen (space through which blood flows)

Partially obstructed lumen

Atherosclerotic plaque

LM about 20x

LM about 20x

Normal artery

Obstructed artery

Biophoto Associates/Science SourceImages

Normal and obstructed arteries

Q What is the role of HDL?

atherosclerotic plaques begins when excess LDLs from the blood accumulate in the inner layer of an artery wall (layer closest to the bloodstream), the lipids and proteins in the LDLs undergo oxidation (removal of electrons), and the proteins bind to sugars. In response, endothelial and smooth muscle cells of the artery secrete substances that attract monocytes from the blood and convert them into macrophages. The macrophages then ingest and become so filled with oxidized LDL particles that they have a foamy appearance when viewed microscopically (**foam cells**). T cells (lymphocytes) follow monocytes into the inner lining of an artery, where they release chemicals that intensify the inflammatory response. Together, the foam cells, macrophages, and T cells form a **fatty streak**, the beginning of an atherosclerotic plaque.

Macrophages secrete chemicals that cause smooth muscle cells of the middle layer of an artery to migrate to the top of the atherosclerotic plaque, forming a cap over it and thus walling it off from the blood.

Because most atherosclerotic plaques expand away from the bloodstream rather than into it, blood can still flow through the affected artery with relative ease, often for decades. Relatively few heart attacks occur when plaque in a coronary artery expands into the bloodstream and restricts blood flow. Most heart attacks occur when the cap over the plaque breaks open in response to chemicals produced by foam cells. In addition, T cells induce foam cells to produce tissue factor (TF), a chemical that begins the cascade of reactions that result in blood clot formation. If the clot in a coronary artery is large enough, it can significantly decrease or stop the flow of blood and result in a heart attack.

A number of other risk factors (all modifiable) have also been identified as significant predictors of CAD when their levels are elevated. **C-reactive proteins (CRPs)** are proteins produced by the liver or present in blood in an inactive form that are converted to an active form during inflammation. CRPs may play a direct role in the development of atherosclerosis by promoting the uptake of LDLs by macrophages. **Lipoprotein (a)** is an LDL-like particle that binds to endothelial cells, macrophages, and blood platelets; may promote the proliferation of smooth muscle fibers; and inhibits the breakdown of blood clots. **Fibrinogen** is a glycoprotein involved in blood clotting that may help regulate cellular proliferation, vasoconstriction, and platelet aggregation. **Homocysteine** (hō′-mō-SIS-tēn) is an amino acid that may induce blood vessel damage by promoting platelet aggregation and smooth muscle fiber proliferation.

Diagnosis of CAD Many procedures may be employed to diagnose CAD; the specific procedure used will depend on the signs and symptoms of the individual.

A resting electrocardiogram (see Section 20.3) is the standard test employed to diagnose CAD. **Stress testing** can also be performed. In an *exercise stress test*, the functioning of the heart is monitored when placed under physical stress by exercising using a treadmill, an exercise bicycle, or arm exercises. During the procedure, ECG recordings are monitored continuously and blood pressure is monitored at intervals. A *nonexercise (pharmacologic) stress test* is used for individuals who cannot exercise due to conditions such as arthritis. A medication is injected that stresses the heart to mimic the effects of exercise. During both exercise and nonexercise stress testing, **radionuclide imaging** may be performed to evaluate blood flow through heart muscle (see **Table 1.3**).

Diagnosis of CAD may also involve **echocardiography** (ek′-ō-kar-dē-OG-ra-fē), a technique that uses ultrasound waves to image the interior of the heart. Echocardiography allows the heart to be seen in motion and can be used to determine the size, shape, and functions of the chambers of the heart; the volume and velocity of blood pumped from the heart; the status of heart valves; the presence of birth defects; and abnormalities of the pericardium. A fairly recent technique for evaluating CAD is **electron beam computerized tomography (EBCT)**, which detects calcium deposits in coronary arteries. These calcium deposits are indicators of atherosclerosis.

Coronary (cardiac) computed tomography radiography (CCTA) is a computer-assisted radiography procedure in which a contrast medium is injected into a vein and a beta blocker is given to decrease heart rate. Then x-ray beams trace an arc around the heart and ultimately produce an image called a *CCTA scan*. This procedure is used primarily to detect blockages such as atherosclerotic plaques or calcium deposits (see **Table 1.3**).

Cardiac catheterization (kath′-e-ter-i-ZĀ-shun) is an invasive procedure used to visualize the heart's chambers, valves, and great vessels in order to diagnose and treat disease not related to abnormalities of the coronary arteries. It may also be used to measure pressure in the heart and great vessels; to assess cardiac output; to measure the flow of blood through the heart and great vessels; to identify the location of septal and valvular defects; and to take tissue and blood samples. The basic procedure involves inserting a long, flexible, radiopaque **catheter** (plastic tube) into a peripheral vein (for right heart catheterization) or a peripheral artery (for *left heart catheterization*) and guiding it under fluoroscopy (x-ray observation).

Coronary angiography (an′-jē-OG-ra-fē; *angio-* = blood vessel; *-grapho* = to write) is an invasive procedure used to obtain information about the coronary arteries. In the procedure, a catheter is inserted into an artery in the groin or wrist and threaded under fluoroscopy toward the heart and then into the coronary arteries. After the tip of the catheter is in place, a radiopaque contrast medium is injected into the coronary arteries. The radiographs of the arteries, called *angiograms*, appear in motion on a monitor and the information is recorded on a videotape or computer disc. Coronary angiography may be used to visualize coronary arteries (see **Table 1.3**) and to inject clot-dissolving drugs, such as streptokinase or tissue plasminogen activator (tPA), into a coronary artery to dissolve an obstructing thrombus.

Treatment of CAD Treatment options for CAD include **drugs** (antihypertensives, nitroglycerin, beta blockers, cholesterol-lowering drugs, and clot-dissolving agents) and various surgical and nonsurgical procedures designed to increase the blood supply to the heart.

Coronary artery bypass grafting (CABG) is a surgical procedure in which a blood vessel from another part of the body is attached ("grafted") to a coronary artery to bypass an area of blockage. A piece of the grafted blood vessel is sutured between the aorta and the unblocked portion of the coronary artery (**Figure SG20.2a**). Sometimes multiple blood vessels have to be grafted.

A nonsurgical procedure used to treat CAD is **percutaneous transluminal coronary angioplasty (PTCA)** (*percutaneous* = through the skin; *trans-* = across; *-lumen* = an opening or channel in a tube; *angio-* = blood vessel; *-plasty* = to mold or to shape). In one variation of this procedure, a balloon catheter is inserted into an artery of an arm or leg and gently guided into a coronary artery (**Figure SG20.2b**). While dye is released, angiograms (videotape x-rays of blood vessels) are taken to locate the plaques. Next, the catheter is advanced to the point of obstruction, and a balloonlike device is inflated with air to squash the plaque against the blood vessel wall. Because 30–50% of PTCA-opened arteries fail due to restenosis (renarrowing) within 6 months after the procedure is done, a stent may be inserted via a catheter. A **stent** is a metallic, fine wire tube that is permanently placed in an artery to keep the artery *patent* (open), permitting blood to circulate (**Figure SG20.2c, d**). Restenosis may be due to damage from the procedure itself, for PTCA may damage the arterial wall, leading to platelet activation, proliferation of smooth muscle fibers, and plaque formation. Recently, *drug-coated (drug-eluting) coronary stents* have been used to prevent restenosis. The stents are coated with one of several antiproliferative drugs (drugs that inhibit the proliferation of smooth muscle fibers of the middle layer of an artery) and anti-inflammatory drugs. It has been shown that drug-coated stents reduce the rate of restenosis when compared with bare-metal (noncoated) stents. In addition to balloon and stent angioplasty, laser-emitting catheters are used to vaporize plaques (excimer laser coronary angioplasty or ELCA) and small blades inside catheters are used to remove part of the plaque (directional coronary atherectomy).

Figure SG20.2 Procedures for reestablishing blood flow in occluded coronary arteries.

Treatment options for CAD include drugs and various nonsurgical and surgical procedures.

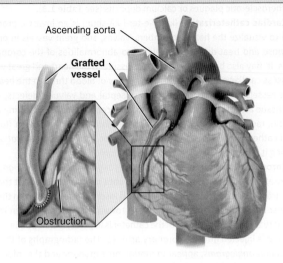

(a) Coronary artery bypass grafting (CABG)

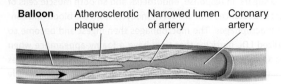
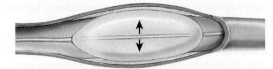

Balloon catheter with uninflated balloon is threaded to obstructed area in artery

When balloon is inflated, it stretches arterial wall and squashes atherosclerotic plaque

After lumen is widened, balloon is deflated and catheter is withdrawn

(b) Percutaneous transluminal coronary angioplasty (PTCA)

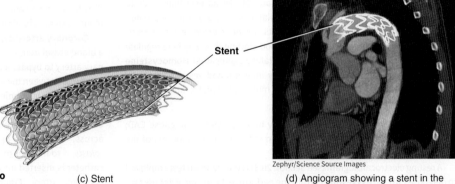

Zephyr/Science Source Images

Q Which diagnostic procedure for CAD is used to visualize coronary blood vessels?

(c) Stent

(d) Angiogram showing a stent in the circumflex artery

One area of current research involves cooling the body's core temperature during procedures such as coronary artery bypass grafting (CABG). There have been some promising results from the application of cold therapy during a cerebral vascular accident (CVA or stroke). This research stemmed from observations of people who had suffered a hypothermic incident (such as cold-water drowning) and recovered with relatively minimal neurologic deficits.

Congenital Heart Defects

A defect that is present at birth, and usually before, is called a **congenital defect** (kon-JEN-i-tal). Many such defects are not serious and may go unnoticed for a lifetime. Others are life-threatening and must be surgically repaired. Among the several congenital defects that affect the heart are the following (**Figure SG20.3**):

- **Coarctation of the aorta** (kō′-ark-TĀ-shun). In this condition, a segment of the aorta is too narrow, and thus the flow of oxygenated blood to the body is reduced, the left ventricle is forced to pump harder, and high blood

pressure develops. Coarctation is usually repaired surgically by removing the area of obstruction. Surgical interventions that are done in childhood may require revisions in adulthood. Another surgical procedure is balloon dilation, insertion and inflation of a device in the aorta to stretch the vessel. A stent can be inserted and left in place to hold the vessel open.

- **Patent ductus arteriosus (PDA)** (PĀ-tent). In some babies, the ductus arteriosus, a temporary blood vessel between the aorta and the pulmonary trunk, remains open rather than closing shortly after birth. As a result, aortic blood flows into the lower-pressure pulmonary trunk, thus increasing the pulmonary trunk blood pressure and overworking both ventricles. In uncomplicated PDA, medication can be used to facilitate the closure of the defect. In more severe cases, surgical intervention may be required.

- **Septal defect.** A septal defect is an opening in the septum that separates the interior of the heart into left and right sides. In an **atrial septal defect** the fetal foramen ovale between the two atria fails to close after birth. A **ventricular septal defect** is caused by incomplete development of the

Figure SG20.3 Congenital heart defects.

A congenital defect is one that is present at birth, and usually before.

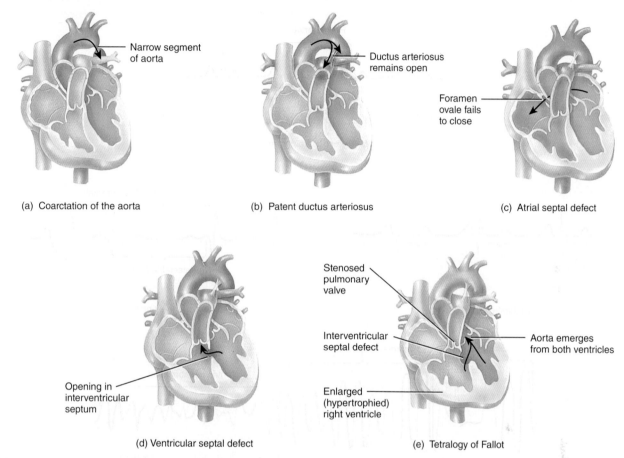

(a) Coarctation of the aorta

Narrow segment of aorta

(b) Patent ductus arteriosus

Ductus arteriosus remains open

(c) Atrial septal defect

Foramen ovale fails to close

(d) Ventricular septal defect

Opening in interventricular septum

(e) Tetralogy of Fallot

Stenosed pulmonary valve

Interventricular septal defect

Enlarged (hypertrophied) right ventricle

Aorta emerges from both ventricles

Q Which four developmental defects occur in tetralogy of Fallot?

interventricular septum. In such cases, oxygenated blood flows directly from the left ventricle into the right ventricle, where it mixes with deoxygenated blood. The condition is treated surgically.

- **Tetralogy of Fallot** (tet-RAL-ō-jē of fal-Ō). This condition is a combination of four developmental defects: an interventricular septal defect, an aorta that emerges from both ventricles instead of from the left ventricle only, a stenosed pulmonary valve, and an enlarged right ventricle. There is a decreased flow of blood to the lungs and mixing of blood from both sides of the heart. This causes cyanosis, the bluish discoloration most easily seen in nail beds and mucous membranes when the level of deoxygenated hemoglobin is high; in infants, this condition is referred to as "blue baby." Despite the apparent complexity of this condition, surgical repair is usually successful.

Arrhythmias

The usual rhythm of heartbeats, established by the SA node, is called **normal sinus rhythm**. The term **arrhythmia** (a-RITH-mē-a) or *dysrhythmia* refers to an abnormal rhythm as a result of a defect in the conduction system of the heart. The heart may beat irregularly, too quickly, or too slowly. Symptoms include chest pain, shortness of breath, lightheadedness, dizziness, and fainting. Arrhythmias may be caused by factors that stimulate the heart such as stress, caffeine, alcohol, nicotine, cocaine, and certain drugs that contain caffeine or other stimulants. Arrhythmias may also be caused by a congenital defect, coronary artery disease, myocardial infarction, hypertension, defective heart valves, rheumatic heart disease, hyperthyroidism, and potassium deficiency.

Arrhythmias are categorized by their speed, rhythm, and origination of the problem. **Bradycardia** (brād'-i-KAR-dē-a; *brady-* = slow) refers to a slow heart rate (below 50 beats per minute); **tachycardia** (tak'-i-KAR-dē-a; *tachy-* = quick) refers to a rapid heart rate (over 100 beats per minute); and **fibrillation** (fi-bri-LĀ-shun) refers to rapid, uncoordinated heartbeats. Arrhythmias that begin in the atria are called **supraventricular** or **atrial arrhythmias**; those that originate in the ventricles are called **ventricular arrhythmias**.

- **Supraventricular tachycardia (SVT)** is a rapid but regular heart rate (160–200 beats per minute) that originates in the atria. The episodes begin and end suddenly and may last from a few minutes to many hours. SVTs can sometimes be stopped by maneuvers that stimulate the vagus (X) nerve and decrease heart rate. These include straining as if having a difficult bowel movement, rubbing the area over the carotid artery in the neck to stimulate the carotid sinus (not recommended for people over 50 since it may cause a stroke), and plunging the face into a bowl of ice-cold water. Treatment may also involve antiarrhythmic drugs and destruction of the abnormal pathway by radiofrequency ablation.

Figure SG20.4 Representative arrhythmias.

An arrhythmia is an abnormal rhythm as a result of a defect in the cardiac conduction system.

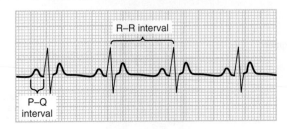

(a) Normal electrocardiogram (ECG)

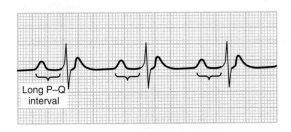

(b) First-degree AV block

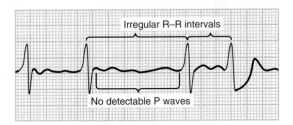

(c) Atrial fibrillation

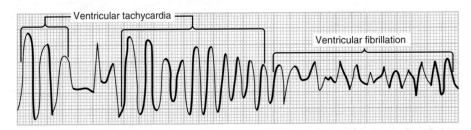

(d) Ventricular tachycardia

(e) Ventricular fibrillation

Q Why is ventricular fibrillation such a serious arrhythmia?

- **Heart block** is an arrhythmia that occurs when the electrical pathways between the atria and ventricles are blocked, slowing the transmission of nerve impulses. The most common site of blockage is the atrioventricular node, a condition called *atrioventricular (AV) block*. In *first-degree AV block*, the P–Q interval is prolonged, usually because conduction through the AV node is slower than normal (**Figure SG20.4b**). In *second-degree AV block*, some of the action potentials from the SA node are not conducted through the AV node. The result is "dropped" beats because excitation doesn't always reach the ventricles. Consequently, there are fewer QRS complexes than P waves on the ECG. In *third-degree (complete) AV block*, no SA node action potentials get through the AV node. Autorhythmic fibers in the atria and ventricles pace the upper and lower chambers separately. With complete AV block, the ventricular contraction rate is less than 40 beats/min.

- **Atrial premature contraction (APC)** is a heartbeat that occurs earlier than expected and briefly interrupts the normal heart rhythm. It often causes a sensation of a skipped heartbeat followed by a more forceful heartbeat. APCs originate in the atrial myocardium and are common in healthy individuals.

- **Atrial flutter** consists of rapid, regular atrial contractions (240–360 beats/min) accompanied by an atrioventricular (AV) block in which some

of the nerve impulses from the SA node are not conducted through the AV node.

- **Atrial fibrillation (AF)** is a common arrhythmia, affecting mostly older adults, in which contraction of the atrial fibers is asynchronous (not in unison) so that atrial pumping ceases altogether. The atria may beat 300–600 beats/min. The ventricles may also speed up, resulting in a rapid heartbeat (up to 160 beats/min). The ECG of an individual with atrial fibrillation typically has no clearly defined P waves and irregularly spaced QRS complexes (and R–R intervals) (**Figure SG20.4c**). Since the atria and ventricles do not beat in rhythm, the heartbeat is irregular in timing and strength. In an otherwise strong heart, atrial fibrillation reduces the pumping effectiveness of the heart by 20–30%. The most dangerous complication of atrial fibrillation is stroke since blood may stagnate in the atria and form blood clots. A stroke occurs when part of a blood clot occludes an artery supplying the brain.

- **Ventricular premature contraction**, another form of arrhythmia, arises when an *ectopic focus* (ek-TOP-ik), a region of the heart other than the conduction system, becomes more excitable than normal and causes an occasional abnormal action potential to occur. As a wave of depolarization spreads outward from the ectopic focus, it causes a ventricular premature contraction (beat). The contraction occurs early in diastole before the SA node is normally scheduled to discharge its action potential. Ventricular premature contractions may be relatively benign and may be caused by emotional stress, excessive intake of stimulants such as caffeine, alcohol, or nicotine, and lack of sleep. In other cases, the premature beats may reflect an underlying pathology.

- **Ventricular tachycardia (VT or V-tach)** is an arrhythmia that originates in the ventricles and is characterized by four or more ventricular premature contractions. It causes the ventricles to beat too fast (at least 120 beats/min) (**Figure SG20.4d**). VT is almost always associated with heart disease or a recent myocardial infarction and may develop into a very serious arrhythmia called ventricular fibrillation (described shortly). Sustained VT is dangerous because the ventricles do not fill properly and thus do not pump sufficient blood. The result may be low blood pressure and heart failure.

- **Ventricular fibrillation (VF or V-fib)** is the most deadly arrhythmia, in which contractions of the ventricular fibers are completely asynchronous so that the ventricles quiver rather than contract in a coordinated way. As a result, ventricular pumping stops, blood ejection ceases, and circulatory failure and death occur unless there is immediate medical intervention. During ventricular fibrillation, the ECG has no clearly defined P waves, QRS complexes, or T waves (**Figure SG20.4e**). The most common cause of ventricular fibrillation is inadequate blood flow to the heart due to coronary artery disease, as occurs during a myocardial infarction. Other causes are cardiovascular shock, electrical shock, drowning, and very low potassium levels. Ventricular fibrillation causes unconsciousness in seconds and, if untreated, seizures occur and irreversible brain damage may occur after 5 minutes. Death soon follows. Treatment involves cardiopulmonary resuscitation (CPR) and defibrillation. In **defibrillation** (dē-fib-re-LĀ-shun), also called **cardioversion** (kar'-dē-ō-VER-shun), a strong, brief electrical current is passed to the heart and often can stop the ventricular fibrillation. The electrical shock is generated by a device

called a **defibrillator** (de-FIB-ri-lā-tor) and applied via two large paddle-shaped electrodes pressed against the skin of the chest. Patients who face a high risk of dying from heart rhythm disorders now can receive an **automatic implantable cardioverter defibrillator (AICD)**, an implanted device that monitors heart rhythm and delivers a small shock directly to the heart when a life-threatening rhythm disturbance occurs. Thousands of patients around the world have AICDs. Also available are **automated external defibrillators (AEDs)** that function like AICDs, except that they are external devices. About the size of a laptop computer, AEDs are used by emergency response teams and are found increasingly in public places such as stadiums, casinos, airports, hotels, and shopping malls. Defibrillation may also be used as an emergency treatment for cardiac arrest.

Congestive Heart Failure

In **congestive heart failure (CHF)**, there is a loss of pumping efficiency by the heart. Causes of CHF include coronary artery disease, congenital defects, long-term high blood pressure (which increases the afterload), myocardial infarctions (regions of dead heart tissue due to a previous heart attack), and valve disorders. As the pump becomes less effective, more blood remains in the ventricles at the end of each cycle, and gradually the end-diastolic volume (preload) increases. Initially, increased preload may promote increased force of contraction (the Frank–Starling law of the heart), but as the preload increases further, the heart is overstretched and contracts less forcefully. The result is a potentially lethal positive feedback loop: Less effective pumping leads to even lower pumping capability.

Often, one side of the heart starts to fail before the other. If the left ventricle fails first, it cannot pump out all the blood it receives. As a result, blood backs up in the lungs and causes *pulmonary edema*, fluid accumulation in the lungs that can cause suffocation if left untreated. If the right ventricle fails first, blood backs up in the systemic veins and, over time, the kidneys cause an increase in blood volume. In this case, the resulting *peripheral edema* usually is most noticeable in the feet and ankles.

MEDICAL TERMINOLOGY

Asystole (ā-SIS-tō-lē; *a-* = without) Failure of the myocardium to contract.

Cardiac arrest (KAR-dē-ak a-REST) Cessation of an effective heartbeat. The heart may be completely stopped or in ventricular fibrillation.

Cardiac rehabilitation (rē-ha-bil-i-TĀ-shun) A supervised program of progressive exercise, psychological support, education, and training to enable a patient to resume normal activities following a myocardial infarction.

Cardiomegaly (kar'-dē-ō-MEG-a-lē; *mega* = large) Heart enlargement.

Cardiomyopathy (kar'-dē-ō-mī-OP-a-thē; *myo-* = muscle; *-pathos* = disease) A progressive disorder in which ventricular structure or function is impaired. In dilated cardiomyopathy, the ventricles enlarge (stretch) and become weaker, reducing the heart's pumping action. In hypertrophic cardiomyopathy, the ventricular walls thicken and the pumping efficiency of the ventricles is reduced.

Commotio cordis (kō-MŌ-shē-ō KOR-dis; *commotio* = disturbance; *cordis* = heart) Damage to the heart, frequently fatal, as a result of a sharp, nonpenetrating blow to the chest while the ventricles are repolarizing.

Cor pulmonale (CP) (KOR pul-mōn-AL-ē; *cor* = heart; *pulmon-* = lung) A term referring to right ventricular hypertrophy from disorders that bring about hypertension (high blood pressure) in the pulmonary circulation.

Ejection fraction The fraction of the end-diastolic volume (EDV) that is ejected during an average heartbeat. Equal to stroke volume (SV) divided by EDV.

Electrophysiological testing (e-lek'-trō-fiz'-ē-ō-LOJ-i-kal) A procedure in which a catheter with an electrode is passed through blood vessels and introduced into the heart. It is used to detect the exact locations of abnormal electrical conduction pathways. Once an abnormal pathway is located, it can be destroyed by sending a current through the electrode, a procedure called *radiofrequency ablation*.

Palpitation (pal'-pi-TĀ-shun) A fluttering of the heart or an abnormal rate or rhythm of the heart about which an individual is aware.

Paroxysmal tachycardia (par'-ok-SIZ-mal tak'-i-KAR-dē-a; *tachy-* = quick) A period of rapid heartbeats that begins and ends suddenly.

Sick sinus syndrome An abnormally functioning SA node that initiates heartbeats too slowly or rapidly, pauses too long between heartbeats, or stops producing heartbeats. Symptoms include lightheadedness, shortness of breath, loss of consciousness, and palpitations. It is caused by degeneration of cells in the SA node and is common in elderly persons. It is also related to coronary artery disease. Treatment consists of drugs to speed up or slow down the heart or implantation of an artificial pacemaker.

Sudden cardiac death The unexpected cessation of circulation and breathing due to an underlying heart disease such as ischemia, myocardial infarction, or a disturbance in cardiac rhythm.

SELF-QUIZ QUESTIONS

Fill in the blanks in the following statements.

1. The chamber of the heart with the thickest myocardium is the _____.

2. The phase of heart contraction is called _____; the phase of relaxation is called _____.

Indicate whether the following statements are true or false.

3. In auscultation, the lubb represents closing of the semilunar valves and the dupp represents closing of the atrioventricular valves.

4. The Frank–Starling law of the heart equalizes the output of the right and left ventricles and keeps the same volume of blood flowing to both the systemic and pulmonary circulations.

Choose the one best answer to the following questions.

5. Which of the following is the correct route of blood through the heart from the systemic circulation to the pulmonary circulation and back to the systemic circulation?
   (a) right atrium, tricuspid valve, right ventricle, pulmonary semilunar valve, left atrium, mitral valve, left ventricle, aortic semilunar valve
   (b) left atrium, tricuspid valve, left ventricle, pulmonary semilunar valve, right atrium, mitral valve, right ventricle, aortic semilunar valve
   (c) left atrium, pulmonary semilunar valve, right atrium, tricuspid valve, left ventricle, aortic semilunar valve, right ventricle, mitral valve
   (d) left ventricle, mitral valve, left atrium, pulmonary semilunar valve, right ventricle, tricuspid valve, right atrium, aortic semilunar valve
   (e) right atrium, mitral valve, right ventricle, pulmonary semilunar valve, left atrium, tricuspid valve, left ventricle, aortic semilunar valve

6. Which of the following represents the correct pathway for conduction of an action potential through the heart?
   (a) AV node, AV bundle, SA node, Purkinje fibers, bundle branches
   (b) AV node, bundle branches, AV bundle, SA node, Purkinje fibers
   (c) SA node, AV node, AV bundle, bundle branches, Purkinje fibers
   (d) SA node, AV bundle, bundle branches, AV node, Purkinje fibers
   (e) SA node, AV node, Purkinje fibers, bundle branches, AV bundle

7. The external boundary between the atria and ventricles is the
   (a) anterior interventricular sulcus.
   (b) interventricular septum.
   (c) interatrial septum.
   (d) coronary sulcus.
   (e) posterior interventricular sulcus.

8. A softball player is found to have a resting cardiac output of 5.0 liters per minute and a heart rate of 50 beats per minute. What is her stroke volume?
   (a) 10 mL   (b) 100 mL
   (c) 1000 mL   (d) 250 mL
   (e) The information given is insufficient to calculate stroke volume.

9. Which of the following are true? (1) ANS regulation of heart rate originates in the cardiovascular center of the medulla oblongata. (2) Proprioceptor input is a major stimulus that accounts for the rapid rise in the heart rate at the onset of physical activity. (3) The vagus nerves release norepinephrine, causing the heart rate to increase. (4) Hormones from the adrenal medulla and the thyroid gland can increase the heart rate. (5) Hypothermia increases the heart rate.
   (a) 1, 2, 3, and 4   (b) 1, 2, and 4
   (c) 2, 3, 4, and 5   (d) 3, 5, and 6
   (e) 1, 2, 4, and 5

10. Which of the following are true concerning action potentials and contraction in the myocardium? (1) The refractory period in a cardiac muscle fiber is very brief. (2) The binding of Ca^{2+} to troponin allows the interaction of actin and myosin filaments, resulting in contraction. (3) Repolarization occurs when the voltage-gated K^+ channels open and calcium channels are closing. (4) Opening of voltage-gated fast Na^+ channels results in depolarization. (5) Opening of voltage-gated slow Ca^{2+} channels results in a period of maintained depolarization, known as the plateau.
    (a) 1, 3, and 5   (b) 2, 3, and 4
    (c) 2 and 5   (d) 3, 4, and 5
    (e) 2, 3, 4, and 5

11. Which of the following would not increase stroke volume?
    (a) increased Ca^{2+} in the interstitial fluid
    (b) epinephrine
    (c) increased K^+ in the interstitial fluid
    (d) increase in venous return
    (e) slow resting heart rate

12. Match the following:
    _____ (a) indicates ventricular repolarization
    _____ (b) represents the time from the beginning of ventricular depolarization to the end of ventricular repolarization
    _____ (c) represents atrial depolarization
    _____ (d) represents the time when the ventricular contractile fibers are fully depolarized; occurs during the plateau phase of the action potential
    _____ (e) represents the onset of ventricular depolarization
    _____ (f) represents the conduction time from the beginning of atrial excitation to the beginning of ventricular excitation

    (1) P wave
    (2) QRS complex
    (3) T wave
    (4) P–Q interval
    (5) S–T segment
    (6) Q–T interval

13. Match the following:

_____ (a) major branch from the ascending aorta; passes inferior to the left auricle

_____ (b) lies in the posterior interventricular sulcus; supplies the walls of the ventricles with oxygenated blood

_____ (c) located in the coronary sulcus on the posterior surface of the heart; receives most of the deoxygenated blood from the myocardium

_____ (d) lies in the coronary sulcus; supplies oxygenated blood to the walls of the right ventricle

_____ (e) lies in the coronary sulcus; drains the right atrium and right ventricle

_____ (f) major branch from the ascending aorta; lies inferior to the right auricle

_____ (g) lies in the posterior interventricular sulcus; drains the right and left ventricles

_____ (h) lies in the anterior interventricular sulcus; supplies oxygenated blood to the walls of both ventricles

_____ (i) lies in the anterior interventricular sulcus; drains the walls of both ventricles and the left atrium

_____ (j) lies in the coronary sulcus; supplies oxygenated blood to the walls of the left ventricle and left atrium

_____ (k) drain the right ventricle and open directly into the right atrium

(1) small cardiac vein
(2) anterior interventricular branch (left anterior descending artery)
(3) anterior cardiac veins
(4) posterior interventricular branch
(5) marginal branch
(6) circumflex branch
(7) middle cardiac vein
(8) left coronary artery
(9) right coronary artery
(10) great cardiac vein
(11) coronary sinus

14. Match the following:

_____ (a) collects oxygenated blood from the pulmonary circulation

_____ (b) pumps deoxygenated blood to the lungs for oxygenation

_____ (c) their contraction pulls on and tightens the chordae tendineae, preventing the valve cusps from everting

_____ (d) cardiac muscle tissue

_____ (e) increase blood-holding capacity of the atria

_____ (f) tendonlike cords connected to the atrioventricular valve cusps which, along with the papillary muscles, prevent valve eversion

_____ (g) the superficial dense irregular connective tissue covering of the heart

_____ (h) outer layer of the serous pericardium; is fused to the fibrous pericardium

_____ (i) endothelial cells lining the interior of the heart; are continuous with the endothelium of the blood vessels

_____ (j) pumps oxygenated blood to all body cells, except the air sacs of the lungs

_____ (k) prevents backflow of blood from the right ventricle into the right atrium

_____ (l) collects deoxygenated blood from the systemic circulation

_____ (m) left atrioventricular valve

_____ (n) the remnant of the foramen ovale, an opening in the interatrial septum of the fetal heart

_____ (o) blood vessels that pierce the heart muscle and supply blood to the cardiac muscle fibers

_____ (p) grooves on the surface of the heart which delineate the external boundaries between the chambers

_____ (q) prevent backflow of blood from the arteries into the ventricles

_____ (r) the gap junction and desmosome connections between individual cardiac muscle fibers

_____ (s) internal wall dividing the chambers of the heart

_____ (t) separate the upper and lower heart chambers, preventing backflow of blood from the ventricles back into the atria

_____ (u) inner visceral layer of the pericardium; adheres tightly to the surface of the heart

_____ (v) ridges formed by raised bundles of cardiac muscle fibers

(1) right atrium
(2) right ventricle
(3) left atrium
(4) left ventricle
(5) tricuspid valve
(6) bicuspid (mitral) valve
(7) chordae tendineae
(8) auricles
(9) papillary muscles
(10) trabeculae carneae
(11) fibrous pericardium
(12) parietal pericardium
(13) epicardium
(14) myocardium
(15) endocardium
(16) atrioventricular valves
(17) semilunar valves
(18) intercalated discs
(19) sulci
(20) septum
(21) fossa ovalis
(22) coronary circulation

15. Match the following:

_____ (a) amount of blood contained in the ventricles at the end of ventricular relaxation

_____ (b) period of time when cardiac muscle fibers are contracting and exerting force but not shortening

_____ (c) amount of blood ejected per beat by each ventricle

_____ (d) amount of blood remaining in the ventricles following ventricular contraction

_____ (e) difference between a person's maximum cardiac output and cardiac output at rest

_____ (f) period of time when semilunar valves are open and blood flows out of the ventricles

_____ (g) period when all four valves are closed and ventricular blood volume does not change

(1) cardiac reserve

(2) stroke volume

(3) end-diastolic volume (EDV)

(4) isovolumetric relaxation

(5) end-systolic volume (ESV)

(6) ventricular ejection

(7) isovolumetric contractiont

21 THE CARDIOVASCULAR SYSTEM: BLOOD VESSELS AND HEMODYNAMICS

21.1 Structure and Function of Blood Vessels

1. Arteries carry blood away from the heart. The wall of an artery consists of a tunica interna, a tunica media (which maintains elasticity and contractility), and a tunica externa. Large arteries are termed elastic (conducting) arteries, and medium-sized arteries are called muscular (distributing) arteries.

2. Many arteries anastomose: The distal ends of two or more vessels unite. An alternative blood route from an anastomosis is called collateral circulation. Arteries that do not anastomose are called end arteries.

3. Arterioles are small arteries that deliver blood to capillaries. Through constriction and dilation, arterioles assume a key role in regulating blood flow from arteries into capillaries and in altering arterial blood pressure.

4. Capillaries are microscopic blood vessels through which materials are exchanged between blood and tissue cells; some capillaries are continuous, and others are fenestrated. Capillaries branch to form an extensive network throughout a tissue. This network increases surface area, allowing a rapid exchange of large quantities of materials.

5. Precapillary sphincters regulate blood flow through capillaries.

6. Microscopic blood vessels in the liver are called sinusoids.

7. Venules are small vessels that continue from capillaries and merge to form veins.

8. Veins consist of the same three tunics as arteries but have a thinner tunica interna and a thinner tunica media. The lumen of a vein is also larger than that of a comparable artery. Veins contain valves to prevent backflow of blood. Weak valves can lead to varicose veins.

9. Vascular (venous) sinuses are veins with very thin walls.

10. Systemic veins are collectively called blood reservoirs because they hold a large volume of blood. If the need arises, this blood can be shifted into other blood vessels through vasoconstriction of veins. The principal blood reservoirs are the veins of the abdominal organs (liver and spleen) and skin.

21.2 Capillary Exchange

1. Substances enter and leave capillaries by diffusion, transcytosis, or bulk flow.

2. The movement of water and solutes (except proteins) through capillary walls depends on hydrostatic and osmotic pressures.

3. The near equilibrium between filtration and reabsorption in capillaries is called Starling's law of the capillaries.

4. Edema is an abnormal increase in interstitial fluid.

21.3 Hemodynamics: Factors Affecting Blood Flow

1. The velocity of blood flow is inversely related to the cross-sectional area of blood vessels; blood flows slowest where cross-sectional area is greatest. The velocity of blood flow decreases from the aorta to arteries to capillaries and increases in venules and veins.

2. Blood pressure and resistance determine blood flow.

3. Blood flows from regions of higher to lower pressure. The higher the resistance, however, the lower the blood flow.

4. Cardiac output equals the mean arterial pressure divided by total resistance ($CO = MAP \div R$).

5. Blood pressure is the pressure exerted on the walls of a blood vessel.

6. Factors that affect blood pressure are cardiac output, blood volume, viscosity, resistance, and the elasticity of arteries.

7. As blood leaves the aorta and flows through the systemic circulation, its pressure progressively falls to 0 mmHg by the time it reaches the right ventricle.

8. Resistance depends on blood vessel diameter, blood viscosity, and total blood vessel length.

9. Venous return depends on pressure differences between the venules and the right ventricle.

10. Blood return to the heart is maintained by several factors, including skeletal muscle contractions, valves in veins (especially in the limbs), and pressure changes associated with breathing.

21.4 Control of Blood Pressure and Blood Flow

1. The cardiovascular (CV) center is a group of neurons in the medulla oblongata that regulates heart rate, contractility, and blood vessel diameter.

2. The cardiovascular center receives input from higher brain regions and sensory receptors (baroreceptors and chemoreceptors).

3. Output from the cardiovascular center flows along sympathetic and parasympathetic axons. Sympathetic impulses propagated along cardioaccelerator nerves increase heart rate and contractility; parasympathetic impulses propagated along vagus nerves decrease heart rate.

4. Baroreceptors monitor blood pressure, and chemoreceptors monitor blood levels of O_2, CO_2, and hydrogen ions. The carotid sinus reflex helps regulate blood pressure in the brain. The aortic reflex regulates general systemic blood pressure.

5. Hormones that help regulate blood pressure are epinephrine, norepinephrine, ADH (antidiuretic hormone), angiotensin II, and ANP (atrial natriuretic peptide).
6. Autoregulation refers to local, automatic adjustments of blood flow in a given region to meet a particular tissue's need.
7. O_2 level is the principal stimulus for autoregulation.

21.5 Checking Circulation

1. Pulse is the alternate expansion and elastic recoil of an artery wall with each heartbeat. It may be felt in any artery that lies near the surface or over a hard tissue.
2. A normal resting pulse (heart) rate is 70–80 beats/min.
3. Blood pressure is the pressure exerted by blood on the wall of an artery when the left ventricle undergoes systole and then diastole. It is measured by the use of a sphygmomanometer.
4. Systolic blood pressure (SBP) is the arterial blood pressure during ventricular contraction. Diastolic blood pressure (DBP) is the arterial blood pressure during ventricular relaxation. Normal blood pressure is less than 120/80.
5. Pulse pressure is the difference between systolic and diastolic blood pressure. It normally is about 40 mmHg.

21.6 Shock and Homeostasis

1. Shock is a failure of the cardiovascular system to deliver enough O_2 and nutrients to meet the metabolic needs of cells.
2. Types of shock include hypovolemic, cardiogenic, vascular, and obstructive.
3. Signs and symptoms of shock include systolic blood pressure less than 90 mmHg; rapid resting heart rate; weak, rapid pulse; clammy, cool, pale skin; sweating; hypotension; altered mental state; decreased urinary output; thirst; and acidosis.

21.7 Circulatory Routes: Systemic Circulation

1. The systemic circulation carries oxygenated blood from the left ventricle through the aorta to all parts of the body, including some lung tissue, but *not* the air sacs (alveoli) of the lungs, and returns the deoxygenated blood to the right atrium.
2. Among the subdivisions of the systemic circulation are the coronary (cardiac) circulation and the hepatic portal circulation.

21.8 The Aorta and Its Branches

1. The aorta is divided into the ascending aorta, arch of the aorta, thoracic aorta, and abdominal aorta.
2. Each section gives off arteries that branch to supply the whole body.

21.9 Ascending Aorta

1. The ascending aorta is the part of the aorta that extends from the aorta valve of the heart.
2. The two branches of the ascending aorta are the right and left coronary arteries.

21.10 The Arch of the Aorta

1. The arch of the aorta is the continuation of the ascending aorta.
2. The three branches of the arch of the aorta are the brachiocephalic trunk, left common carotid artery, and left subclavian artery.

21.11 Thoracic Aorta

1. The thoracic aorta is the continuation of the arch of the aorta.
2. It sends off visceral branches and parietal branches.

21.12 Abdominal Aorta

1. The abdominal aorta is the continuation of the thoracic aorta.
2. It gives rise to unpaired visceral; branches and paired visceral branches.

21.13 Arteries of the Pelvis and Lower Limbs

1. The abdominal aorta ends by dividing into the right and left common iliac arteries.
2. These arteries in turn branch into smaller arteries.

21.14 Veins of the Systemic Circulation

1. Blood returns to the heart through the systemic veins.
2. All veins of the systemic circulation drain into the superior or inferior venae cavae or the coronary sinus, which in turn empty into the right atrium.

21.15 Veins of the Head and Neck

1. The three major veins that drain blood from the head are the internal jugular, external jugular, and vertebral veins.
2. Within the cranial cavity, all veins drain into dural venous sinuses and then into the internal jugular vein.

21.16 Veins of the Upper Limbs

1. Both superficial and deep veins return blood from the upper limbs to the heart.
2. Superficial veins are larger than deep veins and return most of the blood from the upper limbs.

21.17 Veins of the Thorax

1. Most thoracic structures are drained by a network of veins called the azygos system.
2. The azygous system consists of the azygos, hemiazygos, and accessory hemiazygos veins.

21.18 Veins of the Abdomen and Pelvis

1. Many small veins drain blood from the abdomen and pelvis.
2. These veins in turn convey blood into the inferior vena cava.

21.19 Veins of the Lower Limbs

1. Blood from the lower limbs is drained by both superficial and deep veins.
2. The superficial veins often anastomose with one another and with deep veins along their length.

21.20 Circulatory Routes: The Hepatic Portal Circulation

1. The hepatic portal circulation directs venous blood from the gastrointestinal organs and spleen into the hepatic portal veins of the liver before it returns to the heart.
2. It enables the liver to utilize nutrients and detoxify harmful substances in the blood.

21.21 Circulatory Routes: The Pulmonary Circulation

1. The pulmonary circulations takes deoxygenated blood from the right ventricle to the alveoli within the lungs and returns oxygenated blood from the alveoli to the left atrium.
2. The pulmonary circulation includes the pulmonary trunk, pulmonary arteries, and pulmonary veins.

21.22 Circulatory Routes: The Fetal Circulation

1. Fetal circulation exists only in the fetus. It involves the exchange of materials between fetus and mother via the placenta.
2. The fetus derives O_2 and nutrients from and eliminates CO_2 and wastes into maternal blood. At birth, when pulmonary (lung), digestive, and liver functions begin, the special structures of fetal circulation are no longer needed.

21.23 Development of Blood Vessels and Blood

1. Blood vessels develop from mesenchyme (hemangioblasts → angioblasts → blood islands) in mesoderm called blood islands.

2. Blood cells also develop from mesenchyme (hemangioblasts → pluripotent stem cells).
3. The development of blood cells from pluripotent stem cells derived from angioblasts occurs in the walls of blood vessels in the yolk sac, chorion, and allantois at about 3 weeks after fertilization. Within the embryo, blood is produced by the liver at about the fifth week and in the spleen, red bone marrow, and thymus at about the twelfth week.

21.24 Aging and the Cardiovascular System

1. General changes associated with aging include reduced compliance (distensibility) of blood vessels, reduction in cardiac muscle size, reduced cardiac output, and increased systolic blood pressure.
2. The incidence of coronary artery disease (CAD), congestive heart failure (CHF), and atherosclerosis increases with age.

CLINICAL CONNECTIONS

Angiogenesis and Disease (Refer page 637 of Textbook)

Angiogenesis (an'-jē-ō-JEN-e-sis; *angio-* = blood vessel; *-genesis* = production) refers to the growth of new blood vessels. It is an important process in embryonic and fetal development, and in postnatal life serves important functions such as wound healing, formation of a new uterine lining after menstruation, formation of the corpus luteum after ovulation, and development of blood vessels around obstructed arteries in the coronary circulation. Several proteins (peptides) are known to promote and inhibit angiogenesis.

Clinically angiogenesis is important because cells of a malignant tumor secrete proteins called *tumor angiogenesis factors (TAFs)* that stimulate blood vessel growth to provide nourishment for the tumor cells. Scientists are seeking chemicals that would inhibit angiogenesis and thus stop the growth of tumors. In *diabetic retinopathy* (ret-i-NOP-a-thē), angiogenesis may be important in the development of blood vessels that actually cause blindness, so finding inhibitors of angiogenesis may also prevent the blindness associated with diabetes.

Varicose Veins (Refer page 645 of Textbook)

Leaky venous valves can cause veins to become dilated and twisted in appearance, a condition called **varicose veins** (VAR-i-kōs) or *varices* (VAR-i-sēz; *varic-* = a swollen vein). The singular is *varix* (VAR-iks). The condition may occur in the veins of almost any body part, but it is most common in the esophagus, anal canal, and superficial veins of the lower limbs. Those in the lower limbs can range from cosmetic problems to serious medical conditions. The valvular defect may be congenital or may result from mechanical stress (prolonged standing or pregnancy) or aging. The leaking venous valves allow the backflow of blood from the deep veins to the less efficient superficial veins, where the blood pools. This creates pressure that distends the vein and allows fluid to leak into surrounding tissue. As a result, the affected vein and the tissue around it may become inflamed and painfully tender. Veins close to the surface of the legs, especially the saphenous vein, are highly susceptible to varicosities; deeper veins are not as vulnerable because surrounding skeletal muscles prevent their walls from stretching excessively. Varicose veins in the anal canal are referred to as *hemorrhoids* (HEM-o-royds). Esophageal varices result from dilated veins

in the walls of the lower part of the esophagus and sometimes the upper part of the stomach. Bleeding esophageal varices are life-threatening and are usually a result of chronic liver disease.

Several treatment options are available for varicose veins in the lower limbs. *Elastic stockings* (support hose) may be used for individuals with mild symptoms or for whom other options are not recommended. *Sclerotherapy* (skle-rō-THER-a-pē) involves injection of a solution into varicose veins that damages the tunica interna by producing a harmless superficial thrombophlebitis (inflammation involving a blood clot). Healing of the damaged part leads to scar formation that occludes the vein. *Radiofrequency endovenous occlusion* (ō-KLOO-zhun) involves the application of radiofrequency energy to heat up and close off varicose veins. *Laser occlusion* uses laser therapy to shut down veins. In a surgical procedure called *stripping,* veins may be removed. In this procedure, a flexible wire is threaded through the vein and then pulled out to strip (remove) it from the body.

Edema (Refer page 646 of Textbook)

If filtration greatly exceeds reabsorption, the result is **edema** (e-DĒ-ma = swelling), an abnormal increase in interstitial fluid volume. Edema is not usually detectable in tissues until interstitial fluid volume has risen to 30% above normal. Edema can result from either excess filtration or inadequate reabsorption.

Two situations may cause excess filtration:

- *Increased capillary blood pressure* causes more fluid to be filtered from capillaries.
- *Increased permeability of capillaries* raises interstitial fluid osmotic pressure by allowing some plasma proteins to escape. Such leakiness may be caused by the destructive effects of chemical, bacterial, thermal, or mechanical agents on capillary walls.

One situation commonly causes inadequate reabsorption:

- *Decreased concentration of plasma proteins* lowers the blood colloid osmotic pressure. Inadequate synthesis or dietary intake or loss of plasma proteins is associated with liver disease, burns, malnutrition (for example, kwashiorkor; see Disorders: Homeostatic Imbalances in Chapter 25), and kidney disease.

Syncope (Refer page 649 of Textbook)

Syncope (SIN-kō-pē), or fainting, is a sudden, temporary loss of consciousness that is not due to head trauma, followed by spontaneous recovery. It is most commonly due to cerebral ischemia, lack of sufficient blood flow to the brain. Syncope may occur for several reasons:

- *Vasodepressor syncope* is due to sudden emotional stress or real, threatened, or fantasized injury.
- *Situational syncope* is caused by pressure stress associated with urination, defecation, or severe coughing.
- *Drug-induced syncope* may be caused by drugs such as antihypertensives, diuretics, vasodilators, and tranquilizers.
- *Orthostatic hypotension*, an excessive decrease in blood pressure that occurs on standing up, may cause fainting.

Carotid Sinus Massage and Carotid Sinus Syncope (Refer page 653 of Textbook)

Because the carotid sinus is close to the anterior surface of the neck, it is possible to stimulate the baroreceptors there by putting pressure on the neck. Physicians sometimes use **carotid sinus massage**, which involves carefully massaging the neck over the carotid sinus, to slow heart rate in a person who has paroxysmal superventricular tachycardia, a type of tachycardia that originates in the atria. Anything that stretches or puts pressure on the carotid sinus, such as hyperextension of the head, tight collars, or carrying heavy shoulder loads, may also slow heart rate and can cause **carotid sinus syncope**, fainting due to inappropriate stimulation of the carotid sinus baroreceptors.

DISORDERS: HOMEOSTATIC IMBALANCES

Hypertension

About 50 million Americans have **hypertension** (hī′-per-TEN-shun), or persistently high blood pressure. It is the most common disorder affecting the heart and blood vessels and is the major cause of heart failure, kidney disease, and stroke. In May 2003, the Joint National Committee on Prevention, Detection, Evaluation, and Treatment of High Blood Pressure published new guidelines for hypertension because clinical studies have linked what were once considered fairly low blood pressure readings to an increased risk of cardiovascular disease. The new guidelines are as follows:

CATEGORY	SYSTOLIC (mmHg)	DIASTOLIC (mmHg)
Normal	Less than 120 and	Less than 80
Prehypertension	120–139 *or*	80–89
Stage 1 hypertension	140–159 *or*	90–99
Stage 2 hypertension	Greater than 160 or	Greater than 100

Using the new guidelines, the normal classification was previously considered optimal; prehypertension now includes many more individuals previously classified as normal or high-normal; stage 1 hypertension is the same as in previous guidelines; and stage 2 hypertension now combines the previous stage 2 and stage 3 categories since treatment options are the same for the former stages 2 and 3.

TYPES AND CAUSES OF HYPERTENSION Between 90 and 95% of all cases of hypertension are **primary hypertension**, a persistently elevated bloosd pressure that cannot be attributed to any identifiable cause. The remaining 5–10% of cases are **secondary hypertension**, which has an identifiable underlying cause. Several disorders cause secondary hypertension:

- *Obstruction of renal blood flow* or disorders that damage renal tissue may cause the kidneys to release excessive amounts of renin into the blood. The resulting high level of angiotensin II causes vasoconstriction, thus increasing systemic vascular resistance.

- *Hypersecretion of aldosterone*—resulting, for instance, from a tumor of the adrenal cortex—stimulates excess reabsorption of salt and water by the kidneys, which increases the volume of body fluids.

- *Hypersecretion of epinephrine and norepinephrine* may occur by a **pheochromocytoma** (fē-ō-krō′-mō-sī-TŌ-ma), a tumor of the adrenal medulla. Epinephrine and norepinephrine increase heart rate and contractility and increase systemic vascular resistance.

DAMAGING EFFECTS OF UNTREATED HYPERTENSION High blood pressure is known as the "silent killer" because it can cause considerable damage to the blood vessels, heart, brain, and kidneys before it causes pain or other noticeable symptoms. It is a major risk factor for the number-one (heart disease) and number-three (stroke) causes of death in the United States. In blood vessels, hypertension causes thickening of the tunica media, accelerates development of atherosclerosis and coronary artery disease, and increases systemic vascular resistance. In the heart, hypertension increases the afterload, which forces the ventricles to work harder to eject blood.

The normal response to an increased workload due to vigorous and regular exercise is hypertrophy of the myocardium, especially in the wall of the left ventricle. This is a positive effect that makes the heart a more efficient pump. An increased afterload, however, leads to myocardial hypertrophy that is accompanied by muscle damage and fibrosis (a buildup of collagen fibers between the muscle fibers). As a result, the left ventricle enlarges, weakens, and dilates. Because arteries in the brain are usually less protected by surrounding tissues than are the major arteries in other parts of the body, prolonged hypertension can eventually cause them to rupture, resulting in a stroke. Hypertension also damages kidney arterioles, causing them to thicken, which narrows the lumen; because the blood supply to the kidneys is thereby reduced, the kidneys secrete more renin, which elevates the blood pressure even more.

LIFESTYLE CHANGES TO REDUCE HYPERTENSION Although several categories of drugs (described next) can reduce elevated blood pressure, the following lifestyle changes are also effective in managing hypertension:

- *Lose weight.* This is the best treatment for high blood pressure short of using drugs. Loss of even a few pounds helps reduce blood pressure in overweight hypertensive individuals.

- *Limit alcohol intake.* Drinking in moderation may lower the risk of coronary heart disease, mainly among males over 45 and females over 55. Moderation is defined as no more than one 12-oz beer per day for females and no more than two 12-oz beers per day for males.

- *Exercise.* Becoming more physically fit by engaging in moderate activity (such as brisk walking) several times a week for 30 to 45 minutes can lower systolic blood pressure by about 10 mmHg.

- *Reduce intake of sodium (salt).* Roughly half the people with hypertension are "salt sensitive." For them, a high-salt diet appears to promote hypertension, and a low-salt diet can lower their blood pressure.

- *Maintain recommended dietary intake of potassium, calcium, and magnesium.* Higher levels of potassium, calcium, and magnesium in the diet are associated with a lower risk of hypertension.

- *Don't smoke or quit smoking.* Smoking has devastating effects on the heart and can augment the damaging effects of high blood pressure by promoting vasoconstriction.
- *Manage stress.* Various meditation and biofeedback techniques help some people reduce high blood pressure. These methods may work by decreasing the daily release of epinephrine and norepinephrine by the adrenal medulla.

DRUG TREATMENT OF HYPERTENSION Drugs having several different mechanisms of action are effective in lowering blood pressure. Many people are successfully treated with *diuretics* (dī-ū-RET-iks), agents that decrease blood pressure by decreasing blood volume, because they increase elimination of water and salt in the urine. *ACE (angiotensin-converting enzyme) inhibitors* block formation of angiotensin II and thereby promote vasodilation and decrease the secretion of aldosterone. *Beta blockers* (BĀ-ta) reduce blood pressure by inhibiting the secretion of renin and by decreasing heart rate and contractility. *Vasodilators* relax the smooth muscle in arterial walls, causing vasodilation and lowering blood pressure by lowering systemic vascular resistance. An important category of vasodilators are the *calcium channel blockers*, which slow the inflow of Ca^{2+} into vascular smooth muscle cells. They reduce the heart's workload by slowing Ca^{2+} entry into pacemaker cells and regular myocardial fibers, thereby decreasing heart rate and the force of myocardial contraction.

MEDICAL TERMINOLOGY

Aneurysm (AN-ū-rizm) A thin, weakened section of the wall of an artery or a vein that bulges outward, forming a balloonlike sac. Common causes are atherosclerosis, syphilis, congenital blood vessel defects, and trauma. If untreated, the aneurysm enlarges and the blood vessel wall becomes so thin that it bursts. The result is massive hemorrhage with shock, severe pain, stroke, or death. Treatment may involve surgery in which the weakened area of the blood vessel is removed and replaced with a graft of synthetic material.

Aortography (ā′-or-TOG-ra-fē) X-ray examination of the aorta and its main branches after injection of a radiopaque dye.

Carotid endarterectomy (ka-ROT-id end′-ar-ter-EK-tō-mē) The removal of atherosclerotic plaque from the carotid artery to restore greater blood flow to the brain.

Claudication (klaw′-di-KĀ-shun) Pain and lameness or limping caused by defective circulation of the blood in the vessels of the limbs.

Deep vein thrombosis (DVT) The presence of a thrombus (blood clot) in a deep vein of the lower limbs. It may lead to (1) pulmonary embolism, if the thrombus dislodges and then lodges within the pulmonary arterial blood flow, and (2) postphlebitic syndrome, which consists of edema, pain, and skin changes due to destruction of venous valves.

Doppler ultrasound scanning Imaging technique commonly used to measure blood flow. A transducer is placed on the skin and an image is displayed on a monitor that provides the exact position and severity of a blockage.

Femoral angiography An imaging technique in which a contrast medium is injected into the femoral artery and spreads to other arteries in the lower limb, and then a series of radiographs are taken of one or more sites. It is used to diagnose narrowing or blockage of arteries in the lower limbs.

Hypotension (hī-pō-TEN-shun) Low blood pressure; most commonly used to describe an acute drop in blood pressure, as occurs during excessive blood loss.

Normotensive (nor′-mō-TEN-siv) Characterized by normal blood pressure.

Occlusion (ō-KLOO-zhun) The closure or obstruction of the lumen of a structure such as a blood vessel. An example is an atherosclerotic plaque in an artery.

Orthostatic hypotension (or′-thō-STAT-ik; *ortho-* = straight; *-static* = causing to stand) An excessive lowering of systemic blood pressure when a person assumes an erect or semierect posture; it is usually a sign of a disease. May be caused by excessive fluid loss, certain drugs, and cardiovascular or neurogenic factors. Also called **postural hypotension**.

Phlebitis (fle-BĪ-tis; *phleb-* = vein) Inflammation of a vein, often in a leg.

Thrombectomy (throm-BEK-tō-mē; *thrombo-* = clot) An operation to remove a blood clot from a blood vessel.

Thrombophlebitis (throm′-bō-fle-BĪ-tis) Inflammation of a vein involving clot formation. Superficial thrombophlebitis occurs in veins under the skin, especially in the calf.

Venipuncture (VEN-i-punk-chur; *vena-* = vein) The puncture of a vein, usually to withdraw blood for analysis or to introduce a solution, for example, an antibiotic. The median cubital vein is frequently used.

White coat (*office*) **hypertension** A stress-induced syndrome found in patients who have elevated blood pressure when being examined by health-care personnel, but otherwise have normal blood pressure.

SELF-QUIZ QUESTIONS

Fill in the blanks in the following statements.

1. The _____ reflex helps maintain normal blood pressure in the brain; the _____ reflex governs general systemic blood pressure.

2. In addition to the pressure created by contraction of the left ventricle, venous return is aided by the _____ and the _____, both of which depend on the presence of valves in the veins.

Indicate whether the following statements are true or false.

3. Baroreceptors and chemoreceptors are located in the aorta and carotid arteries.

4. The most important method of capillary exchange is simple diffusion.

Choose the one best answer to the following questions.

5. Which of the following are *not* true? (1) Muscular arteries are also known as conducting arteries. (2) Capillaries play a key role in regulating resistance. (3) The flow of blood through true capillaries is controlled by precapillary sphincters. (4) The lumen of an artery is larger than in a comparable vein. (5) Elastic arteries help propel blood. (6) The tunica media of arteries is thicker than the tunica media of veins.

    (a) 2, 3, and 6   (b) 1, 2, and 4   (c) 1, 2, 4, and 6

    (d) 3, 4, and 5   (e) 1, 2, 3, and 4

6. Which of the following are *true* concerning capillary exchange? (1) Large, lipid-insoluble molecules cross capillary walls by transcytosis. (2) The blood hydrostatic pressure promotes reabsorption of fluid into the capillaries. (3) If the pressures that promote filtration are greater than the pressures that promote reabsorption, fluid will move out of a capillary and into interstitial spaces. (4) A negative net filtration pressure results in reabsorption of fluid from interstitial spaces into a capillary. (5) The difference in osmotic pressure across a capillary wall is due primarily to red blood cells.

    (a) 1, 3, and 4   (b) 1, 2, 3, 4, and 5   (c) 1, 2, 3, and 4

    (d) 3 and 4   (e) 2, 4, and 5

7. Which of the following would *not* increase vascular resistance? (1) vasodilation, (2) polycythemia, (3) obesity, (4) dehydration, (5) anemia.
   (a) 1 and 2
   (b) 1, 3, and 4
   (c) 1 and 5
   (d) 1, 4, and 5
   (e) 1 only

8. Capillary exchange is enhanced by (1) the slow rate of flow through the capillaries, (2) a small cross-sectional area, (3) the thinness of capillary walls, (4) the respiratory pump, (5) extensive branching, which increases the surface area.
   (a) 1, 2, 3, 4, and 5
   (b) 1, 2, 3, and 5
   (c) 1 and 3
   (d) 3 and 5
   (e) 1, 3, and 5

9. Systemic vascular resistance depends on which of the following factors? (1) blood viscosity, (2) total blood vessel length, (3) size of the lumen, (4) type of blood vessel, (5) oxygen concentration of the blood.
   (a) 1, 2, and 3
   (b) 2, 3, and 4
   (c) 3, 4, and 5
   (d) 1, 3, and 5
   (e) 2, 4, and 5

10. Which of the following help regulate blood pressure and help control regional blood flow? (1) baroreceptor and chemoreceptor reflexes, (2) hormones, (3) autoregulation, (4) H^+ concentration of blood, (5) oxygen concentration of the blood.
    (a) 1, 2, and 4
    (b) 2, 4, and 5
    (c) 1, 4, and 5
    (d) 1, 2, 3, 4, and 5
    (e) 3, 4, and 5

11. For each of the following, indicate if it causes vasoconstriction or vasodilation. Use D for vasodilation and C for vasoconstriction.
    (a) atrial natriuretic peptide
    (b) ADH
    (c) decrease in body temperature
    (d) lactic acid
    (e) histamine
    (f) hypoxia
    (g) hypercapnia
    (h) angiotensin II
    (i) nitric oxide
    (j) decreased sympathetic impulses
    (k) acidosis

12. Match the following:
    _____ (a) pressure generated by the pumping of the heart; pushes fluids out of capillaries
    _____ (b) pressure created by proteins present in the interstitial fluid; pulls fluid out of capillaries
    _____ (c) balance of pressure; determines whether blood volume and interstitial fluid remain steady or change
    _____ (d) force due to presence of plasma proteins; pulls fluid into capillaries from interstitial spaces
    _____ (e) pressure due to fluid in interstitial spaces; pushes fluid back into capillaries

    (1) net filtration pressure
    (2) blood hydrostatic pressure
    (3) interstitial fluid hydrostatic pressure
    (4) blood colloid osmotic pressure
    (5) interstitial fluid osmotic pressure

13. Match the following:
    _____ (a) supplies blood to the kidney
    _____ (b) drains blood from the small intestine, portions of the large intestine, stomach, and pancreas
    _____ (c) main blood supply to arm; commonly used to measure blood pressure
    _____ (d) supply blood to the free lower limbs
    _____ (e) drain oxygenated blood from the lungs and send it to the left atrium
    _____ (f) supplies blood to the stomach, liver, and pancreas
    _____ (g) supply blood to the brain
    _____ (h) supplies blood to the large intestine
    _____ (i) drain blood from the head
    _____ (j) detours venous blood from the gastrointestinal organs and spleen through the liver before it returns to the heart
    _____ (k) drains most of the thorax and abdominal wall; can serve as a bypass for the inferior vena cava
    _____ (l) a part of the venous circulation of the leg; a vessel used in heart bypass surgery
    _____ (m) carry deoxygenated blood from the right ventricle to the lungs

    (1) superior mesenteric vein
    (2) inferior mesenteric artery
    (3) pulmonary veins
    (4) brachial artery
    (5) hepatic portal circulation
    (6) carotid arteries
    (7) jugular veins
    (8) celiac trunk
    (9) common iliac arteries
    (10) azygos veins
    (11) renal artery
    (12) great saphenous vein
    (13) pulmonary arteries

14. Match the following:
    _____ (a) a traveling pressure wave created by the alternate expansion and recoil of elastic arteries after each systole of the left ventricle
    _____ (b) the lowest blood pressure in arteries during ventricular relaxation
    _____ (c) a slow resting heart rate or pulse rate
    _____ (d) an inadequate cardiac output that results in a failure of the cardiovascular system to deliver enough oxygen and nutrients to meet the metabolic needs of body cells
    _____ (e) a rapid resting heart rate or pulse rate
    _____ (f) the highest force with which blood pushes against arterial walls as a result of ventricular contraction

    (1) shock
    (2) pulse
    (3) tachycardia
    (4) bradycardia
    (5) systolic blood pressure
    (6) diastolic blood pressure

15. Match the following (some answers will be used more than once):

_____ (a) returns oxygenated blood from the placenta to the fetal liver

_____ (b) an opening in the septum between the right and left atria

_____ (c) becomes the ligamentum venosum after birth

_____ (d) pass blood from the fetus to the placenta

_____ (e) bypasses the nonfunctioning lungs; becomes the ligamentum arteriosum at birth

_____ (f) become the medial umbilical ligaments at birth

_____ (g) transports oxygenated blood into the inferior vena cava

_____ (h) becomes the ligamentum teres at birth

_____ (i) becomes the fossa ovalis after birth

(1) ductus venosus

(2) ductus arteriosus

(3) foramen ovale

(4) umbilical arteries

(5) umbilical vein

22 THE LYMPHATIC SYSTEM AND IMMUNITY

22.1 The Concept of Immunity

1. The ability to ward off disease is called immunity (resistance). Lack of resistance is called susceptibility.
2. The two general types of immunity are (a) innate and (b) adaptive.
3. Innate immunity refers to a wide variety of body responses to a wide range of pathogens.
4. Adaptive immunity involves activation of specific lymphocytes to combat a particular foreign substance.

22.2 Overview of the Lymphatic System

1. The lymphatic system carries out immune responses and consists of lymph, lymphatic vessels, and structures and organs that contain lymphatic tissue (specialized reticular tissue containing many lymphocytes).
2. The lymphatic system drains interstitial fluid, transports dietary lipids, and protects against invasion through immune responses.

22.3 Lymphatic Vessels and Lymph Circulation

1. Lymphatic vessels begin as closed-ended lymphatic capillaries in tissue spaces between cells. Interstitial fluid drains into lymphatic capillaries, thus forming lymph. Lymphatic capillaries merge to form larger vessels, called lymphatic vessels, which convey lymph into and out of lymph nodes.
2. The route of lymph flow is from lymphatic capillaries to lymphatic vessels to lymph trunks to the thoracic duct (left lymphatic duct) and right lymphatic duct to the subclavian veins.
3. Lymph flows because of skeletal muscle contractions and respiratory movements. Valves in lymphatic vessels also aid flow of lymph.

22.4 Lymphatic Organs and Tissues

1. The primary lymphatic organs are red bone marrow and the thymus. Secondary lymphatic organs are lymph nodes, the spleen, and lymphatic nodules.
2. The thymus lies between the sternum and the large blood vessels above the heart. It is the site of T cell maturation.
3. Lymph nodes are encapsulated, egg-shaped structures located along lymphatic vessels. Lymph enters lymph nodes through afferent lymphatic vessels, is filtered, and exits through efferent lymphatic vessels. Lymph nodes are the site of proliferation of B cells and T cells.
4. The spleen is the largest single mass of lymphatic tissue in the body. Within the spleen, B cells and T cells carry out immune functions and macrophages destroy blood-borne pathogens and worn-out red blood cells by phagocytosis.
5. Lymphatic nodules are scattered throughout the mucosa of the gastrointestinal, respiratory, urinary, and reproductive tracts. This lymphatic tissue is termed mucosa-associated lymphatic tissue (MALT).

22.5 Development of Lymphatic Tissues

1. Lymphatic vessels develop from lymph sacs, which arise from developing veins. Thus, they are derived from mesoderm.
2. Lymph nodes develop from lymph sacs that become invaded by mesenchymal cells.

22.6 Innate Immunity

1. Innate immunity includes physical factors, chemical factors, antimicrobial proteins, natural killer cells, phagocytes, inflammation, and fever.
2. The skin and mucous membranes are the first line of defense against entry of pathogens.
3. Antimicrobial substances include interferons, the complement system, iron-binding proteins, and antimicrobial proteins.
4. Natural killer cells and phagocytes attack and kill pathogens and defective cells in the body.
5. Inflammation aids disposal of microbes, toxins, or foreign material at the site of an injury, and prepares the site for tissue repair.
6. Fever intensifies the antiviral effects of interferons, inhibits growth of some microbes, and speeds up body reactions that aid repair.
7. **Table 22.1** summarizes the innate defenses.

22.7 Adaptive Immunity

1. Adaptive immunity involves lymphocytes called B cells and T cells. B cells and T cells arise from stem cells in red bone marrow. B cells mature in red bone marrow; T cells mature in the thymus gland.
2. Before B cells leave the red bone marrow or T cells leave the thymus, they develop immunocompetence, the ability to carry out adaptive immune responses. This process involves the insertion of antigen receptors into their plasma membranes. Antigen receptors are molecules that are capable of recognizing specific antigens.
3. Two major types of mature T cells exit the thymus: helper T cells (also known as CD4 T cells) and cytotoxic T cells (also referred to as CD8 T cells).
4. There are two types of adaptive immunity: cell-mediated immunity and antibody-mediated immunity. In cell-mediated

immune responses, cytotoxic T cells directly attack invading antigens; in antibody-mediated immune responses, B cells transform into plasma cells that secrete antibodies.

5. Clonal selection is the process by which a lymphocyte proliferates and differentiates in response to a specific antigen. The result of clonal selection is the formation of a clone of cells that can recognize the same specific antigen as the original lymphocyte.

6. A lymphocyte that undergoes clonal selection gives rise to two major types of cells in the clone: effector cells and memory cells. The effector cells of a lymphocyte clone carry out immune responses that ultimately result in the destruction or inactivation of the antigen. Effector cells include active helper T cells, which are part of a helper T cell clone; active cytotoxic T cells, which are part of a cytotoxic T cell clone; and plasma cells, which are part of a B cell clone. The memory cells of a lymphocyte clone do not actively participate in the initial immune response. However, if the antigen reappears in the body in the future, the memory cells can quickly respond to the antigen by proliferating and differentiating into more effector cells and more memory cells. Memory cells include memory helper T cells, which are part of a helper T cell clone; memory cytotoxic T cells, which are part of a cytotoxic T cell clone; and memory B cells, which are part of a B cell clone.

7. Antigens (Ags) are chemical substances that are recognized as foreign by the immune system. Antigen receptors exhibit great diversity due to genetic recombination.

8. "Self-antigens" called major histocompatibility complex (MHC) antigens are unique to each person's body cells. All cells except red blood cells display MHC-I molecules. Antigen-presenting cells (APCs) display MHC-II molecules. APCs include macrophages, B cells, and dendritic cells.

9. Exogenous antigens (formed outside body cells) are presented with MHC-II molecules; endogenous antigens (formed inside body cells) are presented with MHC-I molecules.

10. Cytokines are small protein hormones that may stimulate or inhibit many normal cell functions such as growth and differentiation. Other cytokines regulate immune responses (see **Table 22.2**).

22.8 Cell-Mediated Immunity

1. A cell-mediated immune response begins with activation of a small number of T cells by a specific antigen.

2. During the activation process, T-cell receptors (TCRs) recognize antigen fragments associated with MHC molecules on the surface of a body cell.

3. Activation of T cells also requires costimulation, either by cytokines such as interleukin-2 or by pairs of plasma membrane molecules.

4. Once a T cell has been activated, it undergoes clonal selection. The result of clonal selection is the formation of a clone of effector cells and memory cells. The effector cells of a T cell clone carry out immune responses that ultimately result in elimination of the antigen.

5. Helper T cells display CD4 protein, recognize antigen fragments associated with MHC-II molecules, and secrete several cytokines, most importantly interleukin-2, which acts as a costimulator for other helper T cells, cytotoxic T cells, and B cells.

6. Cytotoxic T cells display CD8 protein and recognize antigen fragments associated with MHC-I molecules.

7. Active cytotoxic T cells eliminate invaders by (1) releasing granzymes that cause target cell apoptosis (phagocytes then kill the microbes) and (2) releasing perforin, which causes cytolysis, and granulysin that destroys the microbes.

8. Cytotoxic T cells, macrophages, and natural killer cells carry out immunological surveillance, recognizing and destroying cancerous cells that display tumor antigens.

22.9 Antibody-Mediated Immunity

1. An antibody-mediated immune response begins with activation of a B cell by a specific antigen.

2. B cells can respond to unprocessed antigens, but their response is more intense when they process the antigen. Interleukin-2 and other cytokines secreted by helper T cells provide costimulation for activation of B cells.

3. Once activated, a B cell undergoes clonal selection, forming a clone of plasma cells and memory cells. Plasma cells are the effector cells of a B cell clone; they secrete antibodies.

4. An antibody (Ab) is a protein that combines specifically with the antigen that triggered its production.

5. Antibodies consist of heavy and light chains and variable and constant regions.

6. Based on chemistry and structure, antibodies are grouped into five principal classes (IgG, IgA, IgM, IgD, and IgE), each with specific biological roles.

7. Actions of antibodies include neutralization of antigen, immobilization of bacteria, agglutination and precipitation of antigen, activation of complement, and enhancement of phagocytosis.

8. Complement is a group of proteins that complement immune responses and help clear antigens from the body.

9. Immunization against certain microbes is possible because memory B cells and memory T cells remain after a primary response to an antigen. The secondary response provides protection should the same microbe enter the body again.

22.10 Self-Recognition and Self-Tolerance

1. T cells undergo positive selection to ensure that they can recognize self-MHC proteins (self-recognition), and negative selection to ensure that they do not react to other self-proteins (self-tolerance). Negative selection involves both deletion and anergy.

2. B cells develop tolerance through deletion and anergy.

22.11 Stress and Immunity

1. Psychoneuroimmunology (PNI) deals with communication pathways that link the nervous, endocrine, and immune systems. Thoughts, feelings, moods, and beliefs influence health and the course of disease.

2. Under stress, people are less likely to eat well or exercise regularly, two habits that enhance immunity.

22.12 Aging and the Immune System

1. With advancing age, individuals become more susceptible to infections and malignancies, respond less well to vaccines, and produce more autoantibodies.

2. Immune responses also diminish with age.

CLINICAL CONNECTIONS

Metastasis through Lymphatic Vessels (Refer page 711 of Textbook)

Metastasis (me-TAS-ta-sis; *meta-* = beyond; *-stasis* = to stand), the spread of a disease from one part of the body to another, can occur via lymphatic vessels. All malignant tumors eventually metastasize. Cancer cells may travel in the blood or lymph and establish new tumors where they lodge. When metastasis occurs via lymphatic vessels, secondary tumor sites can be predicted according to the direction of lymph flow from the primary tumor site. Cancerous lymph nodes feel enlarged, firm, nontender, and fixed to underlying structures. By contrast, most lymph nodes that are enlarged due to an infection are softer, tender, and movable.

Ruptured Spleen (Refer page 714 of Textbook)

The spleen is the organ most often damaged in cases of abdominal trauma. Severe blows over the inferior left chest or superior abdomen can fracture the protecting ribs. Such crushing injury may result in a **ruptured spleen**, which causes significant hemorrhage and shock. Prompt removal of the spleen, called a **splenectomy** (splē-NEK-tō-mē), is needed to prevent death due to bleeding. Other structures, particularly red bone marrow and the liver, can take over some functions normally carried out by the spleen. Immune functions, however, decrease in the absence of a spleen. The spleen's absence also places the patient at higher risk for sepsis (a blood infection) due to loss of the filtering and phagocytic functions of the spleen. To reduce the risk of sepsis, patients who have undergone a splenectomy take prophylactic (preventive) antibiotics before any invasive procedures.

Tonsillitis (Refer page 715 of Textbook)

Tonsillitis is an infection or inflammation of the tonsils. Most often, it is caused by a virus, but it may also be caused by the same bacteria that cause strep throat. The principal symptom of tonsillitis is a sore throat. Additionally, fever, swollen lymph nodes, nasal congestion, difficulty in swallowing, and headache may also occur. Tonsillitis of viral origin usually resolves on its own. Bacterial tonsillitis is typically treated with antibiotics. **Tonsillectomy** (ton-si-LEK-tō-mē; *ectomy* = incision), the removal of a tonsil, may be indicated for individuals who do not respond to other treatments. Such individuals usually have tonsillitis lasting for more than 3 months (despite medication), obstructed air pathways, and difficulty in swallowing and talking. It appears that tonsillectomy does not interfere with a person's response to subsequent infections.

Microbial Evasion of Phagocytosis (Refer page 717 of Textbook)

Some microbes, such as the bacteria that cause pneumonia, have extracellular structures called capsules that prevent adherence. This makes it physically difficult for phagocytes to engulf the microbes. Other microbes, such as the toxin-producing bacteria that cause one kind of food poisoning, may be ingested but not killed; instead, the toxins they produce (leukocidins) may kill the phagocytes by causing the release of the phagocyte's own lysosomal enzymes into its cytoplasm. Still other microbes—such as the bacteria that cause tuberculosis—inhibit fusion of phagosomes and lysosomes and thus prevent exposure of the microbes to lysosomal enzymes. These bacteria apparently can also use chemicals in their cell walls to counter the effects of lethal oxidants produced by phagocytes. Subsequent multiplication of the microbes within phagosomes may eventually destroy the phagocyte.

Abscesses and Ulcers (Refer page 719 of Textbook)

If pus cannot drain out of an inflamed region, the result is an **abscess**—an excessive accumulation of pus in a confined space. Common examples are pimples and boils. When superficial inflamed tissue sloughs off the surface of an organ or tissue, the resulting open sore is called an **ulcer**. People with poor circulation—for instance, diabetics with advanced atherosclerosis—are susceptible to ulcers in the tissues of their legs. These ulcers, which are called stasis ulcers, develop because of poor oxygen and nutrient supply to tissues that then become very susceptible to a very mild injury or infection.

Cytokine Therapy (Refer page 725 of Textbook)

Cytokine therapy is the use of cytokines to treat medical conditions. Interferons were the first cytokines shown to have limited effects against some human cancers. Alpha-interferon (Intron A®) is approved in the United States for treating Kaposi sarcoma (KAP-ō-sē), a cancer that often occurs in patients infected with HIV, the virus that causes AIDS. Other approved uses for alpha-interferon include treating genital herpes caused by the herpes virus; treating hepatitis B and C, caused by the hepatitis B and C viruses; and treating hairy cell leukemia. A form of beta-interferon (Betaseron®) slows the progression of multiple sclerosis (MS) and lessens the frequency and severity of MS attacks. Of the interleukins, the one most widely used to fight cancer is interleukin-2. Although this treatment is effective in causing tumor regression in some patients, it also can be very toxic. Among the adverse effects are high fever, severe weakness, difficulty breathing due to pulmonary edema, and hypotension leading to shock.

Graft Rejection and Tissue Typing (Refer page 727 of Textbook)

Organ transplantation involves the replacement of an injured or diseased organ, such as the heart, liver, kidney, lungs, or pancreas, with an organ donated by another individual. Usually, the immune system recognizes the proteins in the transplanted organ as foreign and mounts both cell-mediated and antibody-mediated immune responses against them. This phenomenon is known as **graft rejection**.

The success of an organ or tissue transplant depends on **histocompatibility** (his′-tō-kom-pat-i-BIL-i-tē)—that is, the tissue compatibility between the donor and the recipient. The more similar the MHC antigens, the greater the histocompatibility, and thus the greater the probability that the transplant will not be rejected. **Tissue typing (histocompatibility testing)** is done before any organ transplant. In the United States, a nationwide computerized registry helps physicians select the most histocompatible and needy organ transplant recipients whenever donor organs become available. The closer the match between the major histocompatibility complex proteins of the donor and recipient, the weaker is the graft rejection response.

To reduce the risk of graft rejection, organ transplant recipients receive immunosuppressive drugs. One such drug is cyclosporine, derived from a fungus, which inhibits secretion of interleukin-2 by helper T cells but has only a minimal effect on B cells. Thus, the risk of rejection is diminished while resistance to some diseases is maintained.

Monoclonal Antibodies (Refer page 730 of Textbook)

The antibodies produced against a given antigen by plasma cells can be harvested from an individual's blood. However, because an antigen typically has many epitopes, several different clones of plasma cells produce different antibodies against the antigen. If a single plasma cell could be isolated and induced to proliferate into a clone of identical plasma cells, then a large quantity of identical antibodies could be produced. Unfortunately, lymphocytes and plasma cells are difficult to grow in culture, so scientists sidestepped this difficulty by fusing B cells with tumor cells that grow easily and proliferate endlessly. The resulting hybrid cell is called a **hybridoma** (hī-bri-DŌ-ma).

Hybridomas are long-term sources of large quantities of pure, identical antibodies, called **monoclonal antibodies (MAbs)** (mon'-ō-KLŌ-nal) because they come from a single clone of identical cells. One clinical use of monoclonal antibodies is for measuring levels of a drug in a patient's blood. Other uses include the diagnosis of strep throat, pregnancy, allergies, and diseases such as hepatitis, rabies, and some sexually transmitted diseases. MAbs have also been used to detect cancer at an early stage and to ascertain the extent of metastasis. They may also be useful in preparing vaccines to counteract the rejection associated with transplants, to treat autoimmune diseases, and perhaps to treat AIDS.

Cancer Immunology (Refer page 733 of Textbook)

Although the immune system responds to cancerous cells, often immunity provides inadequate protection, as evidenced by the number of people dying each year from cancer. Considerable research is focused on **cancer immunology**, the study of ways to use immune responses for detecting, monitoring, and treating cancer. For example, some tumors of the colon release *carcinoembryonic*

antigen *(CEA)* into the blood, and prostate cancer cells release *prostate-specific antigen (PSA)*. Detecting these antigens in blood does not provide definitive diagnosis of cancer, because both antigens are also released in certain noncancerous conditions. However, high levels of cancer-related antigens in the blood often do indicate the presence of a malignant tumor.

Finding ways to induce our immune system to mount vigorous attacks against cancerous cells has been an elusive goal. Many different techniques have been tried, with only modest success. In one method, inactive lymphocytes are removed in a blood sample and cultured with interleukin-2. The resulting *lymphokine-activated killer (LAK)* cells are then transfused back into the patient's blood. Although LAK cells have produced dramatic improvement in a few cases, severe complications affect most patients. In another method, lymphocytes procured from a small biopsy sample of a tumor are cultured with interleukin-2. After their proliferation in culture, such *tumor-infiltrating lymphocytes (TILs)* are reinjected. About a quarter of patients with malignant melanoma and renal-cell carcinoma who received TIL therapy showed significant improvement. The many studies currently under way provide reason to hope that immune-based methods will eventually lead to cures for cancer.

DISORDERS: HOMEOSTATIC IMBALANCES

AIDS: Acquired Immunodeficiency Syndrome

Acquired immunodeficiency syndrome (AIDS) is a condition in which a person experiences a telltale assortment of infections due to the progressive destruction of immune system cells by the **human immunodeficiency virus (HIV)**. AIDS represents the end stage of infection by HIV. A person who is infected with HIV may be symptom-free for many years, even while the virus is actively attacking the immune system. In the two decades after the first five cases were reported in 1981, 22 million people died of AIDS. Worldwide, 35 to 40 million people are currently infected with HIV.

HIV TRANSMISSION Because HIV is present in the blood and some body fluids, it is most effectively transmitted (spread from one person to another) by actions or practices that involve the exchange of blood or body fluids between people. HIV is transmitted in semen or vaginal fluid during unprotected (without a condom) anal, vaginal, or oral sex. HIV also is transmitted by direct blood-to-blood contact, such as occurs among intravenous drug users who share hypodermic needles or health-care professionals who may be accidentally stuck by HIV-contaminated hypodermic needles. In addition, HIV can be transmitted from an HIV-infected mother to her baby at birth or during breast-feeding.

The chance of transmitting or of being infected by HIV during vaginal or anal intercourse can be greatly reduced—although not entirely eliminated—by the use of latex condoms. Public health programs aimed at encouraging drug users not to share needles have proved effective at checking the increase in new HIV infections in this population. Also, giving certain drugs to pregnant HIV-infected women greatly reduces the risk of transmission of the virus to their babies.

HIV is a very fragile virus; it cannot survive for long outside the human body. The virus is not transmitted by insect bites. One cannot become infected by casual physical contact with an HIV-infected person, such as by hugging or sharing household items. The virus can be eliminated from personal care items and medical equipment by exposing them to heat (135°F for 10 minutes) or by cleaning them with common disinfectants such as hydrogen peroxide, rubbing alcohol, household bleach, or germicidal cleansers such as Betadine or Hibiclens. Standard dishwashing and clothes washing also kill HIV.

HIV: STRUCTURE AND INFECTION HIV consists of an inner core of ribonucleic acid (RNA) covered by a protein coat (capsid). HIV is classified as a **retrovirus** (RET-rō-vī-rus) since its genetic information is carried in RNA instead of DNA.

Surrounding the HIV capsid is an envelope composed of a lipid bilayer that is penetrated by glycoproteins (**Figure SG22.1**).

Outside a living host cell, a virus is unable to replicate. However, when a virus infects and enters a host cell, it uses the host cell's enzymes and ribosomes to make thousands of copies of the virus. New viruses eventually leave and then infect other cells. HIV infection of a host cell begins with the binding of HIV glycoproteins to receptors in the host cell's plasma membrane. This causes the cell to transport the virus into its cytoplasm via receptor-mediated endocytosis. Once inside the host cell, HIV sheds its protein coat, and a viral enzyme called **reverse transcriptase** (tran-SKRIP-tās') reads the viral RNA

Figure SG22.1 Human immunodeficiency virus (HIV), the causative agent of AIDS.

> HIV is most effectively transmitted by practices that involve the exchange of body fluids.

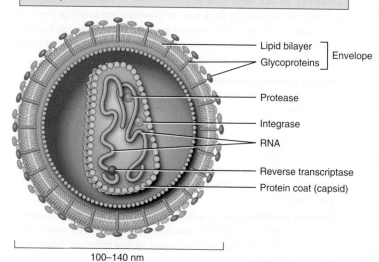

Lipid bilayer ⎤
Glycoproteins ⎦ Envelope

Protease

Integrase

RNA

Reverse transcriptase

Protein coat (capsid)

100–140 nm

Human immunodeficiency virus (HIV)

Q Which cells of the immune system are attacked by HIV?

strand and makes a DNA copy. The viral DNA copy then becomes integrated into the host cell's DNA. Thus, the viral DNA is duplicated along with the host cell's DNA during normal cell division. In addition, the viral DNA can cause the infected cell to begin producing millions of copies of viral RNA and to assemble new protein coats for each copy. The new HIV copies bud off from the cell's plasma membrane and circulate in the blood to infect other cells.

HIV mainly damages helper T cells, and it does so in various ways. Over 10 billion viral copies may be produced each day. The viruses bud so rapidly from an infected cell's plasma membrane that cell lysis eventually occurs. In addition, the body's defenses attack the infected cells, killing them but not all the viruses they harbor. In most HIV-infected individuals, helper T cells are initially replaced as fast as they are destroyed. After several years, however, the body's ability to replace helper T cells is slowly exhausted, and the number of helper T cells in circulation progressively declines.

SIGNS, SYMPTOMS, AND DIAGNOSIS OF HIV INFECTION Soon after being infected with HIV, most people experience a brief flulike illness. Common signs and symptoms are fever, fatigue, rash, headache, joint pain, sore throat, and swollen lymph nodes. About 50% of infected people also experience night sweats. As early as 3 to 4 weeks after HIV infection, plasma cells begin secreting antibodies against HIV. These antibodies are detectable in blood plasma and form the basis for some of the screening tests for HIV. When people test "HIV-positive," it usually means they have antibodies to HIV antigens in their bloodstream.

PROGRESSION TO AIDS After a period of 2 to 10 years, HIV destroys enough helper T cells that most infected people begin to experience symptoms of immunodeficiency. HIV-infected people commonly have enlarged lymph nodes and experience persistent fatigue, involuntary weight loss, night sweats, skin rashes, diarrhea, and various lesions of the mouth and gums. In addition, the virus may begin to infect neurons in the brain, affecting the person's memory and producing visual disturbances.

As the immune system slowly collapses, an HIV-infected person becomes susceptible to a host of *opportunistic infections*. These are diseases caused by microorganisms that are normally held in check but now proliferate because of the defective immune system. AIDS is diagnosed when the helper T cell count drops below 200 cells per microliter (= cubic millimeter) of blood or when opportunistic infections arise, whichever occurs first. In time, opportunistic infections usually are the cause of death.

TREATMENT OF HIV INFECTION At present, infection with HIV cannot be cured. Vaccines designed to block new HIV infections and to reduce the viral load (the number of copies of HIV RNA in a microliter of blood plasma) in those who are already infected are in clinical trials. Meanwhile, three classes of drugs have proved successful in extending the life of many HIV-infected people:

1. **Reverse transcriptase inhibitors** interfere with the action of reverse transcriptase, the enzyme that the virus uses to convert its RNA into a DNA copy. Among the drugs in this category are zidovudine (ZDV, previously called AZT), didanosine (ddI), and stavudine (trade name d4T®). Trizivir, approved in 2000 for treatment of HIV infection, combines three reverse transcriptase inhibitors in one pill.

2. **Integrase inhibitors** block the enzyme integrase, which inserts the HIV DNA copy into host cell DNA. The drug raltegravir is an example of an integrase inhibitor.

3. **Protease inhibitors** interfere with the action of protease, a viral enzyme that cuts proteins into pieces to assemble the protein coat of newly produced HIV particles. Drugs in this category include nelfinavir, saquinavir, ritonavir, and indinavir.

The recommended treatment for HIV-infected patients is *highly active antiretroviral therapy (HAART)*—a combination of three or more antiretroviral medications from at least two differently acting inhibitor drug classes. Most HIV-infected individuals receiving HAART experience a drastic reduction in viral load and an increase in the number of helper T cells in their blood. Not only does HAART delay the progression of HIV infection to AIDS, but many individuals with AIDS have seen the remission or disappearance of opportunistic infections and an apparent return to health. Unfortunately, HAART is very costly (exceeding $10,000 per year), the dosing schedule is grueling, and not all people can tolerate the toxic side effects of these drugs. Although HIV may virtually disappear from the blood with drug treatment (and thus a blood test may be "negative" for HIV), the virus typically still lurks in various lymphatic tissues. In such cases, the infected person can still transmit the virus to another person.

Allergic Reactions

A person who is overly reactive to a substance that is tolerated by most other people is said to be **allergic** or *hypersensitive*. Whenever an allergic reaction takes place, some tissue injury occurs. The antigens that induce an allergic reaction are called **allergens** (AL-er-jens). Common allergens include certain foods (milk, peanuts, shellfish, eggs), antibiotics (penicillin, tetracycline), vaccines (pertussis, typhoid), venoms (honeybee, wasp, snake), cosmetics, chemicals in plants such as poison ivy, pollens, dust, molds, iodine-containing dyes used in certain x-ray procedures, and even microbes.

There are four basic types of **hypersensitivity** reactions: type I (anaphylactic), type II (cytotoxic), type III (immune-complex), and type IV (cell-mediated). The first three are antibody-mediated immune responses; the last is a cell-mediated immune response.

Type I (*anaphylactic*) **reactions** (AN-a-fil-lak′-tik) are the most common and occur within a few minutes after a person sensitized to an allergen is re-exposed to it. In response to the first exposure to certain allergens, some people produce IgE antibodies that bind to the surface of mast cells and basophils. The next time the same allergen enters the body, it attaches to the IgE antibodies already present. In response, the mast cells and basophils release histamine, prostaglandins, leukotrienes, and kinins. Collectively, these mediators cause vasodilation, increased blood capillary permeability, increased smooth muscle contraction in the airways of the lungs, and increased mucus secretion. As a result, a person may experience inflammatory responses, difficulty in breathing through the constricted airways, and a runny nose from excess mucus secretion. In **anaphylactic shock**, which may occur in a susceptible individual who has just received a triggering drug or been stung by a wasp, wheezing and shortness of breath as airways constrict are usually accompanied by shock due to vasodilation and fluid loss from blood. This life-threatening emergency is usually treated by injecting epinephrine to dilate the airways and strengthen the heartbeat.

Type II (*cytotoxic*) **reactions** are caused by antibodies (IgG or IgM) directed against antigens on a person's blood cells (red blood cells, lymphocytes, or platelets) or tissue cells. The reaction of antibodies and antigens usually leads to activation of complement. Type II reactions, which may occur in incompatible blood transfusion reactions, damage cells by causing lysis.

Type III (*immune-complex*) **reactions** involve antigens, antibodies (IgA or IgM), and complement. When certain ratios of antigen to antibody occur, the immune complexes are small enough to escape phagocytosis, but they become trapped in the basement membrane under the endothelium of blood vessels, where they activate complement and cause inflammation. Glomerulonephritis and rheumatoid arthritis (RA) arise in this way.

Type IV (*cell-mediated*) **reactions** or *delayed hypersensitivity reactions* usually appear 12–72 hours after exposure to an allergen. Type IV reactions occur when allergens are taken up by antigen-presenting cells (such as intraepidermal macrophages in the skin) that migrate to lymph nodes and present the allergen to T cells, which then proliferate. Some of the new T cells return to the site of allergen entry into the body, where they produce gamma-interferon, which activates macrophages, and tumor necrosis factor, which stimulates an inflammatory response. Intracellular bacteria such as *Mycobacterium tuberculosis* (mī-kō-bak-TĒ-rē-um too-ber′-ku-LŌ-sis) trigger this type of cell-mediated immune response, as do certain haptens, such as poison ivy toxin. The skin test for tuberculosis also is a delayed hypersensitivity reaction.

Autoimmune Diseases

In an **autoimmune disease** (aw-tō-i-MŪN) or *autoimmunity*, the immune system fails to display self-tolerance and attacks the person's own tissues. Autoimmune diseases usually arise in early adulthood and are common, afflicting an estimated 5% of adults in North America and Europe. Females suffer autoimmune diseases twice as often as males. Recall that self-reactive B cells and T cells normally are deleted or undergo anergy during negative selection (see **Figure 22.22**). Apparently, this process is not 100% effective. Under the influence of unknown environmental triggers and certain genes that make some people more susceptible, self-tolerance breaks down, leading to activation of self-reactive clones of T cells and B cells. These cells then generate cell-mediated or antibody-mediated immune responses against self-antigens.

A variety of mechanisms produce different autoimmune diseases. Some involve production of **autoantibodies**, antibodies that bind to and stimulate or block self-antigens. For example, autoantibodies that mimic TSH (thyroid-stimulating hormone) are present in Graves disease and stimulate secretion of thyroid hormones (thus producing hyperthyroidism); autoantibodies that bind to and block acetylcholine receptors cause the muscle weakness characteristic of myasthenia gravis. Other autoimmune diseases involve activation of cytotoxic T cells that destroy certain body cells. Examples include type 1 diabetes mellitus, in which T cells attack the insulin-producing pancreatic beta cells, and multiple sclerosis (MS), in which T cells attack myelin sheaths around axons of neurons. Inappropriate activation of helper T cells or excessive production of gamma-interferon also occur in certain autoimmune diseases. Other autoimmune disorders include rheumatoid arthritis (RA), systemic lupus erythematosus (SLE), rheumatic fever, hemolytic and pernicious anemias, Addison's disease, Hashimoto's thyroiditis, and ulcerative colitis.

Therapies for various autoimmune diseases include removal of the thymus gland (thymectomy), injections of beta-interferon, immunosuppressive drugs, and plasmapheresis, in which the person's blood plasma is filtered to remove antibodies and antigen–antibody complexes.

Infectious Mononucleosis

Infectious mononucleosis (mon′-ō-noo-klē-Ō-sis) or "mono" is a contagious disease caused by the *Epstein–Barr virus (EBV)*. It occurs mainly in children and young adults, and more often in females than in males. The virus most commonly enters the body through intimate oral contact such as kissing, which accounts for its common name, the "kissing disease." EBV then multiplies in lymphatic tissues and filters into the blood, where it infects and multiplies in B cells, the primary host cells. Because of this infection, the B cells become so enlarged and abnormal in appearance that they resemble monocytes, the primary reason for the term **mononucleosis**. In addition to an elevated white blood cell count with an abnormally high percentage of lymphocytes, signs and symptoms include fatigue, headache, dizziness, sore throat, enlarged and tender lymph nodes, and fever. There is no cure for infectious mononucleosis, but the disease usually runs its course in a few weeks.

Lymphomas

Lymphomas (lim-FŌ-mas; *lymph-* = clear water; *-oma* = tumor) are cancers of the lymphatic organs, especially the lymph nodes. Most have no known cause. The two main types of lymphomas are Hodgkin disease and non-Hodgkin lymphoma.

Hodgkin disease (HD) (HOJ-kin) is characterized by a painless, nontender enlargement of one or more lymph nodes, most commonly in the neck, chest, and axilla. If the disease has metastasized from these sites, fever, night sweats, weight loss, and bone pain also occur. HD primarily affects individuals between ages 15 and 35 and those over 60, and it is more common in males. If diagnosed early, HD has a 90–95% cure rate.

Non-Hodgkin lymphoma (NHL), which is more common than HD, occurs in all age groups, the incidence increasing with age to a maximum between ages 45 and 70. NHL may start the same way as HD but may also include an enlarged spleen, anemia, and general malaise. Up to half of all individuals with NHL are cured or survive for a lengthy period. Treatment options for both HD and NHL include radiation therapy, chemotherapy, and bone marrow transplantation.

Systemic Lupus Erythematosus

Systemic lupus erythematosus (SLE) (er′-e-thē′-ma-TŌ-sus)**,** or simply *lupus* (= wolf) is a chronic autoimmune, inflammatory disease that affects multiple body systems. Lupus is characterized by periods of active disease and remission; symptoms range from mild to life-threatening. Lupus most often develops between ages 15 and 44 and is 10–15 times more common in females than males. It is also 2–3 times more common in African Americans, Hispanics, Asian Americans, and Native Americans than in European Americans. Although the cause of SLE is not known, both a genetic predisposition to the disease and environmental factors (infections, antibiotics, ultraviolet light, stress, and hormones) may trigger it. Sex hormones appear to influence the development of SLE. The disorder often occurs in females who exhibit extremely low levels of androgens.

Signs and symptoms of SLE include joint pain, muscle pain, chest pain with deep breaths, headaches, pale or purple fingers or toes, kidney dysfunction, low blood cell count, nerve or brain dysfunction, slight fever, fatigue, oral ulcers, weight loss, swelling in the legs or around the eyes, enlarged lymph nodes and spleen, photosensitivity, rapid loss of large amounts of scalp hair, and sometimes an eruption across the bridge of the nose and cheeks called a "butterfly rash." The erosive nature of some of the SLE skin lesions was thought to resemble the damage inflicted by the bite of a wolf—thus, the term *lupus*.

Two immunological features of SLE are excessive activation of B cells and inappropriate production of autoantibodies against DNA (anti-DNA antibodies) and other components of cellular nuclei such as histone proteins. Triggers of B cell activation are thought to include various chemicals and drugs, viral and bacterial antigens, and exposure to sunlight. Circulating complexes of abnormal autoantibodies and their "antigens" cause damage in tissues throughout the body. Kidney damage occurs as the complexes become trapped in the basement membrane of kidney capillaries, obstructing blood filtering. Renal failure is the most common cause of death.

There is no cure for lupus, but drug therapy can minimize symptoms, reduce inflammation, and forestall flare-ups. The most commonly used lupus medications are pain relievers (nonsteroidal anti-inflammatory drugs such as aspirin and ibuprofen), antimalarial drugs (hydroxychloroquine), and corticosteroids (prednisone and hydrocortisone).

Severe Combined Immunodeficiency Disease

Severe combined immunodeficiency disease (SCID) (im′-ū-nō-de-FISH-en-sē) is a rare inherited disorder in which both B cells and T cells are missing or inactive. Scientists have now identified mutations in several genes that are responsible for some types of SCID. In some cases, an infusion of red bone marrow cells from a sibling having very similar MHC (HLA) antigens can provide normal stem cells that give rise to normal B and T cells. The result can be a complete cure. Less than 30% of afflicted patients, however, have a compatible sibling who could serve as a donor. The disorder, which occurs more frequently in males, is also known as *bubble boy disease*, named for David Vetter, who was born with the condition and lived behind plastic barriers to protect him from microbes. He died at age 12 in 1984. The chances of a child born with SCID are about 1 in 500,000 and, until recent years, it was always fatal. Children with SCID have virtually no defenses against microbes. Treatment consists of bringing any current infections under control, bolstering nutrition,

bone marrow transplant (provides stem cells to make new B and T cells), enzymatic replacement therapy (injections of polyethylene glycol–linked adenosine deaminase, or PE-ADA), and gene therapy. In this technique, the most common approach is to insert a normal gene into a genome to replace a nonfunctional gene. The normal gene is usually delivered by a virus. Then, the normal gene would produce B and T cells to provide sufficient immunity.

MEDICAL TERMINOLOGY

Adenitis (ad′-e-NĪ-tis; *aden-* = gland; *-itis* = inflammation of) Enlarged, tender, and inflamed lymph nodes resulting from an infection.

Allograft (AL-ō-graft; *allo-* = other) A transplant between genetically distinct individuals of the same species. Skin transplants from other people and blood transfusions are allografts.

Autograft (AW-tō-graft; *auto-* = self) A transplant in which one's own tissue is grafted to another part of the body (such as skin grafts for burn treatment or plastic surgery).

Chronic fatigue syndrome (CFS) A disorder, usually occurring in young adults and primarily in females, characterized by (1) extreme fatigue that impairs normal activities for at least 6 months and (2) the absence of other known diseases (cancer, infections, drug abuse, toxicity, or psychiatric disorders) that might produce similar symptoms.

Gamma globulin (GLOB-ū-lin) Suspension of immunoglobulins from blood consisting of antibodies that react with a specific pathogen. It is prepared by injecting the pathogen into animals, removing blood from the animals after antibodies have been produced, isolating the antibodies, and injecting them into a human to provide short-term immunity.

Hypersplenism (hī-per-SPLĒN-izm; *hyper-* = over) Abnormal splenic activity due to splenic enlargement and associated with an increased rate of destruction of normal blood cells.

Lymphadenopathy (lim-fad′-e-NOP-a-thē; *lymph-* = clear fluid; *-pathy* = disease) Enlarged, sometimes tender lymph nodes as a response to infection; also called swollen glands.

Lymphangitis (lim-fan-JĪ-tis; *-itis* = inflammation of) Inflammation of lymphatic vessels.

Lymphedema (lim′-fe-DĒ-ma; *edema* = swelling) Accumulation of lymph in lymphatic vessels, causing painless swelling of a limb.

Splenomegaly (splē′-nō-MEG-a-lē; *mega-* = large) Enlarged spleen.

Xenograft (ZEN-ō-graft; *xeno-* = strange or foreign) A transplant between animals of different species. Xenografts from porcine (pig) or bovine (cow) tissue may be used in humans as a physiological dressing for severe burns. Other xenografts include pig heart valves and baboon hearts.

SELF-QUIZ QUESTIONS

Fill in the blanks in the following statements.

1. The first line of defense of innate immunity against pathogens consists of the _____ and _____; the second line of defense consists of _____, _____, and _____.

2. Substances that are recognized as foreign and provoke immune responses are known as _____.

Indicate whether the following statements are true or false.

3. Your body's ability to ward off damage or disease through your defenses is known as resistance; vulnerability to disease is susceptibility.

4. Your T cells must be able to recognize your own MHC molecules, a process known as self-recognition, and lack reactivity to peptide fragments from your own proteins, a condition known as self-tolerance.

Choose the one best answer to the following questions.

5. Trace the sequence of fluid from blood vessel to blood vessel by way of the lymphatic system. (1) lymphatic vessels, (2) blood capillaries, (3) subclavian veins, (4) lymphatic capillaries, (5) interstitial spaces, (6) arteries, (7) lymphatic ducts.
   (a) 2, 5, 4, 1, 7, 6, 3   (b) 3, 6, 2, 4, 5, 1, 7   (c) 6, 2, 5, 4, 1, 7, 3
   (d) 6, 2, 5, 4, 7, 1, 3   (e) 2, 5, 4, 7, 1, 3, 6

6. Which of the following describe lymph nodes? (1) Lymph enters the nodes through efferent lymphatic vessels and leaves through afferent lymphatic vessels. (2) The outer cortex consists of lymphatic nodules that contain B cells and are the sites of plasma cell and memory B cell formation. (3) The inner cortex contains lymphatic nodules with mature T cells.

(4) The reticular fibers within the sinuses of the lymph nodes trap foreign substances in the lymph. (5) The sinuses of lymph nodes are known as red pulp.
   (a) 1, 2, 3, and 4   (b) 2, 4, and 5   (c) 1, 2, 3, 4, and 5
   (d) 2 and 4   (e) 1, 2, and 4

7. Which of the following statements are *correct*? (1) Lymphatic vessels are found throughout the body, except in avascular tissues, the CNS, portions of the spleen, and red bone marrow. (2) Lymphatic capillaries allow interstitial fluid to flow into them but not out of them. (3) Anchoring filaments attach lymphatic endothelial cells to surrounding tissues. (4) Lymphatic vessels freely receive all the components of blood, including the formed elements. (5) Lymph ducts directly connect to blood vessels by way of the subclavian veins.
   (a) 1, 3, 4, and 5   (b) 2, 3, 4, and 5   (c) 1, 2, 3, and 4
   (d) 1, 2, 4, and 5   (e) 1, 2, 3, and 5

8. Which of the following are physical factors that help fight pathogens and disease? (1) numerous layers of the epidermis, (2) mucus of mucous membranes, (3) saliva, (4) interferons, (5) complement.
   (a) 1, 3, and 4   (b) 2, 4, and 5   (c) 1, 4, and 5
   (d) 1, 2, and 3   (e) 1, 2, and 4

9. Which of the following are functions of antibodies? (1) neutralization of antigens, (2) immobilization of bacteria, (3) agglutination and precipitation of antigens, (4) activation of complement, (5) enhancement of phagocytosis.
   (a) 1, 3, and 4   (b) 2, 4, and 5   (c) 1, 2, 3, and 4
   (d) 1, 2, 3, and 5   (e) 1, 2, 3, 4, and 5

10. Which of the following are *true*? (1) Lymphatic vessels resemble arteries. (2) Lymph is very similar to interstitial fluid. (3) Lacteals are specialized lymphatic capillaries responsible for transporting dietary lipids. (4) Lymph is normally a cloudy, pale yellow fluid. (5) The thoracic duct drains lymph from the upper right side of the body. (6) Lymph flow is maintained by skeletal muscle contractions, one-way valves, and breathing movements.

(a) 1, 2, 5, and 6   (b) 2, 3, and 6   (c) 2, 3, 4, and 6
(d) 2, 4, and 6   (e) 3, 5, and 6

11. Place the stages of phagocytosis in the correct order of occurrence. (1) formation of phagolysosome, (2) adherence to microbe, (3) destruction of microbe, (4) ingestion to form a phagosome, (5) chemotactic attraction of phagocyte.

(a) 2, 5, 4, 1, 3   (b) 4, 5, 2, 1, 3   (c) 5, 2, 4, 1, 3
(d) 5, 4, 2, 3, 1   (e) 2, 5, 1, 4, 3

12. Place in order the steps involved in cell-mediated immune response to an exogenous antigen.

(a) costimulation and activation of helper T cells
(b) presentation of antigen to helper T cells
(c) elimination of invaders through the release of granzymes, perforin, granulysin, or lymphotoxin or by attraction and activation of phagocytes
(d) proliferation and differentiation of helper T cells to produce a helper T cell clone
(e) antigen processing by dendritic cells, macrophages, or B cells
(f) recognition of antigen fragments associated with MHC-II molecules by T-cell receptors
(g) secretion of cytokines such as interleukin-2 by activated helper T cells
(h) migration of antigen-presenting cells to lymphatic tissue
(i) activation of cytotoxic T cells

13. Match the following:

_____ (a) encapsulated bean-shaped structures located along the length of lymphatic vessels; contain T and B cells, macrophages, and follicular dendritic cells; filter lymph

_____ (b) produces pre-T cells and B cells; found in flat bones and epiphyses of long bones

_____ (c) clusters of lymphatic nodules involved in immune responses against inhaled or ingested foreign substances

_____ (d) the single largest mass of lymphatic tissue in the body; consists of red and white pulp

_____ (e) responsible for the maturation of T cells

_____ (f) lymphatic nodules associated with mucous membranes of the digestive, urinary, reproductive, and respiratory systems

_____ (g) nonencapsulated clusters of lymphocytes

(1) red bone marrow
(2) thymus
(3) lymph nodes
(4) spleen
(5) mucosa-associated lymphatic tissue
(6) lymphatic nodules
(7) tonsils

14. Match the following:

_____ (a) recognize foreign antigens combined with MHC-I molecules on the surface of body cells infected by microbes, some tumor cells, and cells of a tissue transplant; display CD8 proteins

_____ (b) are programmed to recognize the reappearance of a previously encountered antigen

_____ (c) differentiate into plasma cells that secrete specific antibodies

(d) process and present exogenous antigens; include macrophages, B cells, and dendritic cells

_____ (e) secrete cytokines as costimulators; display CD4 proteins

_____ (f) ingest microbes or any foreign particulate matter; include neutrophils and macrophages

_____ (g) lymphocytes that have the ability to kill a wide variety of infectious microbes plus certain spontaneously arising tumor cells; lack antigen receptors

(1) active helper T cells
(2) cytotoxic T cells
(3) memory T cells
(4) B cells
(5) NK cells
(6) phagocytes
(7) antigen-presenting cells

23 | THE RESPIRATORY SYSTEM

23.1 Overview of the Respiratory System

1. Three basic steps are involved in respiration: (1) pulmonary ventilation, (2) external (pulmonary) respiration, and (3) internal (tissue) respiration.
2. The respiratory system consists of the nose, pharynx, larynx, trachea, bronchi, and lungs. They act with the cardiovascular system to supply oxygen (O_2) and remove carbon dioxide (CO_2) from the blood.
3. It is divided into an upper and lower respiratory system.

23.2 The Upper Respiratory System

1. The external portion of the nose is made of cartilage and skin and is lined with a mucous membrane. Openings to the exterior are the external nares. The internal portion of the nose communicates with the paranasal sinuses and nasopharynx through the internal nares. The nasal cavity is divided by a nasal septum. The anterior portion of the cavity is called the vestibule. The nose warms, moistens, and filters air and functions in olfaction and speech.
2. The pharynx (throat) is a muscular tube lined by a mucous membrane. The anatomical regions are the nasopharynx, oropharynx, and laryngopharynx. The nasopharynx functions in respiration. The oropharynx and laryngopharynx function in both breathing and digestion.

23.3 The Lower Respiratory System

1. The larynx (voice box) is a passageway that connects the pharynx with the trachea. It contains the thyroid cartilage (Adam's apple); the epiglottis, which prevents food from entering the larynx; the cricoid cartilage, which connects the larynx and trachea; and the paired arytenoid, corniculate, and cuneiform cartilages. The larynx contains vocal folds, which produce sound as they vibrate. Taut folds produce high pitches, and relaxed ones produce low pitches.
2. The trachea (windpipe) extends from the larynx to the main bronchi. It is composed of C-shaped rings of cartilage and smooth muscle and is lined with ciliated pseudostratified columnar epithelium.
3. The bronchial tree consists of the trachea, main bronchi, lobar bronchi, segmental bronchi, bronchioles, and terminal bronchioles. Walls of bronchi contain rings of cartilage; walls of bronchioles contain increasingly smaller plates of cartilage and increasing amounts of smooth muscle.

4. Lungs are paired organs in the thoracic cavity enclosed by the pleural membrane. The parietal pleura is the superficial layer that lines the thoracic cavity; the visceral pleura is the deep layer that covers the lungs. The right lung has three lobes separated by two fissures; the left lung has two lobes separated by one fissure and a depression, the cardiac notch.
5. Lobar bronchi give rise to branches called segmental bronchi, which supply segments of lung tissue called bronchopulmonary segments. Each bronchopulmonary segment consists of lobules, which contain lymphatics, arterioles, venules, terminal bronchioles, respiratory bronchioles, alveolar ducts, alveolar sacs, and alveoli.
6. Alveolar walls consist of type I alveolar cells, type II alveolar cells, and associated alveolar macrophages.
7. Gas exchange occurs across the respiratory membranes.

23.4 Pulmonary Ventilation

1. Pulmonary ventilation, or breathing, consists of inhalation and exhalation.
2. The movement of air into and out of the lungs depends on pressure changes governed in part by Boyle's law, which states that the volume of a gas varies inversely with pressure, assuming that temperature remains constant.
3. Inhalation occurs when alveolar pressure falls below atmospheric pressure. Contraction of the diaphragm and external intercostals increases the size of the thorax, thereby decreasing the intrapleural pressure so that the lungs expand. Expansion of the lungs decreases alveolar pressure so that air moves down a pressure gradient from the atmosphere into the lungs.
4. During forceful inhalation, accessory muscles of inhalation (sternocleidomastoids, scalenes, and pectoralis minors) are also used.
5. Exhalation occurs when alveolar pressure is higher than atmospheric pressure. Relaxation of the diaphragm and external intercostals results in elastic recoil of the chest wall and lungs, which increases intrapleural pressure; lung volume decreases and alveolar pressure increases, so air moves from the lungs to the atmosphere.
6. Forceful exhalation involves contraction of the internal intercostal and abdominal muscles.
7. The surface tension exerted by alveolar fluid is decreased by the presence of surfactant.
8. Compliance is the ease with which the lungs and thoracic wall can expand.

9. The walls of the airways offer some resistance to breathing.
10. Normal quiet breathing is termed eupnea; other patterns are costal breathing and diaphragmatic breathing. Modified respiratory movements, such as coughing, sneezing, sighing, yawning, sobbing, crying, laughing, and hiccupping, are used to express emotions and to clear the airways. (See **Table 23.2**.)

23.5 Lung Volumes and Capacities

1. Lung volumes exchanged during breathing and the rate of respiration are measured with a spirometer.
2. Lung volumes measured by spirometry include tidal volume, minute ventilation, alveolar ventilation rate, inspiratory reserve volume, expiratory reserve volume, and $FEV_{1.0}$. Other lung volumes include anatomic dead space, residual volume, and minimal volume.
3. Lung capacities, the sum of two or more lung volumes, include inspiratory, functional, residual, vital, and total lung capacities.

23.6 Exchange of Oxygen and Carbon Dioxide

1. The partial pressure of a gas is the pressure exerted by that gas in a mixture of gases. It is symbolized by P_x, where the subscript is the chemical formula of the gas.
2. According to Dalton's law, each gas in a mixture of gases exerts its own pressure as if all the other gases were not present.
3. Henry's law states that the quantity of a gas that will dissolve in a liquid is proportional to the partial pressure of the gas and its solubility (given constant temperature).
4. In internal and external respiration, O_2 and CO_2 diffuse from areas of higher partial pressures to areas of lower partial pressures.
5. External respiration or pulmonary gas exchange is the exchange of gases between alveoli and pulmonary blood capillaries. It depends on partial pressure differences, a large surface area for gas exchange, a small diffusion distance across the respiratory membrane, and the rate of airflow into and out of the lungs.
6. Internal respiration or systemic gas exchange is the exchange of gases between systemic blood capillaries and tissue cells.

23.7 Transport of Oxygen and Carbon Dioxide

1. In each 100 mL of oxygenated blood, 1.5% of the O_2 is dissolved in blood plasma and 98.5% is bound to hemoglobin as oxyhemoglobin ($Hb–O_2$).
2. The binding of O_2 to hemoglobin is affected by P_{O_2}, acidity (pH), P_{CO_2}, temperature, and 2,3-bisphosphoglycerate (BPG).
3. Fetal hemoglobin differs from adult hemoglobin in structure and has a higher affinity for O_2.
4. In each 100 mL of deoxygenated blood, 7% of CO_2 is dissolved in blood plasma, 23% combines with hemoglobin as carbamino-hemoglobin ($Hb–CO_2$), and 70% is converted to bicarbonate ions (HCO_3^-).
5. In an acidic environment, hemoglobin's affinity for O_2 is lower, and O_2 dissociates more readily from it (Bohr effect).
6. In the presence of O_2, less CO_2 binds to hemoglobin (Haldane effect).

23.8 Control of Breathing

1. The respiratory center consists of a medullary respiratory center in the medulla and a pontine respiratory group in the pons.
2. The medullary respiratory center in the medulla is made up of a dorsal respiratory group (DRG), which controls normal quiet breathing, and a ventral respiratory group (VRG), which is used during forceful breathing and controls the rhythm of breathing.
3. The pontine respiratory group in the pons may modify the rhythm of breathing during exercise, speaking, and sleep.
4. The activity of the respiratory center can be modified in response to inputs from various parts of the body in order to maintain the homeostasis of breathing.
5. These include cortical influences; the inflation reflex; chemical stimuli, such as O_2 and CO_2 and H^+ levels; proprioceptor input; blood pressure changes; limbic system stimulation; temperature; pain; and irritation to the airways. (See **Table 23.3**.)

23.9 Exercise and the Respiratory System

1. The rate and depth of breathing change in response to both the intensity and duration of exercise.
2. An increase in pulmonary perfusion and O_2-diffusing capacity occurs during exercise.
3. The abrupt increase in breathing at the start of exercise is due to neural changes that send excitatory impulses to the dorsal respiratory group of the medullary respiratory center in the medulla oblongata. The more gradual increase in breathing during moderate exercise is due to chemical and physical changes in the bloodstream.

23.10 Development of the Respiratory System

1. The respiratory system begins as an outgrowth of endoderm called the respiratory diverticulum.
2. Smooth muscle, cartilage, and connective tissue of the bronchial tubes and pleural sacs develop from mesoderm.

23.11 Aging and the Respiratory System

1. Aging results in decreased vital capacity, decreased blood level of O_2, and diminished alveolar macrophage activity.
2. Elderly people are more susceptible to pneumonia, emphysema, bronchitis, and other pulmonary disorders.

CLINICAL CONNECTIONS

Rhinoplasty (Refer page 743 of Textbook)

Rhinoplasty (Rī-nō-plas′-tē; *thin* = nose; *-plasty* = to mold or to shape), or "nose job," is a surgical procedure in which the shape of the external nose is altered. Although rhinoplasty is often done for cosmetic reasons, it is sometimes performed to repair a fractured nose or a deviated nasal septum. In the procedure, both local and general anesthetics are given. Instruments are then inserted through the nostrils, the nasal cartilage is reshaped, and the nasal bones are fractured and repositioned to achieve the desired shape. An internal packing and splint are inserted to keep the nose in the desired position as it heals.

Tonsillectomy (Refer page 743 of Textbook)

Tonsillectomy (ton-si-LEK-tō-mē-; *-ektome* = excision or to cut out) is surgical removal of the tonsils. The procedure is usually performed under general anesthesia on an outpatient basis. Tonsillectomies are performed in individuals who have frequent *tonsillitis* (ton'-si-LĪ-tis), that is, inflammation of the tonsils; tonsils that develop an abscess or tumor; or tonsils that obstruct breathing during sleep.

Laryngitis and Cancer of the Larynx (Refer page 747 of Textbook)

Laryngitis is an inflammation of the larynx that is most often caused by a respiratory infection or irritants such as cigarette smoke. Inflammation of the vocal folds causes hoarseness or loss of voice by interfering with the contraction of the folds or by causing them to swell to the point where they cannot vibrate freely. Many long-term smokers acquire a permanent hoarseness from the damage done by chronic inflammation. **Cancer of the larynx** is found almost exclusively in individuals who smoke. The condition is characterized by hoarseness, pain on swallowing, or pain radiating to an ear. Treatment consists of radiation therapy and/or surgery.

Tracheotomy and Intubation (Refer page 748 of Textbook)

Several conditions may block airflow by obstructing the trachea. The rings of cartilage that support the trachea may be accidentally crushed, the mucous membrane may become inflamed and swell so much that it closes off the passageway, excess mucus secreted by inflamed membranes may clog the lower respiratory passages, a large object may be aspirated (breathed in), or a cancerous tumor may protrude into the airway. Two methods are used to reestablish airflow past a tracheal obstruction. If the obstruction is above the level of the larynx, a **tracheotomy** (trā-kē-O-tō-mē) may be performed. In this procedure, also called a *tracheostomy*, a skin incision is followed by a short longitudinal incision into the trachea below the cricoid cartilage. A tracheal tube is then inserted to create an emergency air passageway. The second method is **intubation** (in'-too-BĀ-shun), in which a tube is inserted into the mouth or nose and passed inferiorly through the larynx and trachea. The firm wall of the tube pushes aside any flexible obstruction, and the lumen of the tube provides a passageway for air; any mucus clogging the trachea can be suctioned out through the tube.

Pneumothorax and Hemothorax (Refer page 751 of Textbook)

In certain conditions, the pleural cavities may fill with air (**pneumothorax**; noo'-mō-THOR-aks; *pneumo-* = air or breath), blood (**hemothorax**), or pus. Air in the pleural cavities, most commonly introduced in a surgical opening of the chest or as a result of a stab or gunshot wound, may cause the lungs to collapse. This collapse of a part of a lung, or rarely an entire lung, is called **atelectasis** (at'-e-LEK-ta-sis; *ateles-* = incomplete; *-ectasis* = expansion). The goal of treatment is the evacuation of air (or blood) from the pleural space, which allows the lung to reinflate. A small pneumothorax may resolve on its own, but it is often necessary to insert a chest tube to assist in evacuation.

Coryza, Seasonal Influenza, and H1N1 Influenza (Refer page 752 of Textbook)

Hundreds of viruses can cause coryza (ko-RĪ-za), or the **common cold,** but a group of viruses called *rhinoviruses* (RĪ-nō-vī-rus-es) is responsible for about 40% of all colds in adults. Typical symptoms include sneezing, excessive nasal secretion, dry cough, and congestion. The uncomplicated common cold is not usually accompanied by a fever. Complications include sinusitis, asthma, bronchitis, ear infections, and laryngitis. Recent investigations suggest an association between emotional stress and the common cold. The higher the stress level, the greater the frequency and duration of colds. **Seasonal influenza (flu)** is also caused by a virus. Its symptoms include chills, fever (usually higher than 101°F = 39°C), headache, and muscular aches. Seasonal influenza can become life-threatening and may develop into pneumonia. It is important to recognize that influenza is a respiratory disease, not a gastrointestinal (GI) disease. Many people mistakenly report having seasonal flu when they are suffering from a GI illness.

H1N1 influenza (flu), also known as *swine flu*, is a type of influenza caused by a new virus called *influenza H1N1*. The virus is spread in the same way that seasonal flu spreads: from person to person through coughing or sneezing or by touching infected objects and then touching one's mouth or nose. Most individuals infected with the virus have mild disease and recover without medical treatment, but some people have severe disease and have even died. The symptoms of H1N1 flu include fever, cough, runny or stuffy nose, headache, body aches, chills, and fatigue. Some people also have vomiting and diarrhea. Most people who have been hospitalized for H1N1 flu have had one or more preexisting medical conditions such as diabetes, heart disease, asthma, kidney disease, or pregnancy. People infected with the virus can infect others from 1 day before symptoms occur to 5–7 days or more after they occur. Treatment of H1N1 flu involves taking antiviral drugs, such as Tamiflu® and Relenza®. A vaccine is also available, but the H1N1 flu vaccine is not a substitute for seasonal flu vaccines.

Respiratory Distress Syndrome (Refer page 759 of Textbook)

Respiratory distress syndrome (RDS) is a breathing disorder of premature newborns in which the alveoli do not remain open due to a lack of surfactant. Recall that surfactant reduces surface tension and is necessary to prevent the collapse of alveoli during exhalation. The more premature the newborn, the greater the chance that RDS will develop. The condition is also more common in infants whose mothers have diabetes and in males; it also occurs more often in European Americans than African Americans. Symptoms of RDS include labored and irregular breathing, flaring of the nostrils during inhalation, grunting during exhalation, and perhaps a blue skin color. Besides the symptoms, RDS is diagnosed on the basis of chest radiographs and a blood test. A newborn with mild RDS may require only supplemental oxygen administered through an oxygen hood or through a tube placed in the nose. In severe cases oxygen may be delivered by continuous positive airway pressure (CPAP) through tubes in the nostrils or a mask on the face. In such cases surfactant may be administered directly into the lungs.

Hyperbaric Oxygenation (Refer page 762 of Textbook)

A major clinical application of Henry's law is **hyperbaric oxygenation** (*hyper-* = over; *-baros* = pressure), the use of pressure to cause more O_2 to dissolve in the blood. It is an effective technique in treating patients infected by anaerobic bacteria, such as those that cause tetanus and gangrene. (Anaerobic bacteria cannot live in the presence of free O_2.) A person undergoing hyperbaric oxygenation is placed in a hyperbaric chamber, which contains O_2 at a pressure greater than 1 atmosphere (760 mmHg). As body tissues pick up the O_2, the bacteria are killed. Hyperbaric chambers may also be used for treating certain heart disorders, carbon monoxide poisoning, gas embolisms, crush injuries, cerebral edema, certain hard-to-treat bone infections caused by anaerobic bacteria, smoke inhalation, near-drowning, asphyxia, vascular insufficiencies, and burns.

Carbon Monoxide Poisoning (Refer page 768 of Textbook)

Carbon monoxide (CO) is a colorless and odorless gas found in exhaust fumes from automobiles, gas furnaces and space heaters, and in tobacco smoke. It is a by-product of the combustion of carbon-containing materials such as coal, gas, and wood. CO binds to the heme group of hemoglobin, just as O_2 does, except that the binding of carbon monoxide to hemoglobin is over 200 times as strong as the binding of O_2 to hemoglobin. Thus, at a concentration as small as 0.1% (P_{CO} = 0.5 mmHg), CO will combine with half the available hemoglobin molecules and reduce the oxygen-carrying capacity of the blood by 50%. Elevated blood levels of CO cause **carbon monoxide poisoning**, which can cause the lips and oral mucosa to appear bright, cherry-red (the color of hemoglobin with carbon monoxide bound to it). Without prompt treatment, carbon monoxide poisoning is fatal. It is possible to rescue a victim of CO poisoning by administering pure oxygen, which speeds up the separation of carbon monoxide from hemoglobin.

Hypoxia (Refer page 773 of Textbook)

Hypoxia (hī-POK-sē-a; *hypo-* = under) is a deficiency of O_2 at the tissue level. Based on the cause, we can classify hypoxia into four types, as follows:

1. **Hypoxic hypoxia** is caused by a low P_{O_2} in arterial blood as a result of high altitude, airway obstruction, or fluid in the lungs.
2. In **anemic hypoxia**, too little functioning hemoglobin is present in the blood, which reduces O_2 transport to tissue cells. Among the causes are hemorrhage, anemia, and failure of hemoglobin to carry its normal complement of O_2, as in carbon monoxide poisoning.
3. In **ischemic hypoxia** (is-KĒ-mik), blood flow to a tissue is so reduced that too little O_2 is delivered to it, even though P_{O_2} and oxyhemoglobin levels are normal.
4. In **histotoxic hypoxia** (his-tō-TOK-sik), the blood delivers adequate O_2 to tissues, but the tissues are unable to use it properly because of the action of some toxic agent. One cause is cyanide poisoning, in which cyanide blocks an enzyme required for the use of O_2 during ATP synthesis.

Effects of Smoking on the Respiratory System (Refer page 774 of Textbook)

Smoking may cause a person to become easily "winded" during even moderate exercise because several factors decrease respiratory efficiency in smokers: (1) Nicotine constricts terminal bronchioles, which decreases airflow into and out of the lungs. (2) Carbon monoxide in smoke binds to hemoglobin and reduces its oxygen-carrying capability. (3) Irritants in smoke cause increased mucus secretion by the mucosa of the bronchial tree and swelling of the mucosal lining, both of which impede airflow into and out of the lungs. (4) Irritants in smoke also inhibit the movement of cilia and destroy cilia in the lining of the respiratory system. Thus, excess mucus and foreign debris are not easily removed, which further adds to the difficulty in breathing. The irritants can also convert the normal respiratory epithelium into stratified squamous epithelium, which lacks cilia and goblet cells. (5) With time, smoking leads to destruction of elastic fibers in the lungs and is the prime cause of emphysema (described in Disorders: Homeostatic Imbalances at the end of the chapter). These changes cause collapse of small bronchioles and trapping of air in alveoli at the end of exhalation. The result is less efficient gas exchange.

DISORDERS: HOMEOSTATIC IMBALANCES

Asthma

Asthma (AZ-ma = panting) is a disorder characterized by chronic airway inflammation, airway hypersensitivity to a variety of stimuli, and airway obstruction. It is at least partially reversible, either spontaneously or with treatment. Asthma affects 3–5% of the U.S. population and is more common in children than in adults. Airway obstruction may be due to smooth muscle spasms in the walls of smaller bronchi and bronchioles, edema of the mucosa of the airways, increased mucus secretion, and/or damage to the epithelium of the airway.

Individuals with asthma typically react to concentrations of agents too low to cause symptoms in people without asthma. Sometimes the trigger is an allergen such as pollen, house dust mites, molds, or a particular food. Other common triggers of asthma attacks are emotional upset, aspirin, sulfiting agents (used in wine and beer and to keep greens fresh in salad bars), exercise, and breathing cold air or cigarette smoke. In the early phase (acute) response, smooth muscle spasm is accompanied by excessive secretion of mucus that may clog the bronchi and bronchioles and worsen the attack. The late phase (chronic) response is characterized by inflammation, fibrosis, edema, and necrosis (death) of bronchial epithelial cells. A host of mediator chemicals, including leukotrienes, prostaglandins, thromboxane, platelet-activating factor, and histamine, take part.

Symptoms include difficult breathing, coughing, wheezing, chest tightness, tachycardia, fatigue, moist skin, and anxiety. An acute attack is treated by giving an inhaled $beta_2$-adrenergic agonist (albuterol) to help relax smooth muscle in the bronchioles and open up the airways. This drug mimics the effect of sympathetic stimulation, that is, it causes bronchodilation. However, long-term therapy of asthma strives to suppress the underlying inflammation. The anti-inflammatory drugs that are used most often are inhaled corticosteroids (glucocorticoids), cromolyn sodium (Intal®), and leukotriene blockers (Accolate®).

Chronic Obstructive Pulmonary Disease

Chronic obstructive pulmonary disease (COPD) is a type of respiratory disorder characterized by chronic and recurrent obstruction of airflow, which increases airway resistance. COPD affects about 30 million Americans and is the fourth leading cause of death behind heart disease, cancer, and cerebrovascular disease. The principal types of COPD are emphysema and chronic bronchitis. In most cases, COPD is preventable because its most common cause is cigarette smoking or breathing secondhand smoke. Other causes include air pollution, pulmonary infection, occupational exposure to dusts and gases, and genetic factors. Because men, on average, have more years of exposure to cigarette smoke than women, they are twice as likely as women to suffer from COPD; still, the incidence of COPD in women has risen sixfold in the past 50 years, a reflection of increased smoking among women.

EMPHYSEMA Emphysema (em-fi-SE-ma = blown up or full of air) is a disorder characterized by destruction of the walls of the alveoli, producing abnormally large air spaces that remain filled with air during exhalation. With less surface area for gas exchange, O_2 diffusion across the damaged respiratory membrane is reduced. Blood O_2 level is somewhat lowered, and any mild exercise that raises the O_2 requirements of the cells leaves the patient breathless. As increasing numbers of alveolar walls are damaged, lung elastic recoil decreases due to loss of elastic fibers, and an increasing amount of air becomes trapped in the lungs at the end of exhalation. Over several years, added exertion during inhalation increases the size of the chest cage, resulting in a "barrel chest."

Emphysema is generally caused by a long-term irritation; cigarette smoke, air pollution, and occupational exposure to industrial dust are the most common irritants. Some destruction of alveolar sacs may be caused by an enzyme imbalance. Treatment consists of cessation of smoking, removal of other environmental irritants, exercise training under careful medical supervision, breathing exercises, use of bronchodilators, and oxygen therapy.

CHRONIC BRONCHITIS Chronic bronchitis is a disorder characterized by excessive secretion of bronchial mucus accompanied by a productive cough (sputum is raised) that lasts for at least 3 months of the year for two successive years. Cigarette smoking is the leading cause of chronic bronchitis. Inhaled irritants lead to chronic inflammation with an increase in the size and number of mucous glands and goblet cells in the airway epithelium. The thickened and excessive mucus produced narrows the airway and impairs ciliary function. Thus, inhaled pathogens become embedded in airway secretions and multiply rapidly. Besides a productive cough, symptoms of chronic bronchitis are shortness of breath, wheezing, cyanosis, and pulmonary hypertension. Treatment for chronic bronchitis is similar to that for emphysema.

Lung Cancer

In the United States, **lung cancer** is the leading cause of cancer death in both males and females, accounting for 160,000 deaths annually. At the time of diagnosis, lung cancer is usually well advanced, with distant metastases present in about 55% of patients, and regional lymph node involvement in an additional 25%. Most people with lung cancer die within a year of the initial diagnosis; the overall survival rate is only 10–15%. Cigarette smoke is the most common cause of lung cancer. Roughly 85% of lung cancer cases are related to smoking, and the disease is 10 to 30 times more common in smokers than nonsmokers. Exposure to secondhand smoke is also associated with lung cancer and heart disease. In the United States, secondhand smoke causes an estimated 4000 deaths a year from lung cancer, and nearly 40,000 deaths a year from heart disease. Other causes of lung cancer are ionizing radiation and inhaled irritants, such as asbestos and radon gas. Emphysema is a common precursor to the development of lung cancer.

The most common type of lung cancer, **bronchogenic carcinoma** (brong′-kō-JEN-ik), starts in the epithelium of the bronchial tubes. Bronchogenic tumors are named based on where they arise. For example, *adenocarcinomas* (ad-en-ō-kar-si-NŌ-mas; *adeno-* = gland) develop in peripheral areas of the lungs from bronchial glands and alveolar cells, *squamous cell carcinomas* develop from the squamous cells in the epithelium of larger bronchial tubes, and *small (oat) cell carcinomas* develop from epithelial cells in primary bronchi near the hilum of the lungs that get their name due to their flat cell shape with little cytoplasm. They tend to involve the mediastinum early on. Depending on the type, bronchogenic tumors may be aggressive, locally invasive, and undergo widespread metastasis. The tumors begin as epithelial lesions that grow to form masses that obstruct the bronchial tubes or invade adjacent lung tissue. Bronchogenic carcinomas metastasize to lymph nodes, the brain, bones, liver, and other organs.

Symptoms of lung cancer are related to the location of the tumor. These may include a chronic cough, spitting blood from the respiratory tract, wheezing, shortness of breath, chest pain, hoarseness, difficulty swallowing, weight loss, anorexia, fatigue, bone pain, confusion, problems with balance, headache, anemia, thrombocytopenia, and jaundice.

Treatment consists of partial or complete surgical removal of a diseased lung (pulmonectomy), radiation therapy, and chemotherapy.

Malignant Mesothelioma

Malignant mesothelioma (mē-zō-thē-lē-OMA) is a rare form of cancer that affects the mesothelium (simple squamous epithelium) of a serous membrane. The most common form of the disease, about 75% of all cases, affects the pleurae of the lungs (*pleural mesothelioma*). The second most common form of the disease affects the peritoneum (*peritoneal mesothelioma*). Other forms of the disease develop in the pericardium (*pericardial mesothelioma*) and the testes (*testicular mesothelioma*). About 2000–3000 cases of malignant mesothelioma are diagnosed each year in the United States, accounting for about 3% of all cancers. The disease is almost entirely caused by asbestos, which has been widely used in insulation, textiles, cement, brake linings, gaskets, roof shingles, and floor products.

The signs and symptoms of malignant mesothelioma may not appear until 20–50 years or more after exposure to asbestos. With respect to pleural mesothelioma, signs and symptoms include chest pain, shortness of breath, pleural effusion, fatigue, anemia, blood in the sputum (fluid) coughed up, wheezing, hoarseness, and unexplained weight loss. Diagnosis is based on a medical history, physical examination, radiographs, CT scans, and biopsy.

There is usually no cure for malignant mesothelioma unless the tumor is found very early and can be completely removed by surgery. However, the prognosis (chance of recovery) is poor since it is typically diagnosed in its later stages after symptoms have appeared. Chemotherapy, radiation therapy, and/or immunotherapy (using the body's immune system) may be used to help decrease symptoms. Sometimes multimodality therapy (combination of therapies) is used.

Pneumonia

Pneumonia (noo-MŌ-ne-a) is an acute infection or inflammation of the alveoli. It is the most common infectious cause of death in the United States, where an estimated 4 million cases occur annually. When certain microbes enter the lungs of susceptible individuals, they release damaging toxins, stimulating inflammation and immune responses that have damaging side effects. The toxins and immune response damage alveoli and bronchial mucous membranes; inflammation and edema cause the alveoli to fill with fluid, interfering with ventilation and gas exchange.

The most common cause of pneumonia is the pneumococcal bacterium *Streptococcus pneumoniae* (strep′-tō-KOK-us noo-MŌ-nē-ī), but other microbes may also cause pneumonia. Those who are most susceptible to pneumonia are the elderly, infants, immunocompromised individuals (AIDS or cancer patients, or those taking immunosuppressive drugs), cigarette smokers, and individuals with an obstructive lung disease. Most cases of pneumonia are preceded by an upper respiratory infection that often is viral. Individuals then develop fever, chills, productive or dry cough, malaise, chest pain, and sometimes dyspnea (difficult breathing) and hemoptysis (spitting blood).

Treatment may involve antibiotics, bronchodilators, oxygen therapy, increased fluid intake, and chest physiotherapy (percussion, vibration, and postural drainage).

Tuberculosis

The bacterium *Mycobacterium tuberculosis* (mī′-kō-bak-TĒR-ē-um) produces an infectious, communicable disease called **tuberculosis (TB)** that most often affects the lungs and the pleurae but may involve other parts of the body. Once the bacteria are inside the lungs, they multiply and cause inflammation, which stimulates neutrophils and macrophages to migrate to the area and engulf the bacteria to prevent their spread. If the immune system is not impaired, the bacteria remain dormant for life, but impaired immunity may enable the bacteria to escape into blood and lymph to infect other organs. In many people, symptoms—fatigue, weight loss, lethargy, anorexia, a low-grade fever, night sweats, cough, dyspnea, chest pain, and hemoptysis—do not develop until the disease is advanced.

During the past several years, the incidence of TB in the United States has risen dramatically. Perhaps the single most important factor related to this increase is the spread of the human immunodeficiency virus (HIV). People infected with HIV are much more likely to develop tuberculosis because their immune systems are impaired. Among the other factors that have contributed to the increased number of cases are homelessness, increased drug abuse, increased immigration from countries with a high prevalence of tuberculosis, increased crowding in housing among the poor, and airborne transmission of tuberculosis in prisons and shelters. In addition, recent outbreaks of tuberculosis involving multi-drug-resistant strains of Mycobacterium tuberculosis have occurred because patients fail to complete their antibiotic and other treatment regimens. TB is treated with the medication isoniazid.

Pulmonary Edema

Pulmonary edema is an abnormal accumulation of fluid in the interstitial spaces and alveoli of the lungs. The edema may arise from increased permeability of the pulmonary capillaries (pulmonary origin) or increased pressure in the pulmonary capillaries (cardiac origin); the latter cause may coincide with congestive heart failure. The most common symptom is dyspnea. Others include wheezing, tachypnea (rapid breathing rate), restlessness, a feeling of suffocation, cyanosis, pallor (paleness), diaphoresis (excessive perspiration), and pulmonary hypertension. Treatment consists of administering oxygen, drugs that dilate the bronchioles and lower blood pressure, diuretics to rid the body of excess fluid, and drugs that correct acid–base imbalance; suctioning of airways; and mechanical ventilation. One of the recent culprits in the development of pulmonary edema was found to be "phen-fen" diet pills.

Sudden Infant Death Syndrome

Sudden infant death syndrome (SIDS) is the sudden, unexpected death of an apparently healthy infant during sleep. It rarely occurs before 2 weeks or after 6 months of age, with the peak incidence between the second and fourth months. SIDS is more common in premature infants, male babies, low-birth-weight babies, babies of drug users or smokers, babies who have stopped breathing and have had to be resuscitated, babies with upper respiratory tract infections, and babies who have had a sibling die of SIDS. African-American and Native American babies are at higher risk. The exact cause of SIDS is unknown. However, it may be due to an abnormality in the mechanisms

that control respiration or low levels of oxygen in the blood. SIDS may also be linked to hypoxia while sleeping in a prone position (on the stomach) and the rebreathing of exhaled air trapped in a depression of a mattress. It is recommended that for the first 6 months infants be placed on their backs for sleeping ("back to sleep").

Severe Acute Respiratory Syndrome

Severe acute respiratory syndrome (SARS) is an example of an emerging infectious disease, that is, a disease that is new or changing. Other examples of emerging infectious diseases are West Nile encephalitis, mad cow disease, and AIDS. SARS first appeared in southern China in late 2002 and has subsequently spread worldwide. It is a respiratory illness caused by a new variety of coronavirus. Symptoms of SARS include fever, malaise, muscle aches, nonproductive (dry) cough, difficulty in breathing, chills, headache, and diarrhea. About 10–20% of patients require mechanical ventilation and in some cases death may result. The disease is primarily spread through person-to-person contact. There is no effective treatment for SARS and the death rate is 5–10%, usually among the elderly and in persons with other medical problems.

MEDICAL TERMINOLOGY

Abdominal thrust maneuver First-aid procedure designed to clear the airways of obstructing objects. It is performed by applying a quick upward thrust between the navel and costal margin that causes sudden elevation of the diaphragm and forceful, rapid expulsion of air in the lungs; this action forces air out the trachea to eject the obstructing object. The abdominal thrust maneuver is also used to expel water from the lungs of near-drowning victims before resuscitation is begun.

Asphyxia (as-FIK-sē-a; *sphyxia* = pulse) Oxygen starvation due to low atmospheric oxygen or interference with ventilation, external respiration, or internal respiration.

Aspiration (as′-pi-RĀ-shun) Inhalation of a foreign substance such as water, food, or a foreign body into the bronchial tree; also, the drawing of a substance in or out by suction.

Black lung disease A condition in which the lungs appear black instead of pink due to inhalation of coal dust over a period of many years. Most often it affects people who work in the coal industry.

Bronchiectasis (brong-kē-EK-ta-sis; *-ektasis* = stretching) A chronic dilation of the bronchi or bronchioles resulting from damage to the bronchial wall, for example, from respiratory infections.

Bronchoscopy (brong-KOS-ko-pē) Visual examination of the bronchi through a **bronchoscope**, an illuminated, flexible tubular instrument that is passed through the mouth (or nose), larynx, and trachea into the bronchi. The examiner can view the interior of the trachea and bronchi to biopsy a tumor, clear an obstructing object or secretions from an airway, take cultures or smears for microscopic examination, stop bleeding, or deliver drugs.

Cheyne–Stokes respiration (CHĀN STŌKS res′-pi-RĀ-shun) A repeated cycle of irregular breathing that begins with shallow breaths that increase in depth and rapidity and then decrease and cease altogether for 15 to 20 seconds. Cheyne–Stokes is normal in infants; it is also often seen just before death from pulmonary, cerebral, cardiac, and kidney disease.

Dyspnea (DISP-nē-a; *dys-* = painful, difficult) Painful or labored breathing.

Epistaxis (ep′-i-STAK-sis) Loss of blood from the nose due to trauma, infection, allergy, malignant growths, or bleeding disorders. It can be arrested by cautery with silver nitrate, electrocautery, or firm packing. Also called **nosebleed**.

Hypoventilation (*hypo-* = below) Slow and shallow breathing.

Mechanical ventilation The use of an automatically cycling device (ventilator or respirator) to assist breathing. A plastic tube is inserted into the nose or mouth and the tube is attached to a device that forces air into the lungs. Exhalation occurs passively due to the elastic recoil of the lungs.

Rales (RĀLS) Sounds sometimes heard in the lungs that resemble bubbling or rattling. Rales are to the lungs what murmurs are to the heart. Different types are due to the presence of an abnormal type or amount of fluid or mucus within the bronchi or alveoli, or to bronchoconstriction that causes turbulent airflow.

Respirator (RES-pi-rā′-tor) An apparatus fitted to a mask over the nose and mouth, or hooked directly to an endotracheal or tracheotomy tube, that is used to assist or support ventilation or to provide nebulized medication to the air passages.

Respiratory failure A condition in which the respiratory system either cannot supply sufficient O_2 to maintain metabolism or cannot eliminate enough CO_2 to prevent respiratory acidosis (a lower-than-normal pH in interstitial fluid).

Rhinitis (rī-NĪ-tis; *rhin-* = nose) Chronic or acute inflammation of the mucous membrane of the nose due to viruses, bacteria, or irritants. Excessive mucus production produces a runny nose, nasal congestion, and postnasal drip.

Sleep apnea (AP-nē-a; *a-* = without; *-pnea* = breath) A disorder in which a person repeatedly stops breathing for 10 or more seconds while sleeping. Most often, it occurs because loss of muscle tone in pharyngeal muscles allows the airway to collapse.

Sputum (SPŪ-tum = to spit) Mucus and other fluids from the air passages that is expectorated (expelled by coughing).

Strep throat Inflammation of the pharynx caused by the bacterium *Streptococcus pyogenes*. It may also involve the tonsils and middle ear.

Tachypnea (tak′-ip-NĒ-a; *tachy-* = rapid; *-pnea* = breath) Rapid breathing rate.

Wheeze (HWēZ) A whistling, squeaking, or musical high-pitched sound during breathing resulting from a partially obstructed airway.

SELF-QUIZ QUESTIONS

Fill in the blanks in the following statements.

1. Oxygen in blood is carried primarily in the form of _____; carbon dioxide is carried as _____, _____, and _____.

2. Write the equation for the chemical reaction that occurs for the transport of carbon dioxide as bicarbonate ions in blood: _____.

Indicate whether the following statements are true or false.

3. The three basic steps of respiration are pulmonary ventilation, external respiration, and cellular respiration.

4. For inhalation to occur, air pressure in the alveoli must be less than atmospheric pressure; for exhalation to occur, air pressure in the alveoli must be greater than atmospheric pressure.

Choose the one best answer to the following questions.

5. What structural changes occur from primary bronchi to terminal bronchioles? (1) The mucous membrane changes from pseudostratified ciliated columnar epithelium to nonciliated simple cuboidal epithelium. (2) The number of goblet cells increases. (3) The amount of smooth muscle increases. (4) Incomplete rings of cartilage disappear. (5) The amount of branching decreases.
    (a) 1, 2, 3, 4, and 5   (b) 2, 3, and 4   (c) 1, 3, and 4
    (d) 1, 3, 4, and 5   (e) 1, 2, 3, and 4

6. Which of the following would cause oxygen to dissociate more readily from hemoglobin? (1) low P_{O_2}, (2) an increase in H^+ in blood, (3) hypercapnia, (4) hypothermia, (5) low levels of BPG (2,3-bisphosphoglycerate).
    (a) 1 and 2   (b) 2, 3, and 4   (c) 1, 2, 3, and 5
    (d) 1, 3, and 5   (e) 1, 2, and 3

7. Which of the following statements are *correct*? (1) Normal exhalation during quiet breathing is an active process involving intensive muscle contraction. (2) Passive exhalation results from elastic recoil of the chest wall and lungs. (3) Air flow during breathing is due to a pressure gradient between the lungs and the atmospheric air. (4) During normal breathing, the pressure between the two pleural layers (intrapleural pressure) is always subatmospheric. (5) Surface tension of alveolar fluid facilitates inhalation.
    (a) 1, 2, and 3   (b) 2, 3, and 4   (c) 3, 4, and 5
    (d) 1, 3, and 5   (e) 2, 3, and 5

8. Which of the following factors affect the rate of external respiration? (1) partial pressure differences of the gases, (2) surface area for gas exchange, (3) diffusion distance, (4) solubility and molecular weight of the gases, (5) presence of bisphosphoglycerate (BPG).
    (a) 1, 2, and 3   (b) 2, 4, and 5   (c) 1, 2, 4, and 5
    (d) 1, 2, 3, and 4   (e) 2, 3, 4, and 5

9. The most important factor in determining the percent oxygen saturation of hemoglobin is
    (a) the partial pressure of oxygen.
    (b) acidity.
    (c) the partial pressure of carbon dioxide.
    (d) temperature.
    (e) BPG.

10. Which of the following statements are *true*? (1) Central chemoreceptors are stimulated by changes in P_{CO_2}, H^+, and P_{O_2}. (2) Respiratory rate increases during the initial onset of exercise due to input to the inspiratory area from proprioceptors. (3) When baroreceptors in the lungs are stimulated, the expiratory area is activated. (4) Stimulation of the limbic system can result in excitation of the inspiratory area. (5) Sudden severe pain causes brief apnea, while prolonged somatic pain causes an increase in respiratory rate. (6) The respiratory rate increases during fever.
    (a) 1, 2, 3, and 6   (b) 1, 4, and 5   (c) 1, 2, 4, 5, and 6
    (d) 2, 3, 4, 5, and 6   (e) 2, 4, 5, and 6

11. Place the steps for normal inhalation in order.
    (a) decrease in intrapleural pressure to 754 mmHg
    (b) increase in the size of the thoracic cavity
    (c) flow of air from higher to lower pressure
    (d) outward pull of pleurae, resulting in lung expansion
    (e) stimulation of primary breathing muscles by phrenic and intercostal nerves
    (f) decrease in alveolar pressure to 758 mmHg
    (g) contraction of the diaphragm and external intercostals
    (h) increase in the volume of the pleural cavity

12. Match the following:
    _____ (a) total volume of air inhaled and exhaled each minute
    _____ (b) tidal volume + inspiratory reserve volume + expiratory reserve volume
    _____ (c) additional amount of air inhaled beyond tidal volume when taking a very deep breath
    _____ (d) residual volume + expiratory reserve volume
    _____ (e) amount of air remaining in lungs after expiratory reserve volume is expelled
    _____ (f) tidal volume + inspiratory reserve volume
    _____ (g) vital capacity + residual volume
    _____ (h) volume of air in one breath
    _____ (i) amount of air exhaled in forced exhalation following a normal exhalation
    _____ (j) provides a medical and legal tool for determining if a baby was born dead or died after birth

    (1) tidal volume
    (2) residual volume
    (3) minute ventilation
    (4) expiratory reserve volume
    (5) inspiratory reserve volume
    (6) minimal volume
    (7) inspiratory capacity
    (8) vital capacity
    (9) functional residual capacity
    (10) total lung capacity

13. Match the following:

_____ (a) functions as a passageway for air and food, provides a resonating chamber for speech sounds, and houses the tonsils

_____ (b) site of external respiration

_____ (c) connects the laryngopharynx with the trachea; houses the vocal cords

_____ (d) serous membrane that surrounds the lungs

_____ (e) functions in warming, moistening, and filtering air; receives olfactory stimuli; is a resonating chamber for sound

_____ (f) simple squamous epithelial cells that form a continuous lining of the alveolar wall; sites of gas exchange

_____ (g) forms anterior wall of the larynx

_____ (h) a tubular passageway for air connecting the larynx to the bronchi

_____ (i) secrete alveolar fluid and surfactant

_____ (j) forms inferior wall of larynx; landmark for tracheotomy

_____ (k) prevents food or fluid from entering the airways

_____ (l) air passageways entering the lungs

_____ (m) ridge covered by a sensitive mucous membrane; irritation triggers cough reflex

(1) nose
(2) pharynx
(3) larynx
(4) epiglottis
(5) trachea
(6) bronchi
(7) carina
(8) cricoid cartilage
(9) pleura
(10) thyroid cartilage
(11) alveoli
(12) type I alveolar cells
(13) type II alveolar cells

14. Match the following:

_____ (a) a deficiency of oxygen at the tissue level

_____ (b) above-normal partial pressure of carbon dioxide

_____ (c) normal quiet breathing

_____ (d) deep, abdominal breathing

_____ (e) the ease with which the lungs and thoracic wall can be expanded

_____ (f) hypoxia-induced vasoconstriction to divert pulmonary blood from poorly ventilated to well-ventilated regions of the lungs

_____ (g) absence of breathing

_____ (h) rapid and deep breathing

_____ (i) shallow, chest breathing

(1) eupnea
(2) apnea
(3) hyperventilation
(4) costal breathing
(5) diaphragmatic breathing
(6) compliance
(7) hypoxia
(8) hypercapnia
(9) ventilation– perfusion coupling

15. Match the following:

_____ (a) prevents excessive inflation of the lungs

_____ (b) the lower the amount of oxyhemoglobin, the higher the carbon dioxide–carrying capacity of the blood

_____ (c) controls the basic rhythm of respiration

_____ (d) active during normal inhalation; sends nerve impulses to external intercostals and diaphragm

_____ (e) sends stimulatory impulses to the inspiratory area that activate it and prolong inhalation

_____ (f) as acidity increases, the affinity of hemoglobin for oxygen decreases and oxygen dissociates more readily from hemoglobin; shifts oxygen-dissociation curve to the right

_____ (g) active during forceful exhalation

_____ (h) pressure of a gas in a closed container is inversely proportional to the volume of the container

_____ (i) transmits inhibitory impulses to turn off the inspiratory area before the lungs become too full of air

_____ (j) the quantity of a gas that dissolves in a liquid is proportional to the partial pressure of the gas and its solubility

_____ (k) relates to the partial pressure of a gas in a mixture of gases whereby each gas in a mixture exerts its own pressure as if all the other gases were not present

(1) Bohr effect
(2) Dalton's law
(3) medullary rhythmicity area
(4) inspiratory area
(5) expiratory area
(6) apneustic area
(7) pneumotaxic area
(8) Henry's law
(9) inflation (Hering–Breuer) reflex
(10) Boyle's law
(11) Haldane effect

24 | THE DIGESTIVE SYSTEM

CHAPTER REVIEW

Introduction

1. The breaking down of larger food molecules into smaller molecules is called digestion.
2. The organs involved in the breakdown of food are collectively known as the digestive system.

24.1 Overview of the Digestive System

1. The digestive system is composed of two main groups of organs: the gastrointestinal (GI) tract and accessory digestive organs.
2. The GI tract is a continuous tube extending from the mouth to the anus.
3. The accessory digestive organs include the teeth, tongue, salivary glands, liver, gallbladder, and pancreas.
4. Digestion includes six basic processes: ingestion, secretion, mixing and propulsion, mechanical and chemical digestion, absorption, and defecation.
5. Mechanical digestion consists of mastication and movements of the gastrointestinal tract that aid chemical digestion.
6. Chemical digestion is a series of hydrolysis reactions that break down large carbohydrates, lipids, proteins, and nucleic acids in foods into smaller molecules that are usable by body cells.

24.2 Layers of the GI Tract

1. The basic arrangement of layers in most of the gastrointestinal tract, from deep to superficial, is the mucosa, submucosa, muscularis, and serosa.
2. Associated with the lamina propria of the mucosa are extensive patches of lymphatic tissue called mucosa-associated lymphoid tissue (MALT).

24.3 Neural Innervation of the GI Tract

1. The gastrointestinal tract is regulated by an intrinsic set of nerves known as the enteric nervous system (ENS) and by an extrinsic set of nerves that are part of the autonomic nervous system (ANS).
2. The ENS consists of neurons arranged into two plexuses: the myenteric plexus and the submucosal plexus.
3. The myenteric plexus, which is located between the longitudinal and circular smooth muscle layers of the muscularis, regulates GI tract motility.
4. The submucosal plexus, which is located in the submucosa, regulates GI secretion.

5. Although the neurons of the ENS can function independently, they are subject to regulation by the neurons of the ANS.
6. Parasympathetic fibers of the vagus (X) nerves and pelvic splanchnic nerves increase GI tract secretion and motility by increasing the activity of ENS neurons.
7. Sympathetic fibers from the thoracic and upper lumbar regions of the spinal cord decrease GI tract secretion and motility by inhibiting ENS neurons.

24.4 Peritoneum

1. The peritoneum is the largest serous membrane of the body; it lines the wall of the abdominal cavity and covers some abdominal organs.
2. Folds of the peritoneum include the mesentery, mesocolon, falciform ligament, lesser omentum, and greater omentum.

24.5 Mouth

1. The mouth is formed by the cheeks, hard and soft palates, lips, and tongue.
2. The vestibule is the space bounded externally by the cheeks and lips and internally by the teeth and gums.
3. The oral cavity proper extends from the vestibule to the fauces.
4. The tongue, together with its associated muscles, forms the floor of the oral cavity. It is composed of skeletal muscle covered with mucous membrane. The upper surface and sides of the tongue are covered with papillae, some of which contain taste buds. Glands in the tongue secrete lingual lipase, which digests triglycerides into fatty acids and diglycerides once in the acid environment of the stomach.
5. The major portion of saliva is secreted by the major salivary glands, which lie outside the mouth and pour their contents into ducts that empty into the oral cavity. There are three pairs of major salivary glands: parotid, submandibular, and sublingual glands.
6. Saliva lubricates food and starts the chemical digestion of carbohydrates. Salivation is controlled by the nervous system.
7. The teeth (dentes) project into the mouth and are adapted for mechanical digestion.
8. A typical tooth consists of three principal regions: crown, root, and neck. Teeth are composed primarily of dentin and are covered by enamel, the hardest substance in the body. There are two dentitions: deciduous and permanent.

9. Through mastication, food is mixed with saliva and shaped into a soft, flexible mass called a bolus. Salivary amylase then begins the digestion of starches, and lingual lipase acts on triglycerides.

24.6 Pharynx

1. The pharynx is a funnel-shaped tube that extends from the internal nares to the esophagus posteriorly and to the larynx anteriorly.

2. The pharynx has both respiratory and digestive functions.

24.7 Esophagus

1. The esophagus is a collapsible, muscular tube that connects the pharynx to the stomach.

2. It contains an upper and a lower esophageal sphincter.

24.8 Deglutition

1. Deglutition, or swallowing, moves a bolus from the mouth to the stomach.

2. Swallowing consists of voluntary, pharyngeal (involuntary), and esophageal (involuntary) stages.

24.9 Stomach

1. The stomach connects the esophagus to the duodenum.

2. The principal anatomical regions of the stomach are the cardia, fundus, body, and pylorus.

3. Adaptations of the stomach for digestion include rugae; glands that produce mucus, hydrochloric acid, pepsin, gastric lipase, and intrinsic factor; and a three-layered muscularis.

4. Mechanical digestion consists of propulsion and retropulsion.

5. Chemical digestion consists mostly of the conversion of proteins into peptides by pepsin.

6. The stomach wall is impermeable to most substances.

7. Among the substances the stomach can absorb are water, certain ions, drugs, and alcohol.

24.10 Pancreas

1. The pancreas consists of a head, a body, and a tail and is connected to the duodenum via the pancreatic duct and accessory duct.

2. Endocrine pancreatic islets secrete hormones, and exocrine acini secrete pancreatic juice.

3. Pancreatic juice contains enzymes that digest starch (pancreatic amylase), proteins (trypsin, chymotrypsin, carboxypeptidase, and elastase), triglycerides (pancreatic lipase), and nucleic acids (ribonuclease and deoxyribonuclease).

24.11 Liver and Gallbladder

1. The liver has left and right lobes; the left lobe includes a quadrate lobe and a caudate lobe. The gallbladder is a sac located in a depression on the posterior surface of the liver that stores and concentrates bile.

2. The lobes of the liver are made up of lobules that contain hepatocytes (liver cells), sinusoids, stellate reticuloendothelial (Kupffer) cells, and a central vein.

3. Hepatocytes produce bile that is carried by a duct system to the gallbladder for concentration and temporary storage.

4. Bile's contribution to digestion is the emulsification of dietary lipids.

5. The liver also functions in carbohydrate, lipid, and protein metabolism; processing of drugs and hormones; excretion of bilirubin; synthesis of bile salts; storage of vitamins and minerals; phagocytosis; and activation of vitamin D.

24.12 Small Intestine

1. The small intestine extends from the pyloric sphincter to the ileocecal sphincter. It is divided into duodenum, jejunum, and ileum.

2. Its glands secrete fluid and mucus, and the circular folds, villi, and microvilli of its wall provide a large surface area for digestion and absorption.

3. Brush-border enzymes digest α-dextrins, maltose, sucrose, lactose, peptides, and nucleotides at the surface of mucosal epithelial cells.

4. Pancreatic and intestinal brush-border enzymes break down starches into maltose, maltotriose, and α-dextrins (pancreatic amylase), α-dextrins into glucose (α-dextrinase), maltose to glucose (maltase), sucrose to glucose and fructose (sucrase), lactose to glucose and galactose (lactase), and proteins into peptides (trypsin, chymotrypsin, and elastase). Also, enzymes break off amino acids at the carboxyl ends of peptides (carboxypeptidases) and break off amino acids at the amino ends of peptides (aminopeptidases). Finally, enzymes split dipeptides into amino acids (dipeptidases), triglycerides to fatty acids and monoglycerides (lipases), and nucleotides to pentoses and nitrogenous bases (nucleosidases and phosphatases).

5. Mechanical digestion in the small intestine involves segmentation and migrating motility complexes.

6. Absorption occurs via diffusion, facilitated diffusion, osmosis, and active transport; most absorption occurs in the small intestine.

7. Monosaccharides, amino acids, and short-chain fatty acids pass into the blood capillaries.

8. Long-chain fatty acids and monoglycerides are absorbed from micelles, resynthesized to triglycerides, and formed into chylomicrons.

9. Chylomicrons move into lymph in the lacteal of a villus.

10. The small intestine also absorbs electrolytes, vitamins, and water.

24.13 Large Intestine

1. The large intestine extends from the ileocecal sphincter to the anus.

2. Its regions include the cecum, colon, rectum, and anal canal.

3. The mucosa contains many goblet cells, and the muscularis consists of teniae coli and haustra.

4. Mechanical movements of the large intestine include haustral churning, peristalsis, and mass peristalsis.

5. The last stages of chemical digestion occur in the large intestine through bacterial action. Substances are further broken down, and some vitamins are synthesized.

6. The large intestine absorbs water, ions, and vitamins.

7. Feces consist of water, inorganic salts, epithelial cells, bacteria, and undigested foods.

8. The elimination of feces from the rectum is called defecation.

9. Defecation is a reflex action aided by voluntary contractions of the diaphragm and abdominal muscles and relaxation of the external anal sphincter.

24.14 Phases of Digestion

1. Digestive activities occur in three overlapping phases: cephalic, gastric, and intestinal.
2. During the cephalic phase of digestion, salivary glands secrete saliva and gastric glands secrete gastric juice in order to prepare the mouth and stomach for food that is about to be eaten.
3. The presence of food in the stomach causes the gastric phase of digestion, which promotes gastric juice secretion and gastric motility.
4. During the intestinal phase of digestion, food is digested in the small intestine. In addition, gastric motility and gastric secretion decrease in order to slow the exit of chyme from the stomach, which prevents the small intestine from being overloaded with more chyme than it can handle.
5. The activities that occur during the various phases of digestion are coordinated by neural pathways and by hormones. **Table 24.8** summarizes the major hormones that control digestion.

24.15 Development of the Digestive System

1. The endoderm of the primitive gut forms the epithelium and glands of most of the GI tract.
2. The mesoderm of the primitive gut forms the smooth muscle and connective tissue of the GI tract.

24.16 Aging and the Digestive System

1. General changes include decreased secretory mechanisms, decreased motility, and loss of tone.
2. Specific changes may include loss of taste, pyorrhea, hernias, peptic ulcer disease, constipation, hemorrhoids, and diverticular diseases.

CLINICAL CONNECTIONS

Peritonitis (Refer page 784 of Textbook)

A common cause of **peritonitis** (per′-i-tō-NĪ-tis), an acute inflammation of the peritoneum, is contamination of the peritoneum by infectious microbes, which can result from accidental or surgical wounds in the abdominal wall, or from perforation or rupture of microbe–containing abdominal organs. If, for example, bacteria gain access to the peritoneal cavity through an intestinal perforation or rupture of the appendix, they can produce an acute, life-threatening form of peritonitis. A less serious (but still painful) form of peritonitis can result from the rubbing together of inflamed peritoneal surfaces. The increased risk of peritonitis is of particular concern to those who rely on peritoneal dialysis, a procedure in which the peritoneum is used to filter the blood when the kidneys do not function properly (see Clinical Connection: Dialysis in Chapter 26).

Mumps (Refer page 788 of Textbook)

Although any of the salivary glands may be the target of a nasopharyngeal infection, the mumps virus (*paramyxovirus*) typically attacks the parotid glands. **Mumps** is an inflammation and enlargement of the parotid glands accompanied by moderate fever, malaise (general discomfort), and extreme pain in the throat, especially when swallowing sour foods or acidic juices. Swelling occurs on one or both sides of the face, just anterior to the ramus of the mandible. In about 30% of males past puberty, the testes may also become inflamed; sterility rarely occurs because testicular involvement is usually unilateral (one testis only). Since a vaccine became available for mumps in 1967, the incidence of the disease has declined dramatically.

Root Canal Therapy (Refer page 788 of Textbook)

Root canal therapy is a multistep procedure in which all traces of pulp tissue are removed from the pulp cavity and root canals of a badly diseased tooth. After a hole is made in the tooth, the root canals are filed out and irrigated to remove bacteria. Then, the canals are treated with medication and sealed tightly. The damaged crown is then repaired.

Gastroesophageal Reflux Disease (Refer page 794 of Textbook)

If the lower esophageal sphincter fails to close adequately after food has entered the stomach, the stomach contents can reflux (back up) into the inferior portion of the esophagus. This condition is known as **gastroesophageal**

reflux disease (GERD) (gas′-trō-e-sof-a-JĒ-al). Hydrochloric acid (HCl) from the stomach contents can irritate the esophageal wall, resulting in a burning sensation that is called **heartburn** because it is experienced in a region very near the heart; it is unrelated to any cardiac problem. Drinking alcohol and smoking can cause the sphincter to relax, worsening the problem. The symptoms of GERD often can be controlled by avoiding foods that strongly stimulate stomach acid secretion (coffee, chocolate, tomatoes, fatty foods, orange juice, peppermint, spearmint, and onions). Other acid-reducing strategies include taking over-the-counter histamine-2 (H_2) blockers such as Tagamet HB® or Pepcid AC® 30 to 60 minutes before eating to block acid secretion, and neutralizing acid that has already been secreted with antacids such as Tums® or Maalox®. Symptoms are less likely to occur if food is eaten in smaller amounts and if the person does not lie down immediately after a meal. GERD may be associated with cancer of the esophagus.

Pylorospasm and Pyloric Stenosis (Refer page 795 of Textbook)

Two abnormalities of the pyloric sphincter can occur in infants. In **pylorospasm** (pī-LOR-ō-spazm), the smooth muscle fibers of the sphincter fail to relax normally, so food does not pass easily from the stomach to the small intestine, the stomach becomes overly full, and the infant vomits often to relieve the pressure. Pylorospasm is treated by drugs that relax the muscle fibers of the pyloric sphincter. **Pyloric stenosis** (ste-NŌ-sis) is a narrowing of the pyloric sphincter that must be corrected surgically. The hallmark symptom is *projectile vomiting*—the spraying of liquid vomitus some distance from the infant.

Vomiting (Refer page 798 of Textbook)

Vomiting or *emesis* is the forcible expulsion of the contents of the upper GI tract (stomach and sometimes duodenum) through the mouth. The strongest stimuli for vomiting are irritation and distension of the stomach; other stimuli include unpleasant sights, general anesthesia, dizziness, and certain drugs such as morphine and derivatives of digitalis. Nerve impulses are transmitted to the vomiting center in the medulla oblongata, and returning impulses propagate to the upper GI tract organs, diaphragm, and abdominal muscles. Vomiting involves squeezing the stomach between the diaphragm and abdominal muscles and expelling the contents through open esophageal sphincters. Prolonged vomiting, especially in infants and elderly people, can be serious because the loss of acidic gastric juice can lead to alkalosis (higher than normal blood pH), dehydration, and damage to the esophagus and teeth.

Pancreatitis and Pancreatic Cancer (Refer page 801 of Textbook)

Inflammation of the pancreas, as may occur in association with alcohol abuse or chronic gallstones, is called **pancreatitis** (pan'-krē-a-TĪ-tis). In a more severe condition known as **acute pancreatitis**, which is associated with heavy alcohol intake or biliary tract obstruction, the pancreatic cells may release either trypsin instead of trypsinogen or insufficient amounts of trypsin inhibitor, and the trypsin begins to digest the pancreatic cells. Patients with acute pancreatitis usually respond to treatment, but recurrent attacks are the rule. In some people pancreatitis is idiopathic, meaning that the cause is unknown. Other causes of pancreatitis include cystic fibrosis, high levels of calcium in the blood (hypercalcemia), high levels of blood fats (hyperlipidemia or hypertriglyceridemia), some drugs, and certain autoimmune conditions. However, in roughly 70% of adults with pancreatitis, the cause is alcoholism. Often the first episode happens between ages 30 and 40.

Pancreatic cancer usually affects people over 50 years of age and occurs more frequently in males. Typically, there are few symptoms until the disorder reaches an advanced stage and often not until it has metastasized to other parts of the body such as the lymph nodes, liver, or lungs. The disease is nearly always fatal and is the fourth most common cause of death from cancer in the United States. Pancreatic cancer has been linked to fatty foods, high alcohol consumption, genetic factors, smoking, and chronic pancreatitis.

Jaundice (Refer page 803 of Textbook)

Jaundice (JAWN-dis = yellowed) is a yellowish coloration of the sclerae (whites of the eyes), skin, and mucous membranes due to a buildup of a yellow compound called bilirubin. After bilirubin is formed from the breakdown of the heme pigment in aged red blood cells, it is transported to the liver, where it is processed and eventually excreted into bile. The three main categories of jaundice are (1) *prehepatic jaundice*, due to excess production of bilirubin; (2) *hepatic jaundice*, due to congenital liver disease, cirrhosis of the liver, or hepatitis; and (3) *extrahepatic jaundice*, due to blockage of bile drainage by gallstones or cancer of the bowel or the pancreas.

Because the liver of a newborn functions poorly for the first week or so, many babies experience a mild form of jaundice called *neonatal (physiological) jaundice* that disappears as the liver matures. Usually, it is treated by exposing the infant to blue light, which converts bilirubin into substances the kidneys can excrete.

Liver Function Tests (Refer page 804 of Textbook)

Liver function tests are blood tests designed to determine the presence of certain chemicals released by liver cells. These include albumin globulinase, alanine aminotransferase (ALT), aspartate aminotransferase (AST), alkaline phosphatase (ALP), gamma-glutamyl-transpeptidase (GGT), and bilirubin. The tests are used to evaluate and monitor liver disease or damage. Common causes of elevated liver enzymes include nonsteroidal anti-inflammatory drugs, cholesterol-lowering medications, some antibiotics, alcohol, diabetes, infections (viral hepatitis and mononucleosis), gallstones, tumors of the liver, and excessive use of herbal supplements such as kava, comfrey, pennyroyal, dandelion root, skullcap, and ephedra.

Gallstones (Refer page 804 of Textbook)

If bile contains either insufficient bile salts or lecithin or excessive cholesterol, the cholesterol may crystallize to form **gallstones**. As they grow in size and number, gallstones may cause minimal, intermittent, or complete obstruction to the flow of bile from the gallbladder into the duodenum. Treatment consists of using gallstone-dissolving drugs, lithotripsy (shock-wave therapy), or surgery. For people with a history of gallstones or for whom drugs or lithotripsy are not options, **cholecystectomy** (kō'-lē-sis-TEK-tō-mē)—the removal of the gallbladder and its contents—is necessary. More than half a million cholecystectomies are performed each year in the United States. To prevent side effects resulting from a loss of the gallbladder, patients should make lifestyle and dietary changes, including the following: (1) limiting the intake of saturated fat; (2) avoiding the consumption of alcoholic beverages; (3) eating smaller amounts of food during a meal and eating five to six smaller meals per day instead of two to three larger meals; and (4) taking vitamin and mineral supplements.

Lactose Intolerance (Refer page 810 of Textbook)

In some people the absorptive cells of the small intestine fail to produce enough lactase, which, as you just learned, is essential for the digestion of lactose. This results in a condition called **lactose intolerance**, in which undigested lactose in chyme causes fluid to be retained in the feces; bacterial fermentation of the undigested lactose results in the production of gases. Symptoms of lactose intolerance include diarrhea, gas, bloating, and abdominal cramps after consumption of milk and other dairy products. The symptoms can be relatively minor or serious enough to require medical attention. The *hydrogen breath test* is often used to aid in diagnosis of lactose intolerance. Very little hydrogen can be detected in the breath of a normal person, but hydrogen is among the gases produced when undigested lactose in the colon is fermented by bacteria. The hydrogen is absorbed from the intestines and carried through the bloodstream to the lungs, where it is exhaled. Persons with lactose intolerance should select a diet that restricts lactose (but not calcium) and take dietary supplements to aid in the digestion of lactose.

Absorption of Alcohol (Refer page 813 of Textbook)

The intoxicating and incapacitating effects of alcohol depend on the blood alcohol level. Because it is lipid-soluble, alcohol begins to be absorbed in the stomach. However, the surface area available for absorption is much greater in the small intestine than in the stomach, so when alcohol passes into the duodenum, it is absorbed more rapidly. Thus, the longer the alcohol remains in the stomach, the more slowly blood alcohol level rises. Because fatty acids in chyme slow gastric emptying, blood alcohol level will rise more slowly when fat-rich foods, such as pizza, hamburgers, or nachos, are consumed with alcoholic beverages. Also, the enzyme alcohol dehydrogenase, which is present in gastric mucosa cells, breaks down some of the alcohol to acetaldehyde, which is not intoxicating. When the rate of gastric emptying is slower, proportionally more alcohol will be absorbed and converted to acetaldehyde in the stomach, and thus less alcohol will reach the bloodstream. Given identical consumption of alcohol, females often develop higher blood alcohol levels (and therefore experience greater intoxication) than males of comparable size because the activity of gastric alcohol dehydrogenase is up to 60% lower in females than in males. Asian males may also have lower levels of this gastric enzyme.

Appendicitis (Refer page 814 of Textbook)

Inflammation of the appendix, termed **appendicitis**, is preceded by obstruction of the lumen of the appendix by chyme, inflammation, a foreign body, a carcinoma of the cecum, stenosis, or kinking of the organ. It is characterized by high fever, elevated white blood cell count, and a neutrophil count higher than 75%. The infection that follows may result in edema and ischemia and may progress to gangrene and perforation within 24 hours. Typically, appendicitis begins with referred pain in the umbilical region of the abdomen, followed by anorexia (loss of appetite for food), nausea, and vomiting. After several hours the pain localizes in the right lower quadrant (RLQ) and is continuous, dull or severe, and intensified by coughing, sneezing, or body movements. Early appendectomy (removal of the appendix) is recommended because it is safer to operate than to risk rupture, peritonitis, and gangrene. Although it required major abdominal surgery in the past, today appendectomies are usually performed laparoscopically.

Polyps in the Colon (Refer page 816 of Textbook)

Polyps in the colon are generally slow-developing benign growths that arise from the mucosa of the large intestine. Often, they do not cause symptoms. If symptoms do occur, they include diarrhea, blood in the feces, and mucus discharged from the anus. The polyps are removed by colonoscopy or surgery because some of them may become cancerous.

Occult Blood (Refer page 818 of Textbook)

The term **occult blood** refers to blood that is hidden; it is not detectable by the human eye. The main diagnostic value of occult blood testing is to screen for colorectal cancer. Two substances often examined for occult blood are feces and urine. Several types of products are available for at-home testing for hidden blood in feces. The tests are based on color changes when reagents are added to feces. The presence of occult blood in urine may be detected at home by using dip-and-read reagent strips.

Dietary Fiber (Refer page 819 of Textbook)

Dietary fiber consists of indigestible plant carbohydrates—such as cellulose, lignin, and pectin—found in fruits, vegetables, grains, and beans. **Insoluble fiber**, which does not dissolve in water, includes the woody or structural parts of plants such as the skins of fruits and vegetables and the bran coating around wheat and corn kernels. Insoluble fiber passes through the GI tract largely unchanged but speeds up the passage of material through the tract. **Soluble fiber**, which does dissolve in water, forms a gel that slows the passage of material through the tract. It is found in abundance in beans, oats, barley, broccoli, prunes, apples, and citrus fruits.

People who choose a fiber-rich diet may reduce their risk of developing obesity, diabetes, atherosclerosis, gallstones, hemorrhoids, diverticulitis, appendicitis, and colorectal cancer. Soluble fiber also may help lower blood cholesterol. The liver normally converts cholesterol to bile salts, which are released into the small intestine to help fat digestion. Having accomplished their task, the bile salts are reabsorbed by the small intestine and recycled back to the liver. Since soluble fiber binds to bile salts to prevent their reabsorption, the liver makes more bile salts to replace those lost in feces. Thus, the liver uses more cholesterol to make more bile salts and blood cholesterol level is lowered.

DISORDERS: HOMEOSTATIC IMBALANCES

Dental Caries

Dental caries (KĀR-ēz), or tooth decay, involves a gradual demineralization (softening) of the enamel and dentin. If untreated, microorganisms may invade the pulp, causing inflammation and infection, with subsequent death of the pulp and abscess of the alveolar bone surrounding the root's apex, requiring root canal therapy (see Section 24.5).

Dental caries begin when bacteria, acting on sugars, produce acids that demineralize the enamel. **Dextran**, a sticky polysaccharide produced from sucrose, causes the bacteria to stick to the teeth. Masses of bacterial cells, dextran, and other debris adhering to teeth constitute **dental plaque** (PLAK). Saliva cannot reach the tooth surface to buffer the acid because the plaque covers the teeth. Brushing the teeth after eating removes the plaque from flat surfaces before the bacteria can produce acids. Dentists also recommend that the plaque between the teeth be removed every 24 hours with dental floss.

Periodontal Disease

Periodontal disease is a collective term for a variety of conditions characterized by inflammation and degeneration of the gingivae, alveolar bone, periodontal ligament, and cementum. In one such condition, called **pyorrhea**, initial symptoms include enlargement and inflammation of the soft tissue and bleeding of the gums. Without treatment, the soft tissue may deteriorate and the alveolar bone may be resorbed, causing loosening of the teeth and recession of the gums. Periodontal diseases are often caused by poor oral hygiene; by local irritants, such as bacteria, impacted food, and cigarette smoke; or by a poor "bite."

Peptic Ulcer Disease

In the United States, 5–10% of the population develops **peptic ulcer disease (PUD)**. An **ulcer** is a craterlike lesion in a membrane; ulcers that develop in areas of the GI tract exposed to acidic gastric juice are called **peptic ulcers**. The most common complication of peptic ulcers is bleeding, which can lead to anemia if enough blood is lost. In acute cases, peptic ulcers can lead to shock and death. Three distinct causes of PUD are recognized: (1) the bacterium *Helicobacter pylori* (hel-i-kō-BAK-ter pī-LŌ-rē); (2) nonsteroidal anti-inflammatory drugs (NSAIDs) such as aspirin; and (3) hypersecretion of HCl, as occurs in Zollinger–Ellison syndrome (ZOL-in-jer EL-i-son), a gastrin-producing tumor, usually of the pancreas.

Helicobacter pylori (previously named *Campylobacter pylori*) is the most frequent cause of PUD. The bacterium produces an enzyme called urease, which splits urea into ammonia and carbon dioxide. While shielding the bacterium from the acidity of the stomach, the ammonia also damages the protective mucous layer of the stomach and the underlying gastric cells. The microbe also produces catalase, an enzyme that may protect *H. pylori* from phagocytosis by neutrophils, plus several adhesion proteins that allow the bacterium to attach itself to gastric cells.

Several therapeutic approaches are helpful in the treatment of PUD. Cigarette smoke, alcohol, caffeine, and NSAIDs should be avoided because they can impair mucosal defensive mechanisms, which increases mucosal susceptibility to the damaging effects of HCl. In cases associated with *H. pylori*, treatment with an antibiotic drug often resolves the problem. Oral antacids such as Tums® or Maalox® can help temporarily by buffering gastric acid. When hypersecretion of HCl is the cause of PUD, H_2 blockers (such as Tagamet®) or proton pump inhibitors such as omeprazole (Prilosec®), which block secretion of H^+ from parietal cells, may be used.

Diverticular Disease

In **diverticular disease** (dī'-ver-TIK-ū-lar), saclike outpouchings of the wall of the colon, termed **diverticula**, occur in places where the muscularis has weakened and may become inflamed. Development of diverticula is known as **diverticulosis** (dī-ver-tik'-ū-LŌ-sis). Many people who develop diverticulosis have no symptoms and experience no complications. Of those people known to have diverticulosis, 10–25% eventually develop an inflammation known as **diverticulitis** (dī'-ver-tik-ū-Lī-tis). This condition may be characterized by pain, either constipation or increased frequency of defecation, nausea, vomiting, and low-grade fever. Because diets low in fiber contribute to development of diverticulitis, patients who change to high-fiber diets show marked relief of symptoms. In severe cases, affected portions of the colon may require surgical removal. If diverticula rupture, the release of bacteria into the abdominal cavity can cause peritonitis.

Colorectal Cancer

Colorectal cancer is among the deadliest of malignancies, ranking second to lung cancer in males and third after lung cancer and breast cancer in females. Genetics plays a very important role; an inherited predisposition contributes to more than half of all cases of colorectal cancer. Intake of alcohol and diets high in animal fat and protein are associated with increased risk of colorectal cancer; dietary fiber, retinoids, calcium, and selenium may be protective. Signs and symptoms of colorectal cancer include diarrhea, constipation, cramping, abdominal pain, and rectal bleeding, either visible or occult (hidden in feces). Precancerous growths on the mucosal surface, called **polyps**, also increase the risk of developing colorectal cancer. Screening for colorectal cancer includes testing for blood in the feces, digital rectal examination, sigmoidoscopy, colonoscopy, and barium enema. Tumors may be removed endoscopically or surgically.

Hepatitis

Hepatitis is an inflammation of the liver that can be caused by viruses, drugs, and chemicals, including alcohol. Clinically, several types of viral hepatitis are recognized.

Hepatitis A (infectious hepatitis) is caused by the hepatitis A virus (HAV) and is spread via fecal contamination of objects such as food, clothing, toys, and eating utensils (fecal–oral route). It is generally a mild disease of children and young adults characterized by loss of appetite, malaise, nausea, diarrhea, fever, and chills. Eventually, jaundice appears. This type of hepatitis does not cause lasting liver damage. Most people recover in 4 to 6 weeks. A vaccine is available.

Hepatitis B is caused by the hepatitis B virus (HBV) and is spread primarily by sexual contact and contaminated syringes and transfusion equipment. It can also be spread via saliva and tears. Hepatitis B virus can be present for years or even a lifetime, and it can produce cirrhosis and possibly cancer of the liver. Individuals who harbor the active hepatitis B virus also become carriers. A vaccine is available.

Hepatitis C, caused by the hepatitis C virus (HCV), is clinically similar to hepatitis B. Hepatitis C can cause cirrhosis and possibly liver cancer. In developed nations, donated blood is screened for the presence of hepatitis B and C.

Hepatitis D is caused by the hepatitis D virus (HDV). It is transmitted like hepatitis B, and in fact a person must have been co-infected with hepatitis B before contracting hepatitis D. Hepatitis D results in severe liver damage and has a higher fatality rate than infection with hepatitis B virus alone. HBV vaccine is protective.

Hepatitis E is caused by the hepatitis E virus and is spread like hepatitis A. Although it does not cause chronic liver disease, hepatitis E virus has a very high mortality rate among pregnant women.

MEDICAL TERMINOLOGY

Achalasia (ak′-a-LĀ-zē-a; *a-* = without; *-chalasis* = relaxation) A condition caused by malfunction of the myenteric plexus in which the lower esophageal sphincter fails to relax normally as food approaches. A whole meal may become lodged in the esophagus and enter the stomach very slowly. Distension of the esophagus results in chest pain that is often confused with pain originating from the heart.

Bariatric surgery (bar′-ē-AT-rik; *baros-* = weight; *-iatreia* = medical treatment) A surgical procedure that limits the amount of food that can be ingested and absorbed in order to bring about a significant weight loss in obese individuals. The most commonly performed type of bariatric surgery is called *gastric bypass surgery*. In one variation of this procedure, the stomach is reduced in size by making a small pouch at the top of the stomach about the size of a walnut. The pouch, which is only 5–10% of the stomach, is sealed off from the rest of the stomach using surgical staples or a plastic band. The pouch is connected to the jejunum of the small intestine, thus bypassing the rest of the stomach and the duodenum. The result is that smaller amounts of food are ingested and fewer nutrients are absorbed in the small intestine. This leads to weight loss.

Barrett's esophagus A pathological change in the epithelium of the esophagus from nonkeratinized stratified squamous epithelium to columnar epithelium so that the lining resembles that of the stomach or small intestine due to long-term exposure of the esophagus to stomach acid; increases the risk of developing cancer of the esophagus.

Borborygmus (bor′-bō-RIG-mus) A rumbling noise caused by the propulsion of gas through the intestines.

Bulimia (bū-LĒM-ē-a; *bu-* = ox; *limia* = hunger or *binge–purge syndrome*) A disorder that typically affects young, single, middle-class white females, characterized by overeating at least twice a week followed by purging by self-induced vomiting, strict dieting or fasting, vigorous exercise, or use of laxatives or diuretics; it occurs in response to fears of being overweight or to stress, depression, and physiological disorders such as hypothalamic tumors.

Canker sore (KANG-ker) Painful ulcer on the mucous membrane of the mouth that affects females more often than males, usually between ages 10 and 40; may be an autoimmune reaction or a food allergy.

Cirrhosis (si-RŌ-sis) Distorted or scarred liver as a result of chronic inflammation due to hepatitis, chemicals that destroy hepatocytes, parasites that infect the liver, or alcoholism; the hepatocytes are replaced by fibrous or adipose connective tissue. Symptoms include jaundice, edema in the legs, uncontrolled bleeding, and increased sensitivity to drugs.

Colitis (kō-LĪ-tis) Inflammation of the mucosa of the colon and rectum in which absorption of water and salts is reduced, producing watery, bloody feces and, in severe cases, dehydration and salt depletion. Spasms of the irritated muscularis produce cramps. It is thought to be an autoimmune condition.

Colonoscopy (kō-lon-OS-kō-pē; *-skopes* = to view) The visual examination of the lining of the colon using an elongated, flexible, fiber-optic endoscope called a colonoscope. It is used to detect disorders such as polyps, cancer, and diverticulosis; to take tissue samples; and to remove small polyps. Most tumors of the large intestine occur in the rectum.

Colostomy (kō-LOS-tō-mē; *-stomy* = provide an opening) The diversion of feces through an opening in the colon, creating a surgical "stoma" (artificial opening) that is made in the exterior of the abdominal wall. This opening serves as a substitute anus through which feces are eliminated into a bag worn on the abdomen.

Dysphagia (dis-FĀ-jē-a; *dys-* = abnormal; *-phagia* = to eat) Difficulty in swallowing that may be caused by inflammation, paralysis, obstruction, or trauma.

Flatus (FLĀ-tus) Air (gas) in the stomach or intestine, usually expelled through the anus. If the gas is expelled through the mouth, it is called **eructation** or *belching* (burping). Flatus may result from gas released during the breakdown of foods in the stomach or from swallowing air or gas-containing substances such as carbonated drinks.

Food poisoning A sudden illness caused by ingesting food or drink contaminated by an infectious microbe (bacterium, virus, or protozoan) or a toxin (poison). The most common cause of food poisoning is the toxin produced by the bacterium *Staphylococcus aureus*. Most types of food poisoning cause diarrhea and/or vomiting, often associated with abdominal pain.

Gastroenteritis (gas′-trō-en-ter-Ī-tis; *gastro-* = stomach; *-enteron-* = intestine; *-itis* = inflammation) Inflammation of the lining of the stomach and intestine (especially the small intestine). It is usually caused by a viral or bacterial infection that may be acquired by contaminated food or water or by people in close contact. Symptoms include diarrhea, vomiting, fever, loss of appetite, cramps, and abdominal discomfort.

Gastroscopy (gas-TROS-kō-pē; *-scopy* = to view with a lighted instrument) Endoscopic examination of the stomach in which the examiner can view the interior of the stomach directly to evaluate an ulcer, tumor, inflammation, or source of bleeding.

Halitosis (hal′-i-TŌ-sis; *halitus-* = breath; *-osis* = condition) A foul odor from the mouth; also called **bad breath**.

Heartburn A burning sensation in a region near the heart due to irritation of the mucosa of the esophagus from hydrochloric acid in stomach contents. It is caused by failure of the lower esophageal sphincter to close properly, so that the stomach contents enter the inferior esophagus. It is not related to any cardiac problem.

Hemorrhoids (HEM-ō-royds; *hemo-* = blood; *-rhoia* = flow) Varicosed (enlarged and inflamed) superior rectal veins. Hemorrhoids develop when the veins are put under pressure and become engorged with blood. If the pressure continues, the wall of the vein stretches. Such a distended vessel oozes blood; bleeding or itching is usually the first sign that a hemorrhoid has developed. Stretching of a vein also favors clot formation, further aggravating swelling and pain. Hemorrhoids may be caused by constipation, which may be brought on by low-fiber diets. Also, repeated straining during defecation forces blood down into the rectal veins, increasing pressure in those veins and possibly causing hemorrhoids. Also called **piles**.

Hernia (HER-nē-a) Protrusion of all or part of an organ through a membrane or cavity wall, usually the abdominal cavity. *Hiatus (diaphragmatic) hernia* is the protrusion of a part of the stomach into the thoracic cavity through the esophageal hiatus of the diaphragm. *Inguinal hernia* is the protrusion of the hernial sac into the inguinal opening; it may contain a portion of the bowel in an advanced stage and may extend into the scrotal compartment in males, causing strangulation of the herniated part.

Inflammatory bowel disease (in-FLAM-a-tō′-rē BOW-el) Inflammation of the gastrointestinal tract that exists in two forms. (1) *Crohn's disease* is an inflammation of any part of the gastrointestinal tract in which the inflammation extends from the mucosa through the submucosa, muscularis, and serosa. (2) *Ulcerative colitis* is an inflammation of the mucosa of the colon and rectum, usually accompanied by rectal bleeding. Curiously, cigarette smoking increases the risk of Crohn's disease but decreases the risk of ulcerative colitis.

Irritable bowel syndrome (IBS) Disease of the entire gastrointestinal tract in which a person reacts to stress by developing symptoms (such as cramping and abdominal pain) associated with alternating patterns of diarrhea and constipation. Excessive amounts of mucus may appear in feces; other symptoms include flatulence, nausea, and loss of appetite. The condition is also known as **irritable colon** or **spastic colitis**.

Malabsorption (mal-ab-SORP-shun; *mal-* = bad) A number of disorders in which nutrients from food are not absorbed properly. It may be due to disorders that result in the inadequate breakdown of food during digestion (due to inadequate digestive enzymes or juices), damage to the lining of the small intestine (from surgery, infections, and drugs like neomycin and alcohol), and impairment of motility. Symptoms may include diarrhea, weight loss, weakness, vitamin deficiencies, and bone demineralization.

Malocclusion (mal′-ō-KLOO-zhun; *mal-* = bad; *-occlusion* = to fit together) Condition in which the surfaces of the maxillary (upper) and mandibular (lower) teeth fit together poorly.

Nausea (NAW-sē-a; *nausia* = seasickness) Discomfort characterized by a loss of appetite and the sensation of impending vomiting. Its causes include local irritation of the gastrointestinal tract, a systemic disease, brain disease or injury, overexertion, or the effects of medication or drug overdosage.

Traveler's diarrhea Infectious disease of the gastrointestinal tract that results in loose, urgent bowel movements, cramping, abdominal pain, malaise, nausea, and occasionally fever and dehydration. It is acquired through ingestion of food or water contaminated with fecal material typically containing bacteria (especially *Escherichia coli*); viruses or protozoan parasites are less common causes.

SELF-QUIZ QUESTIONS

Fill in the blanks in the following statements.

1. The end products of chemical digestion of carbohydrates are _____, of proteins are _____, of lipids are _____ and _____, and of nucleic acids are _____, _____, and _____.

2. List the mechanisms of absorption of materials in the small intestine: _____, _____, _____, and _____.

Indicate whether the following statements are true or false.

3. The soft palate, uvula, and epiglottis prevent swallowed foods and liquids from entering the respiratory passages.

4. The coordinated contractions and relaxations of the muscularis, which propels materials through the GI tract, is known as peristalsis.

Choose the one best answer to the following questions.

5. Which of the following are mismatched?
   (a) chemical digestion: splitting food molecules into simple substances by hydrolysis with the assistance of digestive enzymes
   (b) motility: mechanical processes that break apart ingested food into small molecules
   (c) ingestion: taking foods and liquids into the mouth
   (d) propulsion: movement of food through GI tract due to smooth muscle contraction
   (e) absorption: passage into blood or lymph of ions, fluids and small molecules via the epithelial lining of the GI tract lumen

6. Which of the following are *true* concerning the peritoneum? (1) The kidneys and pancreas are retroperitoneal. (2) The greater omentum is the largest of the peritoneal folds. (3) The lesser omentum binds the large intestine to the posterior abdominal wall. (4) The falciform ligament attaches the liver to the anterior abdominal wall and diaphragm. (5) The mesentery is associated with the jejunum and ileum.
   (a) 1, 2, 3, and 5   (b) 1, 2, and 5   (c) 2 and 5
   (d) 1, 2, 4, and 5   (e) 3, 4, and 5

7. When a surgeon makes an incision in the small intestine, in what order would the physician encounter these structures? (1) epithelium, (2) submucosa, (3) serosa, (4) muscularis, (5) lamina propria, (6) muscularis mucosae.
   (a) 3, 4, 5, 6, 2, 1   (b) 1, 2, 3, 4, 6, 5   (c) 1, 5, 6, 2, 4, 3
   (d) 5, 1, 2, 6, 4, 3   (e) 3, 4, 2, 6, 5, 1

8. Which of the following are functions of the liver? (1) carbohydrate, lipid, and protein metabolism, (2) nucleic acid metabolism, (3) excretion of bilirubin, (4) synthesis of bile salts, (5) activation of vitamin D.
   (a) 1, 2, 3, and 5   (b) 1, 2, 3, and 4   (c) 1, 3, 4, and 5
   (d) 2, 3, 4, and 5   (e) 1, 2, 4, and 5

9. Which of the following statements regarding the regulation of gastric secretion and motility are *true*? (1) The sight, smell, taste, or thought of food can initiate the cephalic phase of gastric activity. (2) The gastric phase begins when food enters the small intestine. (3) Once activated, stretch receptors and chemoreceptors in the stomach trigger the flow of gastric juice and peristalsis. (4) The intestinal phase reflexes inhibit gastric activity. (5) The enterogastric reflex stimulates gastric emptying.
   (a) 1, 3, and 4   (b) 2, 4, and 5   (c) 1, 3, 4, and 5
   (d) 1, 2, and 5   (e) 1, 2, 3, and 4

10. Which of the following are *true*? (1) Segmentations in the small intestine help propel chyme through the intestinal tract. (2) The migrating motility complex is a type of peristalsis in the small intestine. (3) The large surface area for absorption in the small intestine is due to the presence of circular folds, villi, and microvilli. (4) The mucus-producing cells of the small intestine are paneth cells. (5) Most long-chain fatty acid and monoglyceride absorption in the small intestine requires the presence of bile salts.
    (a) 1, 2, and 3   (b) 2, 3, and 5   (c) 1, 2, 3, 4, and 5
    (d) 1, 3, and 5   (e) 1, 2, 3, and 5

11. The release of feces from the large intestine is dependent on (1) stretching of the rectal walls, (2) voluntary relaxation of the external anal sphincter, (3) involuntary contraction of the diaphragm and abdominal muscles, (4) activity of the intestinal bacteria, (5) sympathetic stimulation of the internal sphincter.
    (a) 2, 4, and 5   (b) 1, 2, and 5   (c) 1, 2, 3, and 5
    (d) 1 and 2   (e) 3, 4, and 5

12. Which of the following is *not* true concerning the liver?
    (a) The left hepatic duct joins the cystic duct from the gallbladder.
    (b) As blood passes through the sinusoids, it is processed by hepatocytes and phagocytes.
    (c) Processed blood returns from the liver to systemic circulation through the hepatic veins.
    (d) The liver receives oxygenated blood through the hepatic artery.
    (e) The hepatic portal vein delivers deoxygenated blood from the GI tract to the liver.

13. Match the following:
    ____ (a) collapsed, muscular tube involved in deglutition and peristalsis
    ____ (b) coiled tube attached to the cecum
    ____ (c) contains duodenal glands in the submucosa
    ____ (d) produces and secretes bile
    ____ (e) contains aggregated lymphatic follicles in the mucosa
    ____ (f) responsible for ingestion, mastication, and deglutition
    ____ (g) responsible for churning, peristalsis, storage, and chemical digestion with the enzyme pepsin
    ____ (h) storage area for bile
    ____ (i) contain acini that release juices containing several digestive enzymes for protein, carbohydrate, lipid, and nucleic acid digestion and sodium bicarbonate to buffer stomach acid
    ____ (j) composed of enamel, dentin, and pulp cavity; used in mastication
    ____ (k) passageway for food, fluid, and air; involved in deglutition
    ____ (l) forms a semisolid waste material through haustral churning and peristalsis
    ____ (m) forces the food to the back of the mouth for swallowing; places food in contact with the teeth
    ____ (n) produce a fluid in the mouth that helps cleanse the mouth and teeth and that lubricates, dissolves, and begins the chemical breakdown of food

    (1) mouth
    (2) teeth
    (3) salivary glands
    (4) pharynx
    (5) esophagus
    (6) tongue
    (7) stomach
    (8) duodenum
    (9) ileum
    (10) colon
    (11) liver
    (12) gallbladder
    (13) appendix
    (14) pancreas

14. Match the following:
    ____ (a) an activating brush-border enzyme that splits off part of the trypsinogen molecule to form trypsin, a protease
    ____ (b) an enzyme that initiates carbohydrate digestion in the mouth
    ____ (c) the principal triglyceride-digesting enzyme in adults
    ____ (d) stimulates secretion of gastric juices and promotes gastric emptying
    ____ (e) secreted by chief cells in the stomach; a proteolytic enzyme
    ____ (f) stimulates the flow of pancreatic juice rich in bicarbonates; decreases gastric secretions
    ____ (g) a nonenzymatic fat-emulsifying agent
    ____ (h) causes contraction of the gallbladder and stimulates the production of pancreatic juice rich in digestive enzymes
    ____ (i) inhibits gastrin release
    ____ (j) stimulates secretion of ions and water by the intestines and inhibits gastric acid secretion
    ____ (k) secreted by glands in the tongue; begins breakdown of triglycerides in the stomach

    (1) gastrin
    (2) cholecystokinin
    (3) secretin
    (4) enterokinase
    (5) pepsin
    (6) salivary amylase
    (7) pancreatic lipase
    (8) lingual lipase
    (9) bile
    (10) vasoactive intestinal polypeptide
    (11) somatostatin

25 METABOLISM AND NUTRITION

Introduction

1. Our only source of energy for performing biological work is the food we eat. Food also provides essential substances that we cannot synthesize.
2. Most food molecules absorbed by the gastrointestinal tract are used to supply energy for life processes, serve as building blocks during synthesis of complex molecules, or are stored for future use.

25.1 Metabolic Reactions

1. Metabolism refers to all chemical reactions of the body and is of two types: catabolism and anabolism.
2. Catabolism is the term for reactions that break down complex organic compounds into simple ones. Overall, catabolic reactions are exergonic; they produce more energy than they consume.
3. Chemical reactions that combine simple molecules into more complex ones that form the body's structural and functional components are collectively known as anabolism. Overall, anabolic reactions are endergonic; they consume more energy than they produce.
4. The coupling of anabolism and catabolism occurs via ATP.

25.2 Energy Transfer

1. Oxidation is the removal of electrons from a substance; reduction is the addition of electrons to a substance.
2. Two coenzymes that carry hydrogen atoms during coupled oxidation–reduction reactions are nicotinamide adenine dinucleotide (NAD) and flavin adenine dinucleotide (FAD).
3. ATP can be generated via substrate-level phosphorylation, oxidative phosphorylation, and photophosphorylation.

25.3 Carbohydrate Metabolism

1. During digestion, polysaccharides and disaccharides are hydrolyzed into the monosaccharides glucose (about 80%), fructose, and galactose; the latter two are then converted to glucose. Some glucose is oxidized by cells to provide ATP. Glucose also can be used to synthesize amino acids, glycogen, and triglycerides.
2. Glucose moves into most body cells via facilitated diffusion through glucose transporters (GluT) and becomes phosphorylated to glucose 6-phosphate. In muscle cells, this process is stimulated by insulin. Glucose entry into neurons and hepatocytes is always "turned on."

3. Cellular respiration, the complete oxidation of glucose to CO_2 and H_2O, involves glycolysis, the Krebs cycle, and the electron transport chain.
4. Glycolysis is the breakdown of glucose into two molecules of pyruvic acid; there is a net production of two molecules of ATP.
5. When oxygen is in short supply, pyruvic acid is reduced to lactic acid; under aerobic conditions, pyruvic acid enters the Krebs cycle. Pyruvic acid is prepared for entrance into the Krebs cycle by conversion to a 2-carbon acetyl group followed by the addition of coenzyme A to form acetyl coenzyme A. The Krebs cycle involves decarboxylations, oxidations, and reductions of various organic acids. Each molecule of pyruvic acid that is converted to acetyl coenzyme A and then enters the Krebs cycle produces three molecules of CO_2, four molecules of NADH and four H^+, one molecule of $FADH_2$, and one molecule of ATP. The energy originally stored in glucose and then in pyruvic acid is transferred primarily to the reduced coenzymes NADH and $FADH_2$.
6. The electron transport chain involves a series of oxidation–reduction reactions in which the energy in NADH and $FADH_2$ is liberated and transferred to ATP. The electron carriers include FMN, cytochromes, iron–sulfur centers, copper atoms, and coenzyme Q. The electron transport chain yields a maximum of 26 or 28 molecules of ATP and six molecules of H_2O.
7. **Table 25.1** summarizes the ATP yield during cellular respiration. The complete oxidation of glucose can be represented as follows:
$$C_6 H_{12} O_6 + 6 O_2 + 30 \text{ or } 32 \text{ ADPs} + 30 \text{ or } 32 \, \text{\textcircled{P}} \longrightarrow$$
$$6 CO_2 + 6 H_2O + 30 \text{ or } 32 \text{ ATPs}$$
8. The conversion of glucose to glycogen for storage in the liver and skeletal muscle is called glycogenesis. It is stimulated by insulin.
9. The conversion of glycogen to glucose is called glycogenolysis. It occurs between meals and is stimulated by glucagon and epinephrine.
10. Gluconeogenesis is the conversion of noncarbohydrate molecules into glucose. It is stimulated by cortisol and glucagon.

25.4 Lipid Metabolism

1. Lipoproteins transport lipids in the bloodstream. Types of lipoproteins include chylomicrons, which carry dietary lipids to adipose tissue; very-low-density lipoproteins (VLDLs), which carry triglycerides from the liver to adipose tissue; low-density lipoproteins (LDLs), which deliver cholesterol to body cells; and high-density lipoproteins (HDLs), which remove excess cholesterol from body cells and transport it to the liver for elimination.

2. Cholesterol in the blood comes from two sources: from food and from synthesis by the liver.

3. Lipids may be oxidized to produce ATP or stored as triglycerides in adipose tissue, mostly in the subcutaneous layer.

4. A few lipids are used as structural molecules or to synthesize essential molecules.

5. Adipose tissue contains lipases that catalyze the deposition of triglycerides from chylomicrons and hydrolyze triglycerides into fatty acids and glycerol.

6. In lipolysis, triglycerides are split into fatty acids and glycerol and released from adipose tissue under the influence of epinephrine, norepinephrine, cortisol, thyroid hormones, and insulinlike growth factors.

7. Glycerol can be converted into glucose by conversion into glyceraldehyde 3-phosphate.

8. In beta oxidation of fatty acids, carbon atoms are removed in pairs from fatty acid chains; the resulting molecules of acetyl coenzyme A enter the Krebs cycle.

9. The conversion of glucose or amino acids into lipids is called lipogenesis; it is stimulated by insulin.

25.5 Protein Metabolism

1. During digestion, proteins are hydrolyzed into amino acids, which enter the liver via the hepatic portal vein.

2. Amino acids, under the influence of insulinlike growth factors and insulin, enter body cells via active transport.

3. Inside cells, amino acids are synthesized into proteins that function as enzymes, hormones, structural elements, and so forth; are stored as fat or glycogen; or are used for energy.

4. Before amino acids can be catabolized, they must be deaminated and converted to substances that can enter the Krebs cycle.

5. Amino acids may also be converted into glucose, fatty acids, and ketone bodies.

6. Protein synthesis is stimulated by insulinlike growth factors, thyroid hormones, insulin, estrogen, and testosterone.

7. **Table 25.2** summarizes carbohydrate, lipid, and protein metabolism.

25.6 Key Molecules at Metabolic Crossroads

1. Three molecules play a key role in metabolism: glucose 6-phosphate, pyruvic acid, and acetyl coenzyme A.

2. Glucose 6-phosphate may be converted to glucose, glycogen, ribose 5-phosphate, and pyruvic acid.

3. When ATP is low and oxygen is plentiful, pyruvic acid is converted to acetyl coenzyme A; when oxygen supply is low, pyruvic acid is converted to lactic acid. Carbohydrate and protein metabolism are linked by pyruvic acid.

4. Acetyl coenzyme A is the molecule that enters the Krebs cycle; it is also used to synthesize fatty acids, ketone bodies, and cholesterol.

25.7 Metabolic Adaptations

1. During the absorptive state, ingested nutrients enter the blood and lymph from the GI tract.

2. During the absorptive state, blood glucose is oxidized to form ATP, and glucose transported to the liver is converted to glycogen or triglycerides. Most triglycerides are stored in adipose tissue. Amino acids in hepatocytes are converted to carbohydrates, fats, and proteins. **Table 25.3** summarizes the hormonal regulation of metabolism during the absorptive state.

3. During the postabsorptive state, absorption is complete and the ATP needs of the body are satisfied by nutrients already present in the body. The major task is to maintain normal blood glucose level by converting glycogen in the liver and skeletal muscle into glucose, converting glycerol into glucose, and converting amino acids into glucose. Fatty acids, ketone bodies, and amino acids are oxidized to supply ATP. **Table 25.4** summarizes the hormonal regulation of metabolism during the postabsorptive state.

4. Fasting is going without food for a few days; starvation implies weeks or months of inadequate food intake. During fasting and starvation, fatty acids and ketone bodies are increasingly utilized for ATP production.

25.8 Energy Balance

1. Energy balance is the precise matching of energy intake to energy expenditure over time.

2. A calorie (cal) is the amount of energy required to raise the temperature of 1 g of water 1°C. Because the calorie is a relatively small unit, the kilocalorie (kcal) or Calorie (Cal) is often used to measure the body's metabolic rate and to express the energy content of foods; a kilocalorie equals 1000 calories.

3. Metabolic rate is the overall rate at which metabolic reactions use energy. Factors that affect metabolic rate include hormones, exercise, the nervous system, body temperature, ingestion of food, age, gender, climate, sleep, and malnutrition.

4. Measurement of the metabolic rate under basal conditions is called the basal metabolic rate (BMR).

5. Total metabolic rate (TMR) is the total energy expenditure by the body per unit time. Three components contribute to the TMR: (1) BMR, (2) physical activity, and (3) food-induced thermogenesis.

6. Adipose tissue is the major site of stored chemical energy.

7. Two nuclei in the hypothalamus that help regulate food intake are the arcuate and paraventricular nuclei. The hormone leptin, released by adipocytes, inhibits release of neuropeptide Y from the arcuate nucleus and thereby decreases food intake. Melanocortin also decreases food intake. Ghrelin, released by the stomach, increases appetite by stimulating the release of neuropeptide Y.

25.9 Regulation of Body Temperature

1. Normal core temperature is maintained by a delicate balance between heat-producing and heat-losing mechanisms.

2. Mechanisms of heat transfer include conduction, convection, radiation, and evaporation. Conduction is the transfer of heat between two substances or objects in contact with each other. Convection is the transfer of heat by movement of air or water between areas of different temperatures. Radiation is the transfer of heat from a warmer object to a cooler object without physical contact. Evaporation is the conversion of a liquid to a vapor; in the process, heat is lost.

3. The hypothalamic thermostat is in the preoptic area.

4. Responses that produce, conserve, or retain heat when core temperature falls include vasoconstriction; release of epinephrine and norepinephrine; shivering; and release of thyroid hormones.

5. Responses that increase heat loss when core temperature increases include vasodilation, decreased metabolic rate, and evaporation of perspiration.

25.10 Nutrition

1. Nutrients include water, carbohydrates, lipids, proteins, minerals, and vitamins.
2. Nutrition experts suggest dietary calories be 50–60% from carbohydrates, 30% or less from fats, and 12–15% from proteins.
3. MyPlate emphasizes proportionality, variety, moderation, and nutrient density. In a healthy diet vegetables and fruits take up half the plate, while protein and grains take up the other half.

Vegetables and grains represent the largest portion. Three servings of dairy per day are also recommended.

4. Minerals known to perform essential functions include calcium, phosphorus, potassium, sulfur, sodium, chloride, magnesium, iron, iodide, manganese, copper, cobalt, zinc, fluoride, selenium, and chromium. Their functions are summarized in **Table 25.9**.
5. Vitamins are organic nutrients that maintain growth and normal metabolism. Many function in enzyme systems.
6. Fat-soluble vitamins are absorbed with fats and include vitamins A, D, E, and K; water-soluble vitamins include the B vitamins and vitamin C.
7. The functions and deficiency disorders of the principal vitamins are summarized in **Table 25.1**.

CLINICAL CONNECTIONS

Carbohydrate Loading (Refer page 837 of Textbook)

The amount of glycogen stored in the liver and skeletal muscles varies and can be completely exhausted during long-term athletic endeavors. Thus, many marathon runners and other endurance athletes follow a precise exercise and dietary regimen that includes eating large amounts of complex carbohydrates, such as pasta and potatoes, in the 3 days before an event. This practice, **called carbohydrate loading**, helps maximize the amount of glycogen available for ATP production in muscles. For athletic events lasting more than an hour, carbohydrate loading has been shown to increase an athlete's endurance. The increased endurance is due to increased glycogenolysis, which results in more glucose that can be catabolized for energy.

Ketosis (Refer page 841 of Textbook)

The level of ketone bodies in the blood normally is very low because other tissues use them for ATP production as fast as they are generated from the breakdown of fatty acids in the liver. During periods of excessive beta oxidation, however, the production of ketone bodies exceeds their uptake and use by body cells. This might occur after a meal rich in triglycerides, or during fasting or starvation, because few carbohydrates are available for catabolism. Excessive beta oxidation may also occur in poorly controlled or untreated diabetes mellitus for two reasons: (1) Because adequate glucose cannot get into cells, triglycerides are used for ATP production, and (2) because insulin normally inhibits lipolysis, a lack of insulin accelerates the pace of lipolysis. When the concentration of ketone bodies in the blood rises above normal—a condition called ketosis—the ketone bodies, most of which are acids, must be buffered. If too many accumulate, they decrease the concentration of buffers, such as bicarbonate ions, and blood pH falls. Extreme or prolonged ketosis can lead to **acidosis (ketoacidosis),** an abnormally low blood pH. The decreased blood pH in turn causes depression of the central nervous system, which can result in disorientation, coma, and even death if the condition is not treated. When a diabetic becomes seriously insulin-deficient, one of the telltale signs is the sweet smell on the breath from the ketone body acetone.

Phenylketonuria (Refer page 842 of Textbook)

Phenylketonuria (PKU) (fen′-il-kē′-tō-NOO-rē-a) is a genetic error of protein metabolism characterized by elevated blood levels of the amino acid phenylalanine. Most children with phenylketonuria have a mutation in the gene that codes for the enzyme phenylalanine hydroxylase, the enzyme needed to convert phenylalanine into the amino acid tyrosine, which can enter the Krebs cycle (**Figure 25.15**). Because the enzyme is deficient, phenylalanine cannot be metabolized, and what is not used in protein synthesis builds up in the blood. If untreated, the disorder causes vomiting, rashes, seizures, growth deficiency, and severe mental

retardation. Newborns are screened for PKU, and mental retardation can be prevented by restricting the affected child to a diet that supplies only the amount of phenylalanine needed for growth, although learning disabilities may still ensue. Because the artificial sweetener aspartame (NutraSweet) contains phenylalanine, its consumption must be restricted in children with PKU.

Emotion Eating (Refer page 852 of Textbook)

In addition to keeping us alive, eating serves countless psychological, social, and cultural purposes. We eat to celebrate, punish, comfort, defy, and deny. Eating in response to emotional drives, such as feeling stressed, bored, or tired, is called **emotional eating**. Emotional eating is so common that, within limits, it is considered well within the range of normal behavior. Who hasn't at one time or another headed for the refrigerator after a bad day? Problems arise when emotional eating becomes so excessive that it interferes with health. Physical health problems include obesity and associated disorders such as hypertension and heart disease. Psychological health problems include poor self-esteem, an inability to cope effectively with feelings of stress, and in extreme cases, eating disorders such as anorexia nervosa, bulimia, and obesity.

Eating provides comfort and solace, numbing pain and "feeding the hungry heart." Eating may provide a biochemical "fix" as well. Emotional eaters typically overeat carbohydrate foods (sweets and starches), which may raise brain serotonin levels and lead to feelings of relaxation. Food becomes a way to self-medicate when negative emotions arise.

Hypothermia (Refer page 853 of Textbook)

Hypothermia (hī′-pō-THER-mē-a) is a lowering of core body temperature to 35°C (95°F) or below. Causes of hypothermia include an overwhelming cold stress (immersion in icy water), metabolic diseases (hypoglycemia, adrenal insufficiency, or hypothyroidism), drugs (alcohol, antidepressants, sedatives, or tranquilizers), burns, and malnutrition. Hypothermia is characterized by the following as core body temperature falls: sensation of cold, shivering, confusion, vasoconstriction, muscle rigidity, bradycardia, acidosis, hypoventilation, hypotension, loss of spontaneous movement, coma, and death (usually caused by cardiac arrhythmias). Because the elderly have reduced metabolic protection against a cold environment coupled with a reduced perception of cold, they are at greater risk for developing hypothermia.

Vitamin and Mineral Supplements (Refer page 857 of Textbook)

Most nutritionists recommend eating a balanced diet that includes a variety of foods rather than taking vitamin or mineral supplements, except in special circumstances. Common examples of necessary supplementations include iron for women who have excessive menstrual bleeding; iron and calcium for women who are pregnant or breast-feeding; folic acid (folate) for all women who may become pregnant, to reduce the risk of fetal neural tube defects; calcium for most adults, because they do not receive the recommended amount in their diets; and vitamin B$_{12}$ for strict vegetarians, who eat no meat. Because high levels of antioxidant vitamins are thought to have beneficial effects, some experts recommend supplementing vitamins C and E. More is not always better, however; larger doses of vitamins or minerals can be very harmful.

Hypervitaminosis (hī-per-vī-ta-mi-NŌ-sis; *hyper-* = too much or above) refers to dietary intake of a vitamin that exceeds the ability of the body to utilize,

store, or excrete the vitamin. Since water-soluble vitamins are not stored in the body, few cause any problems. However, because lipid-soluble vitamins are stored in the body, excessive consumption may cause problems. For example, excess intake of vitamin A can cause drowsiness, general weakness, irritability, headache, vomiting, dry and peeling skin, partial hair loss, joint pain, liver and spleen enlargement, coma, and even death. **Hypovitaminosis** (*hypo-* = too little or below), or vitamin deficiency, is discussed in **Table 25.10** for the various vitamins.

DISORDERS: HOMEOSTATIC IMBALANCES

Anorexia Nervosa

Anorexia nervosa is a chronic disorder characterized by self-induced weight loss, negative perception of body image, and physiological changes that result from nutritional depletion. Patients with anorexia nervosa have a fixation on weight control and often insist on having a bowel movement every day despite inadequate food intake. They often abuse laxatives, which worsens the fluid and electrolyte imbalances and nutrient deficiencies. The disorder is found predominantly in young, single females, and it may be inherited. Abnormal patterns of menstruation, amenorrhea (absence of menstruation), and a lowered basal metabolic rate reflect the depressant effects of starvation. Individuals may become emaciated and may ultimately die of starvation or one of its complications. Also associated with the disorder are osteoporosis, depression, and brain abnormalities coupled with impaired mental performance. Treatment consists of psychotherapy and dietary regulation.

Fever

A **fever** is an elevation of core temperature caused by a resetting of the hypothalamic thermostat. The most common causes of fever are viral or bacterial infections and bacterial toxins; other causes are ovulation, excessive secretion of thyroid hormones, tumors, and reactions to vaccines. When phagocytes ingest certain bacteria, they are stimulated to secrete a **pyrogen** (PĪ-rō-gen; *pyro-* = fire; *-gen* = produce), a fever-producing substance. One pyrogen is interleukin-1. It circulates to the hypothalamus and induces neurons of the preoptic area to secrete prostaglandins. Some prostaglandins can reset the hypothalamic thermostat at a higher temperature, and temperature–regulating reflex mechanisms then act to bring the core body temperature up to this new setting. *Antipyretics* are agents that relieve or reduce fever. Examples include aspirin, acetaminophen (Tylenol), and ibuprofen (Advil), all of which reduce fever by inhibiting synthesis of certain prostaglandins.

Suppose that due to production of pyrogens the thermostat is reset at 39°C (103°F). Now the heat-promoting mechanisms (vasoconstriction, increased metabolism, shivering) are operating at full force. Thus, even though core temperature is climbing higher than normal—say, 38°C (101°F)—the skin remains cold, and shivering occurs. This condition, called a **chill**, is a definite sign that core temperature is rising. After several hours, core temperature reaches the setting of the thermostat, and the chills disappear. But now the body will continue to regulate temperature at 39°C (103°F). When the pyrogens disappear, the thermostat is reset at normal—37.0°C (98.6°F). Because core temperature is high in the beginning, the heat-losing mechanisms (vasodilation and sweating) go into operation to decrease core temperature. The skin becomes warm, and the person begins to sweat. This phase of the fever is called the **crisis**, and it indicates that core temperature is falling.

Although death results if core temperature rises above 44–46°C (112–114°F), up to a point, fever is beneficial. For example, a higher temperature intensifies the effects of interferons and the phagocytic activities of macrophages while hindering replication of some pathogens. Because fever

increases heart rate, infection-fighting white blood cells are delivered to sites of infection more rapidly. In addition, antibody production and T cell proliferation increase. Moreover, heat speeds up the rate of chemical reactions, which may help body cells repair themselves more quickly.

Obesity

Obesity is body weight more than 20% above a desirable standard due to an excessive accumulation of adipose tissue. More than one-third of the adult population in the United States is obese. (An athlete may be *overweight* due to higher-than-normal amounts of muscle tissue without being obese.) Even moderate obesity is hazardous to health; it is a risk factor in cardiovascular disease, hypertension, pulmonary disease, non-insulin-dependent diabetes mellitus, arthritis, certain cancers (breast, uterus, and colon), varicose veins, and gallbladder disease.

In a few cases, obesity may result from trauma of or tumors in the food-regulating centers in the hypothalamus. In most cases of obesity, no specific cause can be identified. Contributing factors include genetic factors, eating habits taught early in life, overeating to relieve tension, and social customs. Studies indicate that some obese people burn fewer calories during digestion and absorption of a meal, a smaller food-induced thermogenesis effect. Additionally, obese people who lose weight require about 15% fewer calories to maintain normal body weight than do people who have never been obese. Interestingly, people who gain weight easily when deliberately fed excess calories exhibit less NEAT (nonexercise activity thermogenesis, such as occurs with fidgeting) than people who resist weight gains in the face of excess calories. Although leptin suppresses appetite and produces satiety in experimental animals, it is not deficient in most obese people.

Most surplus calories in the diet are converted to triglycerides and stored in adipose cells. Initially, the adipocytes increase in size, but at a maximal size, they divide. As a result, proliferation of adipocytes occurs in extreme obesity. The enzyme endothelial lipoprotein lipase regulates triglyceride storage. The enzyme is very active in abdominal fat but less active in hip fat. Accumulation of fat in the abdomen is associated with higher blood cholesterol level and other cardiac risk factors because adipose cells in this area appear to be more metabolically active.

Treatment of obesity is difficult because most people who are successful at losing weight gain it back within 2 years. Yet even modest weight loss is associated with health benefits. Treatments for obesity include behavior modification programs, very-low-calorie diets, drugs, and surgery. Behavior modification programs, offered at many hospitals, strive to alter eating behaviors and increase exercise activity. The nutrition program includes a "heart-healthy" diet that includes abundant vegetables but is low in fats, especially saturated fats. A typical exercise program suggests walking for 30 minutes a day, five to seven times a week. Regular exercise enhances both weight loss and weight-loss maintenance. Very-low-calorie (VLC) diets include 400 to 800 kcal/day in a commercially made liquid mixture. The VLC diet is usually prescribed for 12 weeks, under close medical supervision. Two drugs are available to treat obesity. Sibutramine is an appetite suppressant that works by inhibiting reuptake of serotonin and norepinephrine in brain areas that govern eating behavior. Orlistat works by inhibiting the lipases released into the lumen of the GI tract. With less lipase activity, fewer dietary triglycerides are absorbed.

For those with extreme obesity who have not responded to other treatments, a surgical procedure may be considered. The two operations most commonly performed—gastric bypass and gastroplasty—both greatly reduce the stomach size so that it can hold just a tiny quantity of food.

MEDICAL TERMINOLOGY

Bulimia (boo-LIM-ē-a; *bu-* = ox; *-limia* = hunger) or *binge–purge syndrome* A disorder that typically affects young, single, middle-class white females, characterized by overeating at least twice a week followed by purging by self-induced vomiting, strict dieting or fasting, vigorous exercise, or use of laxatives or diuretics; it occurs in response to fears of being overweight or to stress, depression, and physiological disorders such as hypothalamic tumors.

Heat cramps Cramps that result from profuse sweating. The salt lost in sweat causes painful contractions of muscles; such cramps tend to occur in muscles used while working but do not appear until the person relaxes once the work is done. Drinking salted liquids usually leads to rapid improvement.

Heat exhaustion (heat prostration) A condition in which the core temperature is generally normal, or a little below, and the skin is cool and moist due to profuse perspiration. Heat exhaustion is usually characterized by loss of fluid and electrolytes, especially salt (NaCl). The salt loss results in muscle cramps, dizziness, vomiting, and fainting; fluid loss may cause low blood pressure. Complete rest, rehydration, and electrolyte replacement are recommended.

Heatstroke (sunstroke) A severe and often fatal disorder caused by exposure to high temperatures, especially when the relative humidity is high,

which makes it difficult for the body to lose heat. Blood flow to the skin is decreased, perspiration is greatly reduced, and body temperature rises sharply because of failure of the hypothalamic thermostat. Body temperature may reach 43°C (110°F). Treatment, which must be undertaken immediately, consists of cooling the body by immersing the victim in cool water and by administering fluids and electrolytes.

Kwashiorkor (kwash-ē-OR-kor) A disorder in which protein intake is deficient despite normal or nearly normal caloric intake, characterized by edema of the abdomen, enlarged liver, decreased blood pressure, low pulse rate, lower-than-normal body temperature, and sometimes mental retardation. Because the main protein in corn (zein) lacks two essential amino acids, which are needed for growth and tissue repair, many African children whose diet consists largely of cornmeal develop kwashiorkor.

Malnutrition (*mal-* = bad) An imbalance of total caloric intake or intake of specific nutrients, which can be either inadequate or excessive.

Marasmus (mar-AZ-mus) A type of protein–calorie undernutrition that results from inadequate intake of both protein and calories. Its characteristics include retarded growth, low weight, muscle wasting, emaciation, dry skin, and thin, dry, dull hair.

SELF-QUIZ QUESTIONS

Fill in the blanks in the following statements.

1. The thermostat and food intake regulating center of the body is in the _____ of the brain.

2. The three key molecules of metabolism are _____, _____, and _____.

Indicate whether the following statements are true or false.

3. Foods that we eat are used to supply energy for life processes, serve as building blocks for synthesis reactions, or are stored for future use.

4. Vitamins A, B, D, and K are fat-soluble vitamins.

Choose the one best answer to the following questions.

5. NAD$^+$ and FAD (1) are both derivatives of B vitamins, (2) are used to carry hydrogen atoms released during oxidation reactions, (3) become NADH and FADH$_2$ in their reduced forms, (4) act as coenzymes in the Krebs cycle, (5) are the final electron acceptors in the electron transport chain.
   (a) 1, 2, 3, 4, and 5   (b) 2, 3, and 4   (c) 2 and 4
   (d) 1, 2, and 3   (e) 1, 2, 3, and 4

6. During glycolysis, (1) a 6-carbon glucose is split into two 3-carbon pyruvic acids, (2) there is a net gain of two ATP molecules, (3) two NADH molecules are oxidized, (4) moderately high levels of oxygen are needed, (5) the activity of phosphofructokinase determines the rate of the chemical reactions.
   (a) 1, 2, and 3   (b) 1 and 2   (c) 1, 2, and 5
   (d) 2, 3, 4, and 5   (e) 1, 2, 3, 4, and 5

7. If glucose is not needed for immediate ATP production, it can be used for (1) vitamin synthesis, (2) amino acid synthesis, (3) gluconeogenesis, (4) glycogenesis, (5) lipogenesis.
   (a) 1, 3, and 5   (b) 2, 4, and 5   (c) 2, 3, 4, and 5
   (d) 1, 2, and 3   (e) 2 and 5

8. Which of the following is the correct sequence for the oxidation of glucose to produce ATP?
   (a) electron transport chain, Krebs cycle, glycolysis, formation of acetyl CoA
   (b) Krebs cycle, formation of acetyl CoA, electron transport chain, glycolysis
   (c) glycolysis, electron transport chain, Krebs cycle, formation of acetyl CoA
   (d) glycolysis, formation of acetyl CoA, Krebs cycle, electron transport chain
   (e) formation of acetyl CoA, Krebs cycle, glycolysis, electron transport chain.

9. Which of the following would you not expect to experience during fasting or starvation?
   (a) decrease in plasma fatty acid levels
   (b) increase in ketone body formation
   (c) lipolysis
   (d) increased use of ketones for ATP production in the brain
   (e) depletion of glycogen

10. If core body temperature rises above normal, which of the following would occur to cool the body? (1) dilation of vessels in the skin, (2) increased radiation and conduction of heat to the environment, (3) increased metabolic rate, (4) evaporation of perspiration, (5) increased secretion of thyroid hormones.
   (a) 3, 4, and 5   (b) 1, 2, and 4   (c) 1, 2, and 5
   (d) 1, 2, 3, 4 and 5   (e) 1, 2, 4, and 5

11. In which of the following situations would the metabolic rate increase? (1) sleep, (2) after ingesting food, (3) increased secretion of thyroid hormones, (4) parasympathetic nervous system stimulation, (5) fever.

(a) 3 and 4   (b) 1, 3, and 5   (c) 2 and 3

(d) 2, 3, and 4   (e) 2, 3, and 5

12. Which of the following are absorptive state reactions? (1) aerobic cellular respiration, (2) glycogenesis, (3) glycogenolysis, (4) gluconeogenesis using lactic acid, (5) lipolysis.

(a) 1 and 2   (b) 2 and 3   (c) 3 and 4

(d) 4 and 5   (e) 1 and 5

13. Match the hormones with the reactions they regulate (answers may be used more than once; some reactions have more than one answer):

_____ (a) gluconeogenesis

_____ (b) glycogenesis

_____ (c) glycogenolysis

_____ (d) lipolysis

_____ (e) lipogenesis

_____ (f) protein catabolism

_____ (g) protein anabolism

(1) insulin

(2) cortisol

(3) glucagon

(4) thyroid hormones

(5) epinephrine

(6) insulinlike growth factors

14. Match the following:

_____ (a) deliver cholesterol to body cells for use in repair of membranes and synthesis of steroid hormones and bile salts

_____ (b) remove excess cholesterol from body cells and transport it to the liver for elimination

_____ (c) organic nutrients required in small amounts for growth and normal metabolism

_____ (d) the energy-transferring molecule of the body

_____ (e) nutrient molecules that can be oxidized to produce ATP or stored in adipose tissue

_____ (f) transport endogenous lipids to adipocytes for storage

_____ (g) the body's preferred source for synthesizing ATP

_____ (h) composed of amino acids and are the primary regulatory molecules in the body

_____ (i) acetoacetic acid, beta-hydroxybutyric acid, and acetone

_____ (j) hormone secreted by adipocytes that acts to decrease total body-fat mass

_____ (k) neurotransmitter that stimulates food intake

_____ (l) inorganic substances that perform many vital functions in the body

_____ (m) carriers of electrons in the electron transport chain

(1) leptin

(2) minerals

(3) glucose

(4) lipids

(5) proteins

(6) neuropeptide Y

(7) cytochromes

(8) ketone bodies

(9) low-density lipoproteins

(10) ATP

(11) vitamins

(12) high-density lipoproteins

(13) very-low-density lipoproteins

15. Match the following:

_____ (a) the mechanism of ATP generation that links chemical reactions with pumping of hydrogen ions

_____ (b) the removal of electrons from an atom or molecule resulting in a decrease in potential energy

_____ (c) the transfer of an amino group from an amino acid to a substance such as pyruvic acid

_____ (d) the formation of glucose from noncarbohydrate sources

_____ (e) refers to all the chemical reactions in the body

_____ (f) the oxidation of glucose to produce ATP

_____ (g) the splitting of a triglyceride into glycerol and fatty acids

_____ (h) the synthesis of lipids

_____ (i) the addition of electrons to a molecule resulting in an increase in potential energy content of the molecule

_____ (j) the formation of ketone bodies

_____ (k) the breakdown of glycogen back to glucose

_____ (l) exergonic chemical reactions that break down complex organic molecules into simpler ones

_____ (m) overall rate at which metabolic reactions use energy

_____ (n) the breakdown of glucose into two molecules of pyruvic acid

_____ (o) removal of CO_2 from a molecule

_____ (p) endergonic chemical reactions that combine simple molecules and monomers to make more complex ones

_____ (q) the addition of a phosphate group to a molecule

_____ (r) the removal of the amino group from an amino acid

_____ (s) the cleavage of one pair of carbon atoms at a time from a fatty acid

_____ (t) the conversion of glucose into glycogen

(1) metabolism

(2) catabolism

(3) beta oxidation

(4) lipolysis

(5) phosphorylation

(6) glycolysis

(7) cellular respiration

(8) transamination

(9) anabolism

(10) lipogenesis

(11) glycogenolysis

(12) glycogenesis

(13) metabolic rate

(14) ketogenesis

(15) oxidation

(16) reduction

(17) chemiosmosis

(18) deamination

(19) gluconeogenesis

(20) decarboxylation

26 THE URINARY SYSTEM

26.1 Overview of the Urinary System

1. The organs of the urinary system are the kidneys, ureters, urinary bladder, and urethra.
2. The kidneys excrete wastes; alter blood ionic composition, blood volume, blood pressure, and blood pH; maintain blood osmolarity; produce the hormones calcitriol and erythropoietin; and perform gluconeogenesis.
3. The ureters convey urine from the kidneys to the urinary bladder; the urinary bladder stores urine; and the urethra allows urine to pass from the urinary bladder to the outside environment.

26.2 Anatomy of the Kidneys

1. The kidneys are retroperitoneal organs attached to the posterior abdominal wall.
2. Three layers of tissue surround the kidneys; renal capsule, adipose capsule, and renal fascia.
3. Internally, the kidneys consist of a renal cortex, a renal medulla, renal pyramids, renal papillae, renal columns, major and minor calyees, and a renal pelvis.
4. Blood flows into the kidney through the renal artery and successively into segmental, interlobar, arcuate, and cortical radiate arteries; afferent arterioles; glomerular capillaries; efferent arterioles; peritubular capillaries and vasa recta; and cortical radiate, arcuate, and interlobar veins before flowing out of the kidney through the renal vein.
5. Vasomotor nerves from the sympathetic division of the autonomic nervous system supply kidney blood vessels; they help regulate the flow of blood through the kidney.

26.3 The Nephron

1. The nephron is the functional unit of the kidneys. A nephron consists of a renal corpuscle (glomerulus and glomerular capsule) and a renal tubule.
2. A renal tubule consists of a proximal convoluted tubule, a nephron loop, and a distal convoluted tubule, which drains into a collecting duct (shared by several nephrons). The nephron loop consists of a descending limb and an ascending limb.
3. A cortical nephron has a short loop that dips only into the superficial region of the renal medulla; a juxtamedullary nephron has a long nephron loop that stretches through the renal medulla almost to the renal papilla.
4. The wall of the entire glomerular capsule, renal tubule, and ducts consists of a single layer of epithelial cells. The epithelium has distinctive histological features in different parts of the tubule. **Table 26.1** summarizes the histological features of the renal tubule and collecting duct.
5. The juxtaglomerular apparatus (JGA) consists of the juxtaglomerular cells of an afferent arteriole and the macula densa of the final portion of the ascending limb of the nephron loop.

26.4 Overview of Renal Physiology

1. Nephrons perform three basic tasks: glomerular filtration, tubular secretion, and tubular reabsorption.

26.5 Glomerular Filtration

1. Fluid that is filtered by glomeruli enters the capsular space and is called glomerular filtrate.
2. The filtration membrane consists of the glomerular endothelium, basement membrane, and filtration slits between pedicels of podocytes.
3. Most substances in blood plasma easily pass through the glomerular filter. However, blood cells and most proteins normally are not filtered.
4. Glomerular filtrate amounts to up to 180 liters of fluid per day. This large amount of fluid is filtered because the filter is porous and thin, the glomerular capillaries are long, and the capillary blood pressure is high.
5. Glomerular blood hydrostatic pressure (GBHP) promotes filtration; capsular hydrostatic pressure (CHP) and blood colloid osmotic pressure (BCOP) oppose filtration. Net filtration pressure (NFP) = GBHP − CHP − BCOP. NFP is about 10 mmHg.
6. Glomerular filtration rate (GFR) is the amount of filtrate formed in both kidneys per minute; it is normally 105–125 mL/min.
7. Glomerular filtration rate depends on renal autoregulation, neural regulation, and hormonal regulation. **Table 26.2** summarizes regulation of GFR.

26.6 Tubular Reabsorption and Tubular Secretion

1. Tubular reabsorption is a selective process that reclaims materials from tubular fluid and returns them to the bloodstream. Reabsorbed substances include water, glucose, amino acids, urea, and ions, such as sodium, chloride, potassium, bicarbonate, and phosphate (**Table 26.3**).
2. Some substances not needed by the body are removed from the blood and discharged into the urine via tubular secretion. Included are ions (K^+, H^+, and NH_4^+), urea, creatinine, and certain drugs.

3. Reabsorption routes include both paracellular (between tubule cells) and transcellular (across tubule cells) routes. The maximum amount of a substance that can be reabsorbed per unit time is called the transport maximum (T_m).

4. About 90% of water reabsorption is obligatory; it occurs via osmosis, together with reabsorption of solutes, and is not hormonally regulated. The remaining 10% is facultative water reabsorption, which varies according to body needs and is regulated by antidiuretic hormone (ADH).

5. Sodium ions are reabsorbed throughout the basolateral membrane via primary active transport.

6. In the proximal convoluted tubule, Na^+ ions are reabsorbed through the apical membranes via Na^+–glucose symporters and Na^+–H^+ antiporters; water is reabsorbed via osmosis; Cl^-, K^+, Ca^{2+}, Mg^{2+}, and urea are reabsorbed via passive diffusion; and NH_3 and NH_4^+ are secreted.

7. The nephron loop reabsorbs 20–30% of the filtered Na^+, K^+, Ca^{2+}, and HCO_3^-; 35% of the filtered Cl^-; and 15% of the filtered water.

8. The distal convoluted tubule reabsorbs sodium and chloride ions via Na^+–Cl^- symporters.

9. In the collecting duct, principal cells reabsorb Na^+ and secrete K^+; intercalated cells reabsorb K^+ and HCO_3^- and secrete H^+.

10. Angiotensin II, aldosterone, antidiuretic hormone, atrial natriuretic peptide, and parathyroid hormone regulate solute and water reabsorption, as summarized in **Table 26.4**.

26.7 Production of Dilute and Concentrated Urine

1. In the absence of ADH, the kidneys produce dilute urine; renal tubules absorb more solutes than water.

2. In the presence of ADH, the kidneys produce concentrated urine; large amounts of water are reabsorbed from the tubular fluid into interstitial fluid, increasing solute concentration of the urine.

3. The countercurrent multiplier establishes an osmotic gradient in the interstitial fluid of the renal medulla that enables production of concentrated urine when ADH is present.

26.8 Evaluation of Kidney Function

1. A urinalysis is an analysis of the volume and physical, chemical, and microscopic properties of a urine sample. **Table 26.5** summarizes the principal physical characteristics of normal urine.

2. Chemically, normal urine contains about 95% water and 5% solutes. The solutes normally include urea, creatinine, uric acid, urobilinogen, and various ions.

3. **Table 26.6** lists several abnormal components that can be detected in a urinalysis, including albumin, glucose, red and white blood cells, ketone bodies, bilirubin, excessive urobilinogen, casts, and microbes.

4. Renal clearance refers to the ability of the kidneys to clear (remove) a specific substance from blood.

26.9 Urine Transportation, Storage, and Elimination

1. The ureters are retroperitoneal and consist of a mucosa, muscularis, and adventitia. They transport urine from the renal pelvis to the urinary bladder, primarily via peristalsis.

2. The urinary bladder is located in the pelvic cavity posterior to the pubic symphysis; its function is to store urine before micturition.

3. The urinary bladder consists of a mucosa with rugae, a muscularis (detrusor muscle), and an adventitia (serosa over the superior surface).

4. The micturition reflex discharges urine from the urinary bladder via parasympathetic impulses that cause contraction of the detrusor muscle and relaxation of the internal urethral sphincter muscle and via inhibition of impulses in somatic motor neurons to the external urethral sphincter.

5. The urethra is a tube leading from the floor of the urinary bladder to the exterior. Its anatomy and histology differ in females and males. In both sexes, the urethra functions to discharge urine from the body; in males, it discharges semen as well.

26.10 Waste Management in Other Body Systems

1. Besides the kidneys, several other tissues, organs, and processes temporarily confine wastes, transport waste materials for disposal, recycle materials, and excrete excess or toxic substances.

2. Buffers bind excess H^+, the blood transports wastes, the liver converts toxic substances into less toxic ones, the lungs exhale CO_2, sweat glands help eliminate excess heat, and the gastrointestinal tract eliminates solid wastes.

26.11 Development of the Urinary System

1. The kidneys develop from intermediate mesoderm.

2. The kidneys develop in the following sequence: pronephros, mesonephros, and metanephros. Only the metanephros remains and develops into a functional kidney.

26.12 Aging and the Urinary System

1. With aging, the kidneys shrink in size, have a decreased blood flow, and filter less blood.

2. Common problems related to aging include urinary tract infections, increased frequency of urination, urinary retention or incontinence, and renal calculi.

CLINICAL CONNECTIONS

Nephroptosis (Floating Kidney) (Refer page 862 of Textbook)

Nephroptosis (nef′-rōp-TŌ-sis; -*ptosis* = falling), or *floating kidney*, is an inferior displacement or dropping of the kidney. It occurs when the kidney slips from its normal position because it is not securely held in place by adjacent organs or its covering of fat. Nephroptosis develops most often in very thin people whose adipose capsule or renal fascia is deficient. It is dangerous because the ureter may kink and block urine flow. The resulting backup of urine puts pressure on the kidney, which damages the tissue. Twisting of the ureter also causes pain. Nephroptosis is very common; about one in four people has some degree of weakening of the fibrous bands that hold the kidney in place. It is 10 times more common in females than males.

Loss of Plasma Proteins in Urine Causes Edema (Refer page 873 of Textbook)

In some kidney diseases, glomerular capillaries are damaged and become so permeable that plasma proteins enter glomerular filtrate. As a result, the filtrate exerts a colloid osmotic pressure that draws water out of the blood. In this situation, the NFP increases, which means more fluid is filtered. At the same time, blood colloid osmotic pressure decreases because plasma proteins are being lost in the urine. Because more fluid filters out of blood capillaries into tissues throughout the body than returns via reabsorption, blood volume decreases and interstitial fluid volume increases. Thus, loss of plasma proteins in urine causes **edema**, an abnormally high volume of interstitial fluid.

Glucosuria (Refer page 878 of Textbook)

When the blood concentration of glucose is above 200 mg/mL, the renal symporters cannot work fast enough to reabsorb all the glucose that enters the glomerular filtrate. As a result, some glucose remains in the urine, a condition called **glucosuria** (gloo′-kō-SOO-rē-a). The most common cause of glucosuria is diabetes mellitus, in which the blood glucose level may rise far above normal because insulin activity is deficient. Excessive glucose in the glomerular filtrate inhibits water reabsorption by kidney tubules. This leads to increased urinary output (polyuria), decreased blood volume, and dehydration.

Diuretics (Refer page 887 of Textbook)

Diuretics (dī-ū-RET-iks) are substances that slow renal reabsorption of water and thereby cause diuresis, an elevated urine flow rate, which in turn reduces blood volume. Diuretic drugs often are prescribed to treat hypertension (high blood pressure) because lowering blood volume usually reduces blood pressure. Naturally occurring diuretics include caffeine in coffee, tea, and sodas, which inhibits Na^+ reabsorption, and alcohol in beer, wine, and mixed drinks, which inhibits secretion of ADH. Most diuretic drugs act by interfering with a mechanism for reabsorption of filtered Na^+. For example, loop diuretics, such as furosemide (Lasix®), selectively inhibit the Na^+–K^+–$2Cl^-$ symporters in the thick ascending limb of the nephron loop (see **Figure 26.15**). The thiazide diuretics, such as chlorothiazide (Diuril®), act in the distal convoluted tubule, where they promote loss of Na^+ and Cl^- in the urine by inhibiting Na^+–Cl^- symporters.

Dialysis (Refer page 889 of Textbook)

If a person's kidneys are so impaired by disease or injury that he or she is unable to function adequately, then blood must be cleansed artificially by **dialysis** (dī-AL-i-sis; *dialyo* = to separate), the separation of large solutes from smaller ones by diffusion through a selectively permeable membrane. One method of dialysis is **hemodialysis** (hē-mō-dī-AL-i-sis; *hemo-* = blood), which directly filters the patient's blood by removing wastes and excess electrolytes and fluid and then returning the cleansed blood to the patient. Blood removed from the body is delivered to a *hemodialyzer* (artificial kidney). Inside the hemodialyzer, blood flows through a *dialysis membrane,* which contains pores large enough to permit the diffusion of small solutes. A special solution, called the *dialysate* (dī-AL-i-sāt), is pumped into the hemodialyzer so that it surrounds the dialysis membrane. The dialysate is specially formulated to maintain diffusion gradients that remove wastes from the blood (such as urea, creatinine, uric acid, excess phosphate, potassium, and sulfate ions) and add needed substances (such as glucose and bicarbonate ions) to it. The cleansed blood is passed through an air embolus detector to remove air and then returned to the body. An anticoagulant (heparin) is added to prevent blood from clotting in the hemodialyzer. As a rule, most people on hemodialysis require about 6–12 hours a week, typically divided into three sessions.

Another method of dialysis, called **peritoneal dialysis** (per′-i-tō-NĒ-al), uses the peritoneum of the abdominal cavity as the dialysis membrane to filter the blood. The peritoneum has a large surface area and numerous blood vessels, and is a very effective filter. A catheter is inserted into the peritoneal cavity and connected to a bag of dialysate. The fluid flows into the peritoneal cavity by gravity and is left there for sufficient time to permit wastes and excess electrolytes and fluids to diffuse into the dialysate. Then the dialysate is drained out into a bag, discarded, and replaced with fresh dialysate.

Each cycle is called an *exchange*. One variation of peritoneal dialysis, called **continuous ambulatory peritoneal dialysis (CAPD)**, can be performed at home. Usually, the dialysate is drained and replenished four times a day and once at night during sleep. Between exchanges the person can move about freely with the dialysate in the peritoneal cavity.

Urinary Incontinence (Refer page 892 of Textbook)

A lack of voluntary control over micturition is called **urinary incontinence** (in-KON-ti-nens). In infants and children under 2–3 years old, incontinence is normal because neurons to the external urethral sphincter muscle are not completely developed; voiding occurs whenever the urinary bladder is sufficiently distended to stimulate the micturition reflex. Urinary incontinence also occurs in adults. There are four types of urinary incontinence—stress, urge, overflow, and functional. **Stress incontinence** is the most common type of incontinence in young and middle-aged females, and results from weakness of the deep muscles of the pelvic floor. As a result, any physical stress that increases abdominal pressure, such as coughing, sneezing, laughing, exercising, straining, lifting heavy objects, and pregnancy, causes leakage of urine from the urinary bladder. **Urge incontinence** is most common in older people and is characterized by an abrupt and intense urge to urinate followed by an involuntary loss of urine. It may be caused by irritation of the urinary bladder wall by infection or kidney stones, stroke, multiple sclerosis, spinal cord injury, or anxiety. **Overflow incontinence** refers to the involuntary leakage of small amounts of urine caused by some type of blockage or weak contractions of the musculature of the urinary bladder. When urine flow is blocked (for example, from an enlarged prostate or kidney stones) or when the urinary bladder muscles can no longer contract, the urinary bladder becomes overfilled and the pressure inside increases until small amounts of urine dribble out. **Functional incontinence** is urine loss resulting from the inability to get to a toilet facility in time as a result of conditions such as stroke, severe arthritis, or Alzheimer's disease. Choosing the right treatment option depends on correct diagnosis of the type of incontinence. Treatments include Kegel exercises (see Clinical Connection: Injury of Levator Ani and Urinary Stress Incontinence in Chapter 11), urinary bladder training, medication, and possibly even surgery.

DISORDERS: HOMEOSTATIC IMBALANCES

Renal Calculi

The crystals of salts present in urine occasionally precipitate and solidify into insoluble stones called **renal calculi** (KAL-kū-lī = pebbles) or **kidney stones**. They commonly contain crystals of calcium oxalate, uric acid, or calcium phosphate. Conditions leading to calculus formation include the ingestion of excessive calcium, low water intake, abnormally alkaline or acidic urine, and overactivity of the parathyroid glands. When a stone lodges in a narrow passage, such as a ureter, the pain can be intense. **Shock-wave lithotripsy**

(LITH-ō-trip'-sē; *litho-* = stone) is a procedure that uses high-energy shock waves to disintegrate kidney stones and offers an alternative to surgical removal. Once the kidney stone is located using x-rays, a device called a *lithotripter* delivers brief, high-intensity sound waves through a water- or gel-filled cushion placed under the back. Over a period of 30 to 60 minutes, 1000 or more shock waves pulverize the stone, creating fragments that are small enough to wash out in the urine.

Urinary Tract Infections

The term **urinary tract infection (UTI)** is used to describe either an infection of a part of the urinary system or the presence of large numbers of microbes in urine. UTIs are more common in females due to the shorter length of the urethra. Symptoms include painful or burning urination, urgent and frequent urination, low back pain, and bed-wetting. UTIs include *urethritis* (ū-rē-THRĪ-tis), inflammation of the urethra; *cystitis* (sis-TĪ-tis), inflammation of the urinary bladder; and *pyelonephritis* (pī-e-lō-ne-FRĪ-tis), inflammation of the kidneys. If pyelonephritis becomes chronic, scar tissue can form in the kidneys and severely impair their function. Drinking cranberry juice can prevent the attachment of *E. coli* bacteria to the lining of the urinary bladder so that they are more readily flushed away during urination.

Glomerular Diseases

A variety of conditions may damage the kidney glomeruli, either directly or indirectly because of disease elsewhere in the body. Typically, the filtration membrane sustains damage, and its permeability increases.

Glomerulonephritis (glō-mer'-ū-lō-ne-FRĪ-tis) is an inflammation of the kidney that involves the glomeruli. One of the most common causes is an allergic reaction to the toxins produced by streptococcal bacteria that have recently infected another part of the body, especially the throat. The glomeruli become so inflamed, swollen, and engorged with blood that the filtration membranes allow blood cells and plasma proteins to enter the filtrate. As a result, the urine contains many erythrocytes (hematuria) and a lot of protein. The glomeruli may be permanently damaged, leading to chronic renal failure.

Nephrotic syndrome (nef-ROT-ik) is a condition characterized by *proteinuria* (prō-tēn-OO-rē-a), protein in the urine, and *hyperlipidemia* (hī'-per-lip-i-DĒ-mē-a), high blood levels of cholesterol, phospholipids, and triglycerides. The proteinuria is due to an increased permeability of the filtration membrane, which permits proteins, especially albumin, to escape from blood into urine. Loss of albumin results in *hypoalbuminemia* (hī'-pō-al-bū-mi-NĒ-mē-a), low blood albumin level, once liver production of albumin fails to meet increased urinary losses. Edema, usually seen around the eyes, ankles, feet, and abdomen, occurs in nephrotic syndrome because loss of albumin from the blood decreases blood colloid osmotic pressure. Nephrotic syndrome is associated with several glomerular diseases of unknown cause, as well as with systemic disorders such as diabetes mellitus, systemic lupus erythematosus (SLE), a variety of cancers, and AIDS.

Renal Failure

Renal failure is a decrease or cessation of glomerular filtration. In **acute renal failure** (ARF), the kidneys abruptly stop working entirely (or almost entirely). The main feature of ARF is the suppression of urine flow, usually characterized either by *oliguria* (ol'-i-GŪ-rē-a), daily urine output between 50 mL and 250 mL, or by *anuria* (an-Ū-rē-a), daily urine output less than 50 mL. Causes include low blood volume (for example, due to hemorrhage), decreased cardiac output, damaged renal tubules, kidney stones, the dyes used to visualize blood vessels in angiograms, nonsteroidal anti-inflammatory d rugs, and some antibiotic drugs. It is also common in people who suffer a devastating illness or overwhelming traumatic injury; in such cases it may be related to a more general organ failure known as *multiple organ dysfunction syndrome (MODS)*.

Renal failure causes a multitude of problems. There is edema due to salt and water retention and metabolic acidosis due to an inability of the kidneys to excrete acidic substances. In the blood, urea builds up due to impaired renal excretion of metabolic waste products and potassium level rises, which can lead to cardiac arrest. Often, there is anemia because the kidneys no longer produce enough erythropoietin for adequate red blood cell production. Because the kidneys are no longer able to convert vitamin D to calcitriol, which is needed for adequate calcium absorption from the small intestine, osteomalacia also may occur.

Chronic renal failure (CRF) refers to a progressive and usually irreversible decline in glomerular filtration rate (GFR). CRF may result from chronic glomerulonephritis, pyelonephritis, polycystic kidney disease, or traumatic loss of kidney tissue. CRF develops in three stages. In the first stage, *diminished renal reserve,* nephrons are destroyed until about 75% of the functioning nephrons are lost. At this stage, a person may have no signs or symptoms because the remaining nephrons enlarge and take over the function of those that have been lost. Once 75% of the nephrons are lost, the person enters the second stage, called *renal insufficiency,* characterized by a decrease in GFR and increased blood levels of nitrogen-containing wastes and creatinine. Also, the kidneys cannot effectively concentrate or dilute the urine. The final stage, called *end-stage renal failure,* occurs when about 90% of the nephrons have been lost. At this stage, GFR diminishes to 10–15% of normal, oliguria is present, and blood levels of nitrogen-containing wastes and creatinine increase further. People with end-stage renal failure need dialysis therapy and are possible candidates for a kidney transplant operation.

Polycystic Kidney Disease

Polycystic kidney disease (PKD) (pol'-ē-SIS-tik) is one of the most common inherited disorders. In PKD, the kidney tubules become riddled with hundreds or thousands of cysts (fluid-filled cavities). In addition, inappropriate apoptosis (programmed cell death) of cells in noncystic tubules leads to progressive impairment of renal function and eventually to end-stage renal failure.

People with PKD also may have cysts and apoptosis in the liver, pancreas, spleen, and gonads; increased risk of cerebral aneurysms; heart valve defects; and diverticula in the colon. Typically, symptoms are not noticed until adulthood, when patients may have back pain, urinary tract infections, blood in the urine, hypertension, and large abdominal masses. Using drugs to restore normal blood pressure, restricting protein and salt in the diet, and controlling urinary tract infections may slow progression to renal failure.

Urinary Bladder Cancer

Each year, nearly 12,000 Americans die from **urinary bladder cancer**. It generally strikes people over 50 years of age and is three times more likely to develop in males than females. The disease is typically painless as it develops, but in most cases blood in the urine is a primary sign of the disease. Less often, people experience painful and/or frequent urination.

As long as the disease is identified early and treated promptly, the prognosis is favorable. Fortunately, about 75% of urinary bladder cancers are confined to the epithelium of the urinary bladder and are easily removed by surgery. The lesions tend to be low-grade, meaning that they have only a small potential for metastasis.

Urinary bladder cancer is frequently the result of a carcinogen. About half of all cases occur in people who smoke or have at some time smoked cigarettes. The cancer also tends to develop in people who are exposed to chemicals called aromatic amines. Workers in the leather, dye, rubber, and aluminum industries, as well as painters, are often exposed to these chemicals.

Kidney Transplant

A **kidney transplant** is the transfer of a kidney from a donor to a recipient whose kidneys no longer function. In the procedure, the donor kidney is placed in the pelvis of the recipient through an abdominal incision. The renal artery and vein of the transplanted kidney are attached to a nearby artery or vein in the pelvis of the recipient and the ureter of the transplanted kidney is then attached to the urinary bladder. During a kidney transplant, the patient receives only one donor kidney, since only one kidney is needed to maintain sufficient renal function. The nonfunctioning diseased kidneys are usually left in place. As with all organ transplants, kidney transplant recipients must be ever vigilant for signs of infection or organ rejection. The transplant recipient will take immunosuppressive drugs for the rest of his or her life to avoid rejection of the "foreign" organ.

Cystoscopy

Cystoscopy (sis-TOS-kō-pē; *cysto-* = bladder; *-scopy* = to examine) is a very important procedure for direct examination of the mucosa of the urethra and urinary bladder and prostate in males. In the procedure, a *cystoscope* (a flexible narrow tube with a light) is inserted into the urethra to examine the structures through which it passes. With special attachments, tissue samples can be removed for examination (biopsy) and small stones can be removed. Cystoscopy is useful for evaluating urinary bladder problems such as cancer and infections. It can also evaluate the degree of obstruction resulting from an enlarged prostate.

MEDICAL TERMINOLOGY

Azotemia (az-ō-TĒ-mē-a; *azot-* = nitrogen; *-emia* = condition of blood) Presence of urea or other nitrogen-containing substances in the blood.

Cystocele (SIS-tō-sēl; *cysto-* = bladder; *-cele* = hernia or rupture) Hernia of the urinary bladder.

Diabetic kidney disease A disorder caused by diabetes mellitus in which glomeruli are damaged. The result is the leakage of proteins into the urine and a reduction in the ability of the kidney to remove water and waste.

Dysuria (dis-Ū-rē-a; *dys-* = painful; *-uria* = urine) Painful urination.

Enuresis (en′-ū-RĒ-sis = to void urine) Involuntary voiding of urine after the age at which voluntary control has typically been attained.

Hydronephrosis (hī′-drō-ne-FRŌ-sis; *hydro-* = water; *-nephros-* = kidney; *-osis* = condition) Swelling of the kidney due to dilation of the renal pelvis and calyces as a result of an obstruction to the flow of urine. It may be due to a congenital abnormality, a narrowing of the ureter, a kidney stone, or an enlarged prostate.

Intravenous pyelogram (IVP) (in′-tra-VĒ-nus PĪ-e-lō-gram′; *intra-* = within; *-veno-* = vein; *pyelo-* = pelvis of kidney; *-gram* = record) Radiograph (x-ray) of the kidneys, ureters, and urinary bladder after venous injection of a radiopaque contrast medium.

Nephropathy (ne-FROP-a-thē; *nephro-* = kidney; *-pathos* = suffering) Any disease of the kidneys. Types include analgesic (from long-term and excessive use of drugs such as ibuprofen), lead (from ingestion of lead-based paint), and solvent (from carbon tetrachloride and other solvents).

Nocturnal enuresis (nokt-Ū-rē-a en′-ū-RĒ-sis) Discharge of urine during sleep, resulting in bed-wetting; occurs in about 15% of 5-year-old children and generally resolves spontaneously, afflicting only about 1% of adults. It may have a genetic basis, as bed-wetting occurs more often in identical twins than in fraternal twins and more often in children whose parents or siblings were bed-wetters. Possible causes include smaller than normal bladder capacity, failure to awaken in response to a full bladder, and above-normal production of urine at night. Also referred to as **nocturia** (nok-too-rē-a).

Polyuria (pol′-ē-Ū-rē-a; *poly-* = too much) Excessive urine formation. It may occur in conditions such as diabetes mellitus and glomerulonephritis.

Stricture (STRIK-chur) Narrowing of the lumen of a canal or hollow organ, as may occur in the ureter, urethra, or any other tubular structure in the body.

Uremia (ū-RĒ-mē-a; *-emia* = condition of blood) Toxic levels of urea in the blood resulting from severe malfunction of the kidneys.

Urinary retention A failure to completely or normally void urine; may be due to an obstruction in the urethra or neck of the urinary bladder, to nervous contraction of the urethra, or to lack of urge to urinate. In men, an enlarged prostate may constrict the urethra and cause urinary retention. If urinary retention is prolonged, a catheter (slender rubber drainage tube) must be placed into the urethra to drain the urine.

SELF-QUIZ QUESTIONS

Fill in the blanks in the following statements.

1. The renal corpuscle consists of the _____ and _____.

2. Discharge of urine from the urinary bladder is called _____.

Indicate whether the following statements are true or false.

3. The most superficial region of the internal kidney is the renal medulla.

4. When dilute urine is being formed, the osmolarity of the fluid in the tubular lumen increases as it flows down the descending limb of the loop of Henle, decreases as it flows up the ascending limb, and continues to decrease as it flows through the rest of the nephron and collecting duct.

Choose the one best answer to the following questions.

5. Which of the following statements are *correct?* (1) Glomerular filtration rate (GFR) is directly related to the pressures that determine net filtration pressure. (2) Angiotensin II and atrial natriuretic peptide help regulate GFR. (3) Mechanisms that regulate GFR work by adjusting blood flow into and out of the glomerulus and by altering the glomerular capillary surface area available for filtration. (4) GFR increases when blood flow into glomerular capillaries decreases. (5) Normally, GFR increases very little when systemic blood pressure rises.

    (a) 1, 2, and 3   (b) 2, 3, and 4   (c) 3, 4, and 5

    (d) 1, 2, 3, and 5   (e) 2, 3, 4, and 5

6. Which of the following hormones affect Na^+, Cl^-, Ca^{2+}, and/or water reabsorption and/or K^+ secretion by the renal tubules? (1) angiotensin II, (2) aldosterone, (3) ADH, (4) atrial natriuretic peptide, (5) thyroid hormone, (6) parathyroid hormone.

    (a) 1, 3, and 5   (b) 2, 3, and 6   (c) 2, 4, and 5

    (d) 1, 2, 4, and 5   (e) 1, 2, 3, 4, and 6

7. Which of the following are features of the renal corpuscle that enhance its filtering capacity? (1) large glomerular capillary surface area, (2) thick, selectively permeable filtration membrane, (3) high capsular hydrostatic pressure, (4) high glomerular capillary pressure, (5) mesangial cells regulating the filtering surface area.

    (a) 1, 2, and 3   (b) 2, 4, and 5   (c) 1, 4, and 5

    (d) 2, 3, and 4   (e) 2, 3, and 5

8. Given the following values, calculate the net filtration pressure: (1) glomerular blood hydrostatic pressure = 40 mmHg, (2) capsular hydrostatic pressure = 10 mmHg, (3) blood colloid osmotic pressure = 30 mmHg.
   (a) −20 mmHg   (b) 0 mmHg   (c) 20 mmHg
   (d) 60 mmHg   (e) 80 mmHg

9. The micturition reflex (1) is initiated by stretch receptors in the ureters, (2) relies on parasympathetic impulses from the micturition center in S2 and S3, (3) results in contraction of the detrusor muscle, (4) results in contraction of the internal urethral sphincter muscle, (5) inhibits motor neurons in the external urethral sphincter.
   (a) 1, 2, 3, 4, and 5   (b) 1, 3, and 4   (c) 2, 3, 4, and 5
   (d) 2 and 5   (e) 2, 3, and 5

10. Which of the following are mechanisms that control GFR? (1) renal autoregulation, (2) neural regulation, (3) hormonal regulation, (4) chemical regulation of ions, (5) presence or absence of a transporter.
    (a) 1, 2, and 3   (b) 2, 3, and 4   (c) 3, 4, and 5
    (d) 1, 3, and 5   (e) 1, 3, and 4

11. Place the route of blood flow through the kidney in the correct order:
    (a) segmental arteries   (b) vasa recta
    (c) arcuate arteries   (d) peritubular venules
    (e) interlobular veins   (f) renal vein
    (g) renal artery   (h) interlobar arteries
    (i) peritubular capillaries (j) efferent arterioles
    (k) interlobar veins   (l) glomeruli
    (m) arcuate veins   (n) afferent arterioles
    (o) interlobular arteries

12. Place the route of filtrate flow in the correct order from its origin to the ureter:
    (a) minor calyx
    (b) ascending limb of loop of Henle
    (c) papillary duct
    (d) distal convoluted tubule
    (e) major calyx
    (f) descending limb of loop of Henle
    (g) proximal convoluted tubule
    (h) collecting duct
    (i) renal pelvis

13. Match the following:
    ____ (a) cells in the last portion of the distal convoluted tubule and in the collecting ducts; regulated by ADH and aldosterone
    ____ (b) a capillary network lying in the glomerular capsule and functioning in filtration
    ____ (c) the functional unit of the kidney
    ____ (d) drains into a collecting duct
    ____ (e) combined glomerulus and glomerular capsule; where plasma is filtered
    ____ (f) the visceral layer of the glomerular capsule consisting of modified simple squamous epithelial cells
    ____ (g) cells of the final portion of the ascending limb of the loop of Henle that make contact with the afferent arteriole
    ____ (h) site of obligatory water reabsorption
    ____ (i) pores in the glomerular endothelial cells that allow filtration of blood solutes but not blood cells and platelets
    ____ (j) can secrete H^+ against a concentration gradient
    ____ (k) modified smooth muscle cells in the wall of the afferent arteriole

    (1) podocytes
    (2) glomerulus
    (3) renal corpuscle
    (4) proximal convoluted tubule
    (5) distal convoluted tubule
    (6) juxtaglomerular cells
    (7) macula densa
    (8) principal cells
    (9) intercalated cells
    (10) nephron
    (11) fenestrations

14. Match the following:
    ____ (a) measure of blood nitrogen resulting from the catabolism and deamination of amino acids
    ____ (b) produced from the catabolism of creatine phosphate in skeletal muscle
    ____ (c) volume of blood that is cleared of a substance per unit of time
    ____ (d) can result from diabetes mellitus
    ____ (e) insoluble stones of crystallized salts
    ____ (f) usually indicates a pathological condition
    ____ (g) lack of voluntary control of micturition
    ____ (h) can be caused by damage to the filtration membranes

    (1) incontinence
    (2) renal calculi
    (3) plasma creatinine
    (4) BUN test
    (5) albuminuria
    (6) glucosuria
    (7) renal plasma clearance
    (8) hematuria

15. Match the following:

_____ (a) membrane proteins that function as water channels

_____ (b) a secondary active transport process that achieves Na^+ reabsorption, returns filtered HCO_3^- and water to the peritubular capillaries, and secretes H^+

_____ (c) stimulates principal cells to secrete more K^+ into tubular fluid and reabsorb more Na^+ and Cl^- from tubular fluid

_____ (d) enzyme secreted by juxtaglomerular cells

_____ (e) reduces glomerular filtration rate; increases blood volume and pressure

_____ (f) inhibits Na^+ and H_2O reabsorption in the proximal convoluted tubules and collecting ducts

_____ (g) regulates facultative water reabsorption by increasing the water permeability of principal cells in the distal convoluted tubules and collecting ducts

_____ (h) reabsorb Na^+ together with a variety of other solutes

_____ (i) stimulates cells in the distal convoluted tubule to reabsorb more calcium into the blood

(1) angiotensin II

(2) atrial natriuretic peptide

(3) Na^+ symporters

(4) Na^+/H^+ antiporters

(5) aquaporins

(6) aldosterone

(7) ADH

(8) renin

(9) parathyroid hormone

27 | FLUID, ELECTROLYTE, AND ACID–BASE HOMEOSTASIS

27.1 Fluid Compartments and Fluid Homeostasis

1. Body fluid includes water and dissolved solutes. About two-thirds of the body's fluid is located within cells and is called intracellular fluid (ICF). The other one-third, called extracellular fluid (ECF), includes interstitial fluid; blood plasma and lymph; cerebrospinal fluid; gastrointestinal tract fluids; synovial fluid; fluids of the eyes and ears; pleural, pericardial, and peritoneal fluids; and glomerular filtrate.

2. Fluid balance means that the required amounts of water and solutes are present and are correctly proportioned among the various compartments.

3. An inorganic substance that dissociates into ions in solution is called an electrolyte.

4. Water is the largest single constituent in the body. It makes up 45–75% of total body mass, depending on age, gender, and the amount of adipose tissue present.

5. Daily water gain and loss are each about 2500 mL. Sources of water gain are ingested liquids and foods, and water produced by cellular respiration and dehydration synthesis reactions (metabolic water). Water is lost from the body via urination, evaporation from the skin surface, exhalation of water vapor, and defecation. In women, menstrual flow is an additional route for loss of body water.

6. Body water gain is regulated by adjusting the volume of water intake, mainly by drinking more or less fluid. The thirst center in the hypothalamus governs the urge to drink. Although increased amounts of water and solutes are lost through sweating and exhalation during exercise, loss of excess body water or excess solutes depends mainly on regulating excretion in the urine. The extent of urinary NaCl loss is the main determinant of body fluid volume; the extent of urinary water loss is the main determinant of body fluid osmolarity. **Table 27.1** summarizes the factors that regulate water gain and water loss in the body.

7. Angiotensin II and aldosterone reduce urinary loss of Na^+ and thereby increase the volume of body fluids. ANP promotes natriuresis, elevated excretion of Na^+, which decreases blood volume.

8. The major hormone that regulates water loss and thus body fluid osmolarity is antidiuretic hormone (ADH).

9. An increase in the osmolarity of interstitial fluid draws water out of cells, and they shrink slightly. A decrease in the osmolarity of interstitial fluid causes cells to swell. Most often a change in osmolarity is due to a change in the concentration of Na^+, the dominant solute in interstitial fluid.

10. When a person consumes water faster than the kidneys can excrete it or when renal function is poor, the result may be water intoxication, in which cells swell dangerously.

27.2 Electrolytes in Body Fluids

1. Ions formed when electrolytes dissolve in body fluids control the osmosis of water between fluid compartments, help maintain acid–base balance, and carry electrical current.

2. The concentrations of cations and anions are expressed in units of milliequivalents/liter (mEq/liter). Blood plasma, interstitial fluid, and intracellular fluid contain varying types and amounts of ions.

3. Sodium ions (Na^+) are the most abundant extracellular ions. They are involved in impulse transmission, muscle contraction, and fluid and electrolyte balance. Na^+ level is controlled by aldosterone, antidiuretic hormone, and atrial natriuretic peptide.

4. Chloride ions (Cl^-) are the major extracellular anions. They play a role in regulating osmotic pressure and forming HCl in gastric juice. Cl^- level is controlled indirectly by antidiuretic hormone and by processes that increase or decrease renal reabsorption of Na^+.

5. Potassium ions (K^+) are the most abundant cations in intracellular fluid. They play a key role in the resting membrane potential and action potential of neurons and muscle fibers; help maintain intracellular fluid volume; and contribute to regulation of pH. K^+ level is controlled by aldosterone.

6. Bicarbonate ions (HCO_3^-) are the second most abundant anions in extracellular fluid. They are the most important buffer in blood plasma.

7. Calcium is the most abundant mineral in the body. Calcium salts are structural components of bones and teeth. Ca^{2+}, which are principally extracellular cations, function in blood clotting, neurotransmitter release, and contraction of muscle. Ca^{2+} level is controlled mainly by parathyroid hormone and calcitriol.

8. Phosphate ions (H_2PO_4-, HPO_4^{2-}, and PO_4^{3-}) are principally intracellular anions, and their salts are structural components of bones and teeth. They are also required for the synthesis of nucleic acids and ATP and participate in buffer reactions. Their level is controlled by parathyroid hormone and calcitriol.

9. Magnesium ions (Mg^{2+}) are primarily intracellular cations. They act as cofactors in several enzyme systems.

10. **Table 27.2** describes the imbalances that result from deficiency or excess of important body electrolytes.

27.3 Acid–Base Balance

1. The overall acid–base balance of the body is maintained by controlling the H^+ concentration of body fluids, especially extracellular fluid.
2. The normal pH of systemic arterial blood is 7.35–7.45.
3. Homeostasis of pH is maintained by buffer systems, via exhalation of carbon dioxide, and via kidney excretion of H^+ and reabsorption of HCO_3^-. The important buffer systems include proteins, carbonic acid–bicarbonate buffers, and phosphates.
4. An increase in exhalation of carbon dioxide increases blood pH; a decrease in exhalation of CO_2 decreases blood pH.
5. In the proximal convoluted tubules of the kidneys, Na^+–H^+ antiporters secrete H^+ as they reabsorb Na^+. In the collecting ducts of the kidneys, some intercalated cells reabsorb K^+ and HCO_3^- and secrete H^+; other intercalated cells secrete HCO_3^-. In these ways, the kidneys can increase or decrease the pH of body fluids.
6. **Table 27.3** summarizes the mechanisms that maintain pH of body fluids.
7. Acidosis is a systemic arterial blood pH below 7.35; its principal effect is depression of the central nervous system (CNS). Alkalosis

is a systemic arterial blood pH above 7.45; its principal effect is overexcitability of the CNS.

8. Respiratory acidosis and alkalosis are disorders due to changes in blood P_{CO_2}; metabolic acidosis and alkalosis are disorders associated with changes in blood HCO_3^- concentration.
9. Metabolic acidosis or alkalosis can be compensated by respiratory mechanisms (respiratory compensation); respiratory acidosis or alkalosis can be compensated by renal mechanisms (renal compensation). **Table 27.4** summarizes the effects of respiratory and metabolic acidosis and alkalosis.
10. By examining systemic arterial blood pH, HCO_3^-, and P_{CO_2} values, it is possible to pinpoint the cause of an acid–base imbalance.

27.4 Aging and Fluid, Electrolyte, and Acid–Base Homeostasis

1. With increasing age, there is decreased intracellular fluid volume and decreased K^+ due to declining skeletal muscle mass.
2. Decreased kidney function with aging adversely affects fluid and electrolyte balance.

CLINICAL CONNECTIONS

Enemas and Fluid Balance (Refer page 902 of Textbook)

An **enema** (EN-e-ma) is the introduction of a solution into the rectum to draw water (and electrolytes) into the colon osmotically. The increased volume increases peristalsis, which evacuates feces. Enemas are used to treat constipation. Repeated enemas, especially in young children, increase the risk of fluid and electrolyte imbalances.

Indicators of Na⁺ Imbalance (Refer page 904 of Textbook)

If excess sodium ions remain in the body because the kidneys fail to excrete enough of them, water is also osmotically retained. The result is increased blood volume, increased blood pressure, and **edema**, an abnormal accumulation of interstitial fluid. Renal failure and hyperaldosteronism (excessive aldosterone secretion) are two causes of Na^+ retention. Excessive urinary loss of Na^+, by contrast, causes excessive water loss, which results in **hypovolemia** (hī'-pō-vō-LĒ-mē-a), an abnormally low blood volume. Hypovolemia related to Na^+ loss is most frequently due to the inadequate secretion of aldosterone associated with adrenal insufficiency or overly vigorous therapy with diuretic drugs.

Diagnosis of Acid–Base Imbalances (Refer page 911 of Textbook)

The cause of an acid–base imbalance can often be pinpointed by careful evaluation of three factors in a sample of systemic arterial blood: pH, concentration of HCO_3^-, and P_{CO_2}. These three blood chemistry values are examined in the following four-step sequence:

1. Note whether the pH is high (alkalosis) or low (acidosis).
2. Decide which value—P_{CO_2} or HCO_3^-—is out of the normal range and could be the *cause* of the pH change. For example, elevated pH could be caused by low P_{CO_2} or high HCO_3^-.
3. If the cause is a *change in P_{CO_2}*, the problem is *respiratory*; if the cause is a *change in HCO_3^-*, the problem is *metabolic*.
4. Now look at the value that doesn't correspond with the observed pH change. If it is within its normal range, there is no compensation. If it is outside the normal range, compensation is occurring and partially correcting the pH imbalance.

SELF-QUIZ QUESTIONS

Fill in the blanks in the following statements.

1. The source of water that is derived from aerobic cellular respiration and dehydration synthesis reactions is _____ water.
2. In the carbonic acid–bicarbonate buffer system, the _____ acts as a weak base, and _____ acts as a weak acid.

Indicate whether the following statements are true or false.

3. The phosphate buffer system is an important regulator of pH in the cytosol.
4. The two compartments in which water can be found are plasma and cytosol.

Choose the one best answer to the following questions.

5. Which of the following statements are *true*? (1) Buffers prevent rapid, drastic changes in pH of a body fluid. (2) Buffers work slowly. (3) Strong acids lower pH more than weak acids because strong acids contribute fewer H^+. (4) Most buffers consist of a weak acid and the salt of that acid, which acts as a weak base. (5) Hemoglobin is an important buffer.
   (a) 1, 2, 3, and 5   (b) 1, 3, 4, and 5   (c) 1, 3, and 5
   (d) 1, 4, and 5   (e) 2, 3, and 5

6. Which of the following are *true* concerning ions in the body? (1) They control osmosis of water between fluid compartments. (2) They help maintain acid–base balance. (3) They carry electrical current. (4) They serve as

cofactors for enzyme activity. (5) They serve as neurotransmitters under special circumstances.

(a) 1, 3, and 5   (b) 2, 4, and 5   (c) 1, 4, and 5

(d) 1, 2, and 4   (e) 1, 2, 3, and 4

7. Which of the following statements are *true*? (1) An increase in the carbon dioxide concentration in body fluids increases H^+ concentration and thus lowers pH. (2) Breath holding results in a decline in blood pH. (3) The respiratory buffer mechanism can eliminate a single volatile acid: carbonic acid. (4) The only way to eliminate nonvolatile acids is to excrete H^+ in the urine. (5) When the diet contains a large amount of protein, normal metabolism produces more acids than bases.

(a) 1, 2, 3, 4, and 5   (b) 1, 3, 4, and 5   (c) 1, 2, 3, and 4

(d) 1, 2, 4, and 5   (e) 1, 3, and 4

8. Concerning acid–base imbalances: (1) Acidosis can cause depression of the central nervous system through depression of synaptic transmission. (2) Renal compensation can resolve respiratory alkalosis or acidosis. (3) A major physiological effect of alkalosis is lack of excitability in the central nervous system and peripheral nerves. (4) Resolution of metabolic acidosis and alkalosis occurs through renal compensation. (5) In adjusting blood pH, renal compensation occurs quickly, but respiratory compensation takes days.

(a) 1, 2, and 5   (b) 1 and 2   (c) 2, 3, and 4

(d) 2, 3, and 5   (e) 1, 2, 3, and 5

9. Match the following:

____ (a) the most abundant cation in intracellular fluid; plays a key role in establishing the resting membrane potential

____ (b) the most abundant mineral in the body; plays important roles in blood clotting, neurotransmitter release, maintenance of muscle tone, and excitability of nervous and muscle tissue

____ (c) second most common intracellular cation; is a cofactor for enzymes involved in carbohydrate, protein, and Na^+/K^+ ATPase metabolism

____ (d) the most abundant extracellular cation; essential in fluid and electrolyte balance

____ (e) ions that are mostly combined with lipids, proteins, carbohydrates, nucleic acids, and ATP inside cells

____ (f) most prevalent extracellular anion; can help balance the level of anions in different fluid compartments

____ (g) second most prevalent extracellular anion; mainly regulated by the kidneys; important for acid–base balance

____ (h) substances that act to prevent rapid, drastic changes in the pH of a body fluid

____ (i) inorganic substances that dissociate into ions when in solution

(1) sodium
(2) chloride
(3) electrolytes
(4) bicarbonate
(5) buffers
(6) phosphate
(7) magnesium
(8) potassium
(9) calcium

10. Match the following:

____ (a) an abnormal increase in the volume of interstitial fluid

____ (b) can occur during renal failure or destruction of body cells, which releases phosphates into the blood

____ (c) the swelling of cells due to water moving from plasma into interstitial fluid and then into cells

____ (d) occurs when water loss is greater than water gain

____ (e) can be caused by excessive sodium in diet or with dehydration

____ (f) condition that can occur as water moves out of plasma into interstitial fluid and blood volume decreases

____ (g) can be caused by decreased potassium intake or kidney disease; results in muscle fatigue, increased urine output, changes in electrocardiogram

____ (h) can occur from hypoparathyroidism

____ (i) can be caused by emphysema, pulmonary edema, injury to the respiratory center of the medulla oblongata, airway destruction, or disorders of the muscles involved in breathing

____ (j) can be caused by excessive water intake, excessive vomiting, or aldosterone deficiency

____ (k) can be caused by actual loss of bicarbonate ions, ketosis, or failure of kidneys to excrete H^+

____ (l) can be caused by excessive vomiting of gastric contents, gastric suctioning, use of certain diuretics, severe dehydration, or excessive intake of alkaline drugs

____ (m) can be caused by oxygen deficiency at high altitude, stroke, or severe anxiety

(1) respiratory acidosis
(2) respiratory alkalosis
(3) metabolic acidosis
(4) metabolic alkalosis
(5) dehydration
(6) hypovolemia
(7) water intoxication
(8) edema
(9) hypokalemia
(10) hypernatremia
(11) hyponatremia
(12) hyperphosphatemia
(13) hypocalcemia

28 | THE REPRODUCTIVE SYSTEMS

28.1 Male Reproductive System

1. The male structures of reproduction include the testes (2), epididymidis (2), ductus (vas) deferens (2), ejaculatory ducts (2), seminal vesicles (2), urethra (1), prostate (1), bulbourethral (Cowper's) glands (2), and penis (1). The scrotum is a sac that hangs from the root of the penis and consists of loose skin and underlying subcutaneous layer; it supports the testes. The temperature of the testes is regulated by the cremaster muscles, which either contract to elevate the testes and move them closer to the pelvic cavity or relax and move them farther from the pelvic cavity. The dartos muscle causes the scrotum to become tight and wrinkled.

2. The testes are paired oval glands (gonads) in the scrotum containing seminiferous tubules, in which sperm cells are made; sustentacular cells, which nourish sperm cells and secrete inhibin; and interstitial (Leydig) cells, which produce the male sex hormone testosterone. The testes descend into the scrotum through the inguinal canals during the seventh month of fetal development. Failure of the testes to descend is called cryptorchidism.

3. Secondary oocytes and sperm, both of which are called gametes, are produced in the gonads. Spermatogenesis, which occurs in the testes, is the process whereby immature spermatogonia develop into sperm. The spermatogenesis sequence, which includes meiosis I, meiosis II, and spermiogenesis, results in the formation of four haploid sperm (spermatozoa) from each primary spermatocyte. Mature sperm consist of a head and a tail. Their function is to fertilize a secondary oocyte.

4. At puberty, gonadotropin-releasing hormone (GnRH) stimulates anterior pituitary secretion of FSH and LH. LH stimulates production of testosterone; FSH and testosterone stimulate spermatogenesis. Sertoli cells secrete androgen-binding protein (ABP), which binds to testosterone and keeps its concentration high in the seminiferous tubule. Testosterone controls the growth, development, and maintenance of sex organs; stimulates bone growth, protein anabolism, and sperm maturation; and stimulates development of masculine secondary sex characteristics. Inhibin is produced by sustentacular cells; its inhibition of FSH helps regulate the rate of spermatogenesis.

5. The duct system of the testes includes the seminiferous tubules, straight tubules, and rete testis. Sperm flow out of the testes through the efferent ducts. The ductus epididymis is the site of sperm maturation and storage. The ductus (vas) deferens stores sperm and propels them toward the urethra during ejaculation.

6. Each ejaculatory duct, formed by the union of the duct from the seminal vesicle and ampulla of the ductus (vas) deferens, is the passageway for ejection of sperm and secretions of the seminal vesicles into the first portion of the urethra, the prostatic urethra.

7. The urethra in males is subdivided into three portions: the prostatic, intermediate, and spongy urethra.

8. The seminal vesicles secrete an alkaline, viscous fluid that contains fructose (used by sperm for ATP production). Seminal fluid constitutes about 60% of the volume of semen and contributes to sperm viability. The prostate secretes a slightly acidic fluid that constitutes about 25% of the volume of semen and contributes to sperm motility. The bulbourethral (Cowper's) glands secrete mucus for lubrication and an alkaline substance that neutralizes acid. Semen is a mixture of sperm and seminal fluid; it provides the fluid in which sperm are transported, supplies nutrients, and neutralizes the acidity of the male urethra and the vagina.

9. The penis consists of a root, a body, and a glans penis. Engorgement of the penile blood sinuses under the influence of sexual excitation is called erection.

28.2 Female Reproductive System

1. The female organs of reproduction include the ovaries (gonads), uterine (fallopian) tubes or oviducts, uterus, vagina, and vulva. The mammary glands are part of the integumentary system and also are considered part of the reproductive system in females.

2. The ovaries, the female gonads, are located in the superior portion of the pelvic cavity, lateral to the uterus. Ovaries produce secondary oocytes, discharge secondary oocytes (the process of ovulation), and secrete estrogens, progesterone, relaxin, and inhibin.

3. Oogenesis (the production of haploid secondary oocytes) begins in the ovaries. The oogenesis sequence includes meiosis I and meiosis II, which goes to completion only after an ovulated secondary oocyte is fertilized by a sperm cell.

4. The uterine (fallopian) tubes transport secondary oocytes from the ovaries to the uterus and are the normal sites of fertilization. Ciliated cells and peristaltic contractions help move a secondary oocyte or fertilized ovum toward the uterus.

5. The uterus is an organ the size and shape of an inverted pear that functions in menstruation, implantation of a fertilized ovum, development of a fetus during pregnancy, and labor. It also is part of the pathway for sperm to reach the uterine tubes to fertilize a secondary oocyte. Normally, the uterus is held in position by a series of ligaments. Histologically, the layers of the uterus are an outer perimetrium (serosa), a middle myometrium, and an inner endometrium.

6. The vagina is a passageway for sperm and the menstrual flow, the receptacle of the penis during sexual intercourse, and the

inferior portion of the birth canal. It is capable of considerable stretching.

7. The vulva, a collective term for the external genitals of the female, consists of the mons pubis, labia majora, labia minora, clitoris, vestibule, vaginal and urethral orifices, hymen, and bulb of the vestibule, as well as three sets of glands: the paraurethral (Skene's), greater vestibular (Bartholin's), and lesser vestibular glands.

8. The perineum is a diamond-shaped area at the inferior end of the trunk medial to the thighs and buttocks.

9. The mammary glands are modified sweat glands lying superficial to the pectoralis major muscles. Their function is to synthesize, secrete, and eject milk (lactation).

10. Mammary gland development depends on estrogens and progesterone. Milk production is stimulated by prolactin, estrogens, and progesterone; milk ejection is stimulated by oxytocin.

28.3 The Female Reproductive Cycle

1. The function of the ovarian cycle is to develop a secondary oocyte; the function of the uterine (menstrual) cycle is to prepare the endometrium each month to receive a fertilized egg. The female reproductive cycle includes both the ovarian and uterine cycles.

2. The uterine and ovarian cycles are controlled by GnRH from the hypothalamus, which stimulates the release of FSH and LH by the anterior pituitary. FSH and LH stimulate development of follicles and secretion of estrogens by the follicles. LH also stimulates ovulation, formation of the corpus luteum, and the secretion of progesterone and estrogens by the corpus luteum.

3. Estrogens stimulate the growth, development, and maintenance of female reproductive structures; stimulate the development of secondary sex characteristics; and stimulate protein synthesis. Progesterone works with estrogens to prepare the endometrium for implantation and the mammary glands for milk synthesis.

4. Relaxin relaxes the myometrium at the time of possible implantation. At the end of a pregnancy, relaxin increases the flexibility of the pubic symphysis and helps dilate the uterine cervix to facilitate delivery.

5. During the menstrual phase, the stratum functionalis of the endometrium is shed, discharging blood, tissue fluid, mucus, and epithelial cells.

6. During the preovulatory phase, a group of follicles in the ovaries begins to undergo final maturation. One follicle outgrows the others and becomes dominant while the others degenerate. At the same time, endometrial repair occurs in the uterus. Estrogens are the dominant ovarian hormones during the preovulatory phase.

7. Ovulation is the rupture of the mature (graafian) follicle and the release of a secondary oocyte into the pelvic cavity. It is brought about by a surge of LH. Signs and symptoms of ovulation include increased basal body temperature; clear, stretchy cervical mucus; changes in the uterine cervix; and abdominal pain.

8. During the postovulatory phase, both progesterone and estrogens are secreted in large quantity by the corpus luteum of the ovary, and the uterine endometrium thickens in readiness for implantation.

9. If fertilization and implantation do not occur, the corpus luteum degenerates, and the resulting low levels of progesterone and estrogens allow discharge of the endometrium followed by the initiation of another reproductive cycle.

10. If fertilization and implantation do occur, the corpus luteum is maintained by hCG. The corpus luteum and later the placenta secrete progesterone and estrogens to support pregnancy and breast development for lactation.

28.4 The Human Sexual Response

1. The similar sequence of changes experienced by both males and females before, during, and after intercourse is termed the human sexual response; it occurs in four phases; excitement, plateau, orgasm, and resolution.

2. During excitement, there is vasocongestion (engorgement with blood) of genital tissues. Other changes that occur during this phase include increased heart rate and blood pressure, increased skeletal muscle tone throughout the body, and hyperventilation.

3. During the plateau phase, the changes that began during the excitement phase are sustained at an intense level.

4. During orgasm, there are several rhythmic muscular contractions, accompanied by pleasurable sensations and a further increase in blood pressure, heart rate, and respiration rate.

5. During the resolution phase, genital tissues, heart rate, blood pressure, breathing, and muscle tone return to the unaroused state.

28.5 Birth Control Methods and Abortion

1. Birth control methods include complete abstinence, surgical sterilization (vasectomy, tubal ligation), non-incisional sterilization, hormonal methods (combined pill, extended cycle pill, minipill, contraceptive skin patch, vaginal contraceptive ring, emergency contraception, hormonal injections), intrauterine devices, spermicides, barrier methods (male condom, vaginal pouch, diaphragm, cervical cap), and periodic abstinence (rhythm and sympto-thermal methods).

2. Contraceptive pills of the combination type contain progestin and estrogens in concentrations that decrease the secretion of FSH and LH and thereby inhibit development of ovarian follicles and ovulation, inhibit transport of ova and sperm in the uterine tubes, and block implantation in the uterus.

3. An abortion is the premature expulsion from the uterus of the products of conception; it may be spontaneous or induced.

28.6 Development of the Reproductive Systems

1. The gonads develop from gonadal ridges that arise from growth of intermediate mesoderm. In the presence of the *SRY* gene, the gonads begin to differentiate into testes during the seventh week. The gonads differentiate into ovaries when the *SRY* gene is absent.

2. In males, testosterone stimulates development of each mesonephric duct into an epididymis, ductus (vas) deferens, ejaculatory duct, and seminal vesicle, and Müllerian-inhibiting substance (MIS) causes the paramesonephric duct cells to die. In females, testosterone and MIS are absent; the paramesonephric ducts develop into the uterine tubes, uterus, and vagina and the mesonephric ducts degenerate.

3. The external genitals develop from the genital tubercle and are stimulated to develop into typical male structures by the hormone dihydrotestosterone (DHT). The external genitals develop into female structures when DHT is not produced, the normal situation in female embryos.

28.7 Aging and the Reproductive Systems

1. Puberty is the period when secondary sex characteristics begin to develop and the potential for sexual reproduction is reached.
2. The onset of puberty is marked by pulses or bursts of LH and FSH secretion, each triggered by a pulse of GnRH. The hormone leptin, released by adipose tissue, may signal the hypothalamus that long-term energy stores (triglycerides in adipose tissue) are adequate for reproductive functions to begin.
3. In females, the reproductive cycle normally occurs once each month from menarche, the first menses, to menopause, the permanent cessation of menses.
4. Between the ages of 40 and 50, the pool of remaining ovarian follicles becomes exhausted and levels of progesterone and estrogens decline. Most women experience a decline in bone mineral density after menopause, together with some atrophy of the ovaries, uterine tubes, uterus, vagina, external genitalia, and breasts. Uterine and breast cancer increase in incidence with age.
5. In older males, decreased levels of testosterone are associated with decreased muscle strength, waning sexual desire, and fewer viable sperm; prostate disorders are common.

CLINICAL CONNECTIONS

Cryptorchidism (Refer page 918 of Textbook)

The condition in which the testes do not descend into the scrotum is called **cryptorchidism** (krip-TOR-ki-dizm; *crypt-* = hidden; *-orchid* = testis); it occurs in about 3% of full-term infants and about 30% of premature infants. Untreated bilateral cryptorchidism results in sterility because the cells involved in the initial stages of spermatogenesis are destroyed by the higher temperature of the pelvic cavity. The chance of testicular cancer is 30–50 times greater in cryptorchid testes. The testes of about 80% of boys with cryptorchidism will descend spontaneously during the first year of life. When the testes remain undescended, the condition can be corrected surgically, ideally before 18 months of age.

Circumcision (Refer page 924 of Textbook)

Circumcision (= to cut around) is a surgical procedure in which part of or the entire prepuce is removed. It is usually performed several days after birth, and is done for social, cultural, religious, and (more rarely) medical reasons. Although most health-care professionals find no medical justification for circumcision, some feel that it has benefits, such as a lower risk of urinary tract infections, protection against penile cancer, and possibly a lower risk for sexually transmitted diseases. Indeed, studies in several African villages have found lower rates of HIV infection among circumcised men.

Premature Ejaculation (Refer page 926 of Textbook)

A **premature ejaculation** is ejaculation that occurs too early, for example, during foreplay or on or shortly after penetration. It is usually caused by anxiety, other psychological causes, or an unusually sensitive foreskin or glans penis. For most males, premature ejaculation can be overcome by various techniques (such as squeezing the penis between the glans penis and shaft as ejaculation approaches), behavioral therapy, or medication.

Ovarian Cysts (Refer page 933 of Textbook)

Ovarian cysts are fluid-filled sacs in or on an ovary. Such cysts are relatively common, are usually noncancerous, and frequently disappear on their own. Cancerous cysts are more likely to occur in women over 40. Ovarian cysts may cause pain, pressure, a dull ache, or fullness in the abdomen; pain during sexual intercourse; delayed, painful, or irregular menstrual periods; abrupt onset of sharp pain in the lower abdomen; and/or vaginal bleeding. Most ovarian cysts require no treatment, but larger ones (more than 5 cm or 2 in.) may be removed surgically.

Uterine Prolapse (Refer page 935 of Textbook)

A condition called **uterine prolapse** (*prolapse* = falling down or downward displacement) may result from weakening of supporting ligaments and pelvic musculature associated with age or disease, traumatic vaginal delivery, chronic straining from coughing or difficult bowel movements, or pelvic tumors. The prolapse may be characterized as *first degree (mild),* in which the cervix remains within the vagina; *second degree (marked),* in which the cervix protrudes through the vagina to the exterior; and *third degree (complete),* in which the entire uterus is outside the vagina. Depending on the degree of prolapse, treatment may involve pelvic exercises, dieting if a patient is overweight, a stool softener to minimize straining during defecation, pessary therapy (placement of a rubber device around the uterine cervix that helps prop up the uterus), or surgery.

Hysterectomy (Refer page 937 of Textbook)

Hysterectomy (his-ter-EK-tō-mē; *hyster-* = uterus), the surgical removal of the uterus, is the most common gynecological operation. It may be indicated in conditions such as fibroids, which are noncancerous tumors composed of muscular and fibrous tissue; endometriosis; pelvic inflammatory disease; recurrent ovarian cysts; excessive uterine bleeding; and cancer of the cervix, uterus, or ovaries. In a *partial (subtotal) hysterectomy,* the body of the uterus is removed but the cervix is left in place. A *complete hysterectomy* is the removal of both the body and cervix of the uterus. A *radical hysterectomy* includes removal of the body and cervix of the uterus, uterine tubes, possibly the ovaries, the superior portion of the vagina, pelvic lymph nodes, and supporting structures, such as ligaments. A hysterectomy can be performed either through an incision in the abdominal wall or through the vagina.

Episiotomy (Refer page 939 of Textbook)

During childbirth, the emerging fetus normally stretches the perineal region. However, if it appears that the stretching could be excessive, a physician may elect to perform an **episiotomy** (e-piz-ē-OT-ō-mē; *episi-* = vulva or pubic region; *-otomy* = incision), a perineal cut between the vagina and anus made with surgical scissors to widen the birth canal. The cut is made along the midline or at about a 45 degree angle to the midline. Reasons for an episiotomy include a very large fetus, breech presentation (buttocks or lower limbs coming first), fetal distress (such as an abnormal heart rate), forceps delivery, or a short perineum. Following delivery, the incision is closed in layers with sutures that are absorbed within a few weeks.

Breast Augmentation and Reduction (Refer page 942 of Textbook)

Breast augmentation (awg-men-TĀ-shun = enlargement), technically called *augmentation mammaplasty* (mam-a-PLAS-tē), is a surgical procedure to increase breast size and shape. It may be done to enhance breast size for females who feel that their breasts are too small, to restore breast volume due to weight loss or following pregnancy, to improve the shape of breasts that are sagging, and to improve breast appearance following surgery, trauma, or congenital abnormalities. The most commonly used implants are filled with either a saline solution or silicone gel. The incision for the implant is made under the breast, around the areola, in the armpit, or in the navel. Then a pocket is made to place the implant either directly behind the breast tissue or beneath the pectoralis major muscle.

Breast reduction or *reduction mammaplasty* is a surgical procedure that involves decreasing breast size by removing fat, skin, and glandular tissue. This procedure is done because of chronic back, neck, and shoulder pain; poor posture; circulation or breathing problems; a skin rash under the breasts; restricted levels of activity; self-esteem problems; deep grooves in the shoulders from bra strap pressure; and difficulty wearing or fitting into certain bras and clothing. The most common procedure involves an incision around the areola, down the breast toward the crease between the breast and abdomen, and then along the crease. The surgeon removes excess tissue through the incision. In most cases, the nipple and areola remain attached to the breast. However, if the breasts are extremely large, the nipple and areola may have to be reattached at a higher position.

Fibrocystic Disease of the Breasts (Refer page 942 of Textbook)

The breasts of females are highly susceptible to cysts and tumors. In **fibrocystic disease** (fī-brō-SIS-tik), the most common cause of breast lumps in females, one or more cysts (fluid-filled sacs) and thickenings of alveoli develop. The condition, which occurs mainly in females between the ages of 30 and 50, is probably due to a relative excess of estrogens or a deficiency of progesterone in the postovulatory (luteal) phase of the reproductive cycle (discussed shortly). Fibrocystic disease usually causes one or both breasts to become lumpy, swollen, and tender a week or so before menstruation begins.

Female Athlete Triad: Disordered Eating, Amenorrhea, and Premature Osteoporosis (Refer page 946 of Textbook)

The female reproductive cycle can be disrupted by many factors, including weight loss, low body weight, disordered eating, and vigorous physical activity. The observation that three conditions—disordered eating, amenorrhea, and osteoporosis—often occur together in female athletes led researchers to coin the term **female athlete triad**.

Many athletes experience intense pressure from coaches, parents, peers, and themselves to lose weight to improve performance. Hence, they may develop disordered eating behaviors and engage in other harmful weight-loss practices in a struggle to maintain a very low body weight. **Amenorrhea** (a-men-ō-RĒ-a; *a-* = without; *-men-* = month; *-rrhea* = a flow) is the absence of menstruation. The most common causes of amenorrhea are pregnancy and menopause. In female athletes, amenorrhea results from reduced secretion of gonadotropin-releasing hormone, which decreases the release of LH and FSH. As a result, ovarian follicles fail to develop, ovulation does not occur, synthesis of estrogens and progesterone wanes, and monthly menstrual bleeding ceases. Most cases of the female athlete triad occur in young women with very low amounts of body fat. Low levels of the hormone leptin, secreted by adipose cells, may be a contributing factor.

Because estrogens help bones retain calcium and other minerals, chronically low levels of estrogens are associated with loss of bone mineral density. The female athlete triad causes "old bones in young women." In one study, amenorrheic runners in their twenties had low bone mineral densities, similar to those of postmenopausal women 50 to 70 years old! Short periods of amenorrhea in young athletes may cause no lasting harm. However, long-term cessation of the reproductive cycle may be accompanied by a loss of bone mass, and adolescent athletes may fail to achieve an adequate bone mass; both of these situations can lead to premature osteoporosis and irreversible bone damage.

DISORDERS: HOMEOSTATIC IMBALANCES

Reproductive System Disorders in Males

TESTICULAR CANCER **Testicular cancer** is the most common cancer in males between the ages of 20 and 35. More than 95% of testicular cancers arise from spermatogenic cells within the seminiferous tubules. An early sign of testicular cancer is a mass in the testis, often associated with a sensation of testicular heaviness or a dull ache in the lower abdomen; pain usually does not occur. To increase the chance for early detection of a testicular cancer, all males should perform regular self-examinations of the testes. The examination should be done starting in the teen years and once each month thereafter. After a warm bath or shower (when the scrotal skin is loose and relaxed) each testis should be examined as follows. The testis is grasped and gently rolled between the index finger and thumb, feeling for lumps, swellings, hardness, or other changes. If a lump or other change is detected, a physician should be consulted as soon as possible.

PROSTATE DISORDERS Because the prostate surrounds part of the urethra, any infection, enlargement, or tumor can obstruct the flow of urine. Acute and chronic infections of the prostate are common in postpubescent males, often in association with inflammation of the urethra. Symptoms may include fever, chills, urinary frequency, frequent urination at night, difficulty in urinating, burning or painful urination, low back pain, joint and muscle pain, blood in the urine, or painful ejaculation. However, often there are no symptoms. Antibiotics are used to treat most cases that result from a bacterial infection. In **acute prostatitis**, the prostate becomes swollen and tender. **Chronic prostatitis** is one of the most common chronic infections in men of the middle and later years. On examination, the prostate feels enlarged, soft, and very tender, and its surface outline is irregular.

Prostate cancer is the leading cause of death from cancer in men in the United States, having surpassed lung cancer in 1991. Each year it is diagnosed in almost 200,000 U.S. men and causes nearly 40,000 deaths. The amount of PSA (prostate-specific antigen), which is produced only by prostate epithelial cells, increases with enlargement of the prostate and may indicate infection, benign enlargement, or prostate cancer. A blood test can measure the level of PSA in the blood. Males over the age of 40 should have an annual examination of the prostate gland. In a **digital rectal exam**, a physician palpates the gland through the rectum with the fingers (digits). Many physicians also recommend an annual PSA test for males over age 50. Treatment for prostate cancer may involve surgery, cryotherapy, radiation, hormonal therapy, and chemotherapy. Because many prostate cancers grow very slowly, some urologists recommend "watchful waiting" before treating small tumors in men over age 70.

ERECTILE DYSFUNCTION **Erectile dysfunction (ED)**, previously termed *impotence,* is the consistent inability of an adult male to ejaculate or to attain or hold an erection long enough for sexual intercourse. Many cases of impotence are caused by insufficient release of nitric oxide (NO), which relaxes the smooth muscle of the penile arterioles and erectile tissue. The drug *Viagra®* (sildenafil) enhances smooth muscle relaxation by nitric oxide in the penis. Other causes of erectile dysfunction include diabetes mellitus, physical abnormalities of the penis, systemic disorders such as syphilis, vascular disturbances (arterial or venous obstructions), neurological disorders, surgery, testosterone deficiency, and drugs (alcohol, antidepressants, antihistamines, antihypertensives, narcotics, nicotine, and tranquilizers). Psychological factors such as anxiety or depression, fear of causing pregnancy, fear of sexually transmitted diseases, religious inhibitions, and emotional immaturity may also cause ED.

Reproductive System Disorders in Females

PREMENSTRUAL SYNDROME AND PREMENSTRUAL DYSPHORIC DISORDER

Premenstrual syndrome (PMS) is a cyclical disorder of severe physical and emotional distress. It appears during the postovulatory (luteal) phase of the female reproductive cycle and dramatically disappears when menstruation begins. The signs and symptoms are highly variable from one woman to another. They may include edema, weight gain, breast swelling and tenderness, abdominal distension, backache, joint pain, constipation, skin eruptions, fatigue and lethargy, greater need for sleep, depression or anxiety, irritability, mood swings, headache, poor coordination and clumsiness, and cravings for sweet or salty foods. The cause of PMS is unknown. For some women, getting regular exercise; avoiding caffeine, salt, and alcohol; and eating a diet that is high in complex carbohydrates and lean proteins can bring considerable relief.

Premenstrual dysphoric disorder (PMDD) is a more severe syndrome in which PMS-like signs and symptoms do not resolve after the onset of menstruation. Clinical research studies have found that suppression of the reproductive cycle by a drug that interferes with GnRH (leuprolide) decreases symptoms significantly. Because symptoms reappear when estradiol or progesterone is given together with leuprolide, researchers propose that PMDD is caused by abnormal responses to normal levels of these ovarian hormones. *SSRIs* (selective serotonin reuptake inhibitors) have shown promise in treating both PMS and PMDD.

ENDOMETRIOSIS

Endometriosis (en'-dō-ME-trē-o'-sis; *endo-* = within; *metri-* = uterus; *-osis* = condition) is characterized by the growth of endometrial tissue outside the uterus. The tissue enters the pelvic cavity via the open uterine tubes and may be found in any of several sites—on the ovaries, the rectouterine pouch, the outer surface of the uterus, the sigmoid colon, pelvic and abdominal lymph nodes, the cervix, the abdominal wall, the kidneys, and the urinary bladder. Endometrial tissue responds to hormonal fluctuations, whether it is inside or outside the uterus. With each reproductive cycle, the tissue proliferates and then breaks down and bleeds. When this occurs outside the uterus, it can cause inflammation, pain, scarring, and infertility. Symptoms include premenstrual pain or unusually severe menstrual pain.

BREAST CANCER

One in eight women in the United States faces the prospect of **breast cancer**. After lung cancer, it is the second-leading cause of death from cancer in U.S. women. Breast cancer can occur in males but is rare. In females, breast cancer is seldom seen before age 30; its incidence rises rapidly after menopause. An estimated 5% of the 180,000 cases diagnosed each year in the United States, particularly those that arise in younger women, stem from inherited genetic mutations (changes in the DNA). Researchers have now identified two genes that increase susceptibility to breast cancer: *BRCA1* (*breast cancer 1*) and *BRCA2*. Mutation of *BRCA1* also confers a high risk for ovarian cancer. In addition, mutations of the *p53* gene increase the risk of breast cancer in both males and females, and mutations of the androgen receptor gene are associated with the occurrence of breast cancer in some males. Because breast cancer generally is not painful until it becomes quite advanced, any lump, no matter how small, should be reported to a physician at once. Early detection—by breast self-examination and mammograms—is the best way to increase the chance of survival.

The most effective technique for detecting tumors less than 1 cm (0.4 in.) in diameter is **mammography** (mam-OG-ra-fē; *-graphy* = to record), a type of radiography using very sensitive x-ray film. The image of the breast, called a mammogram (see **Table 1.3**), is best obtained by compressing the breasts, one at a time, using flat plates. A supplementary procedure for evaluating breast abnormalities is **ultrasound**. Although ultrasound cannot detect tumors smaller than 1 cm in diameter (which mammography can detect), it can be used to determine whether a lump is a benign, fluid-filled cyst or a solid (and therefore possibly malignant) tumor.

Among the factors that increase the risk of developing breast cancer are (1) a family history of breast cancer, especially in a mother or sister; (2) nulliparity (never having borne a child) or having a first child after age 35; (3) previous cancer in one breast; (4) exposure to ionizing radiation, such as x-rays; (5) excessive alcohol intake; and (6) cigarette smoking.

The American Cancer Society recommends the following steps to help diagnose breast cancer as early as possible:

- All women over 20 should develop the habit of monthly breast self-examination.
- A physician should examine the breasts every 3 years when a woman is between the ages of 20 and 40, and every year after age 40.
- A mammogram should be taken in women between the ages of 35 and 39, to be used later for comparison (baseline mammogram).
- Women with no symptoms should have a mammogram every year after age 40.
- Women of any age with a history of breast cancer, a strong family history of the disease, or other risk factors should consult a physician to determine a schedule for mammography.

In November 2009, the United States Preventive Services Task Force (USPSTF) issued a series of recommendations relative to breast cancer screening for females at normal risk for breast cancer, that is, for females who have no signs or symptoms of breast cancer and who are not at a higher risk for breast cancer (for example, no family history). These recommendations are as follows:

- Women aged 50–74 should have a mammogram every 2 years.
- Women over 75 should not have mammograms.
- Breast self-examination is not required.

Treatment for breast cancer may involve hormone therapy, chemotherapy, radiation therapy, **lumpectomy** (lump-EK-tō-mē) (removal of the tumor and the immediate surrounding tissue), a modified or radical mastectomy, or a combination of these approaches. A **radical mastectomy** (mas-TEK-tō-mē; *mast-* = breast) involves removal of the affected breast along with the underlying pectoral muscles and the axillary lymph nodes. (Lymph nodes are removed because metastasis of cancerous cells usually occurs through lymphatic or blood vessels.) Radiation treatment and chemotherapy may follow the surgery to ensure the destruction of any stray cancer cells.

Several types of chemotherapeutic drugs are used to decrease the risk of relapse or disease progression. Tamoxifen (*Nolvadex®*) is an antagonist to estrogens that binds to and blocks receptors for estrogens, thus decreasing the stimulating effect of estrogens on breast cancer cells. Tamoxifen has been used for 20 years and greatly reduces the risk of cancer recurrence. *Herceptin®*, a monoclonal antibody drug, targets an antigen on the surface of breast cancer cells. It is effective in causing regression of tumors and retarding progression of the disease. The early data from clinical trials of two new drugs, *Femara®* and *Amimidex®*, show relapse rates that are lower than those for tamoxifen. These drugs are inhibitors of aromatase, the enzyme needed for the final step in synthesis of estrogens. Finally, two drugs—tamoxifen and *Evista®* (*raloxifene*)—are being marketed for breast cancer *prevention*. Interestingly, raloxifene blocks estrogen receptors in the breasts and uterus but activates estrogen receptors in bone. Thus, it can be used to treat osteoporosis without increasing a woman's risk of breast or endometrial (uterine) cancer.

OVARIAN AND CERVICAL CANCER Even though **ovarian cancer** is the sixth most common form of cancer in females, it is the leading cause of death from all gynecological malignancies (excluding breast cancer) because it is difficult to detect before it metastasizes (spreads) beyond the ovaries. Risk factors associated with ovarian cancer include age (usually over age 50); race (whites are at highest risk); family history of ovarian cancer; more than 40 years of active ovulation; nulliparity or first pregnancy after age 30; a high-fat, low-fiber, vitamin A–deficient diet; and prolonged exposure to asbestos or talc. Early ovarian cancer has no symptoms or only mild ones associated with other common problems, such as abdominal discomfort, heartburn, nausea, loss of appetite, bloating, and flatulence. Later-stage signs and symptoms include an enlarged abdomen, abdominal and/or pelvic pain, persistent gastrointestinal disturbances, urinary complications, menstrual irregularities, and heavy menstrual bleeding.

Cervical cancer is a carcinoma of the cervix of the uterus that affects about 12,000 females a year in the United States with a mortality rate of about 4,000 annually. It begins as a precancerous condition called **cervical dysplasia** (dis-PLĀ-zē-a), a change in the number, shape, and growth of cervical cells, usually the squamous cells. Sometimes the abnormal cells revert to normal; other times they progress to cancer, which usually develops slowly. In most cases, cervical cancer can be detected in its earliest stages by a Pap test (see Clinical Connection: Papanicolaou Test in Chapter 4). Almost all cervical cancers are caused by several types of human papillomavirus (HPV); other types of HPV cause genital warts (described later). It is estimated that about 20 million Americans are currently affected with HPV. In most cases, the body fights off HPV through its immune responses, but sometimes it causes cancer, which can take years to develop. HPV is transmitted via vaginal, anal, and oral sex; the infected partner may not have any signs or symptoms. The signs and symptoms of cervical cancer include abnormal vaginal bleeding (bleeding between periods, after intercourse, or after menopause, heavier and longer than normal periods, or a continuous vaginal discharge that may be pale or tinged with blood). There are several ways to decrease the risk of HPV infection. These include avoiding risky sexual practices (unprotected sex, sex at an early age, multiple sex partners, or partners who engage in high-risk sexual activities), a weakened immune system, and not getting the HPV vaccine. Two vaccines are available to protect males and females against the types of HPV that cause most types of cervical cancer (Gardasil® and Ceravix®). Treatment options for cervical cancer include *loop electrosurgical excision procedure (LEEP); cryotherapy*, freezing abnormal cells; *laser therapy*, the use of light to burn abnormal tissue; *hysterectomy, radical hysterectomy; pelvic exteneration*, the removal of all pelvic organs; *radiation*; and *chemotherapy*.

VULVOVAGINAL CANDIDIASIS *Candida albicans* is a yeastlike fungus that commonly grows on mucous membranes of the gastrointestinal and genitourinary tracts. The organism is responsible for **vulvovaginal candidiasis** (vul-vō-VAJ-i-nal can-di-DĪ-a-sis), the most common form of **vaginitis** (vaj-i-NĪ-tis), inflammation of the vagina. Candidiasis is characterized by severe itching; a thick, yellow, cheesy discharge; a yeasty odor; and pain. The disorder, experienced at least once by about 75% of females, is usually a result of proliferation of the fungus following antibiotic therapy for another condition. Predisposing conditions include the use of oral contraceptives or cortisone-like medications, pregnancy, and diabetes.

Sexually Transmitted Diseases

A **sexually transmitted disease (STD)** is one that is spread by sexual contact. In most developed countries of the world, such as those of Western Europe, Japan, Australia, and New Zealand, the incidence of STDs has declined markedly during the past 25 years. In the United States, by contrast, STDs have been rising to near-epidemic proportions; they currently affect more than 65 million people. AIDS and hepatitis B, which are sexually transmitted diseases that also may be contracted in other ways, are discussed in Chapters 22 and 24, respectively.

CHLAMYDIA Chlamydia (kla-MID-ē-a) is a sexually transmitted disease caused by the bacterium *Chlamydia trachomatis* (*chlamy-* = cloak). This unusual bacterium cannot reproduce outside body cells; it "cloaks" itself inside cells, where it divides. At present, chlamydia is the most prevalent sexually transmitted disease in the United States. In most cases, the initial infection is asymptomatic and thus difficult to recognize clinically. In males, urethritis is the principal result, causing a clear discharge, burning on urination, frequent urination, and painful urination. Without treatment, the epididymides may also become inflamed, leading to sterility. In 70% of females with chlamydia, symptoms are absent, but chlamydia is the leading cause of pelvic inflammatory disease. The uterine tubes may also become inflamed, which increases the risk of ectopic pregnancy (implantation of a fertilized ovum outside the uterus) and infertility due to the formation of scar tissue in the tubes.

TRICHOMONIASIS Trichomoniasis (trik′-ō-mō-NĪ-a-sis) is a very common STD and is considered the most curable. It is caused by the protozoan *Trichomonas vaginalis*, which is a normal inhabitant of the vagina in females and urethra in males. Most infected people do not have any signs or symptoms. When symptoms are present, they include itching, burning, genital soreness, discomfort with urination, and an unusual smelling discharge in females. Males experience itching or irritations in the penis, burning after urination or ejaculation, or some discharge. Trichomoniasis can increase the risk of infection with other STDs, such as HIV and gonorrhea.

GONORRHEA Gonorrhea (gon-ō-RĒ-a) or "*the clap*" is caused by the bacterium *Neisseria gonorrhoeae*. In the United States, 1 million to 2 million new cases of gonorrhea appear each year, most among individuals aged 15–29 years. Discharges from infected mucous membranes are the source of transmission of the bacteria either during sexual contact or during the passage of a newborn through the birth canal. The infection site can be in the mouth and throat after oral–genital contact, in the vagina and penis after genital intercourse, or in the rectum after recto–genital contact.

Males usually experience urethritis with profuse pus drainage and painful urination. The prostate and epididymis may also become infected. In females, infection typically occurs in the vagina, often with a discharge of pus. Both infected males and females may harbor the disease without any symptoms, however, until it has progressed to a more advanced stage; about 5–10% of males and 50% of females are asymptomatic. In females, the infection and consequent inflammation can proceed from the vagina into the uterus, uterine tubes, and pelvic cavity. An estimated 50,000 to 80,000 women in the United States are made infertile by gonorrhea every year as a result of scar tissue formation that closes the uterine tubes. If bacteria in the birth canal are transmitted to the eyes of a newborn, blindness can result. Administration of a 1% silver nitrate solution in the infant's eyes prevents infection.

SYPHILIS Syphilis, caused by the bacterium *Treponema pallidum* (trep-o-NĒ-ma PAL-i-dum), is transmitted through sexual contact or exchange of blood, or through the placenta to a fetus. The disease progresses through several stages. During the *primary stage,* the chief sign is a painless open sore, called a **chancre** (SHANG-ker), at the point of contact. The chancre heals within 1 to 5 weeks. From 6 to 24 weeks later, signs and symptoms such as a skin rash, fever, and aches in the joints and muscles usher in the *secondary stage,* which is systemic—the infection spreads to all major body systems. When signs of organ degeneration appear, the disease is said to be in the *tertiary stage.* If the nervous system is involved, the tertiary stage is called **neurosyphilis**. As motor areas become damaged extensively, victims may be unable to control urine and bowel movements. Eventually they may become bedridden and unable even to feed themselves. In addition, damage to the cerebral cortex produces memory loss and personality changes that range from irritability to hallucinations.

GENITAL HERPES   **Genital herpes** is an incurable STD. Type II herpes simplex virus (HSV-2) causes genital infections, producing painful blisters on the prepuce, glans penis, and penile shaft in males and on the vulva or sometimes high up in the vagina in females. The blisters disappear and reappear in most patients, but the virus itself remains in the body. A related virus, type I herpes simplex virus (HSV-1), causes cold sores on the mouth and lips. Infected individuals typically experience recurrences of symptoms several times a year.

GENITAL WARTS   **Genital warts** typically appear as single or multiple bumps in the genital area and are caused by several types of human papillomavirus (HPV). The lesions can be flat or raised, small or large, or shaped like a cauliflower with multiple fingerlike projections. Nearly 1 million people in the United States develop genital warts annually. Genital warts can be transmitted sexually and may appear weeks or months after sexual contact, even if an infected partner has no signs or symptoms of the disease. In most cases, the immune system defends against HPV and the infected cells revert to normal within two years. When immunity is ineffective, lesions appear. There is no cure for genital warts, although topical gels are often useful treatments. As noted earlier, the vaccine Gardasil® is available to protect against most genital warts.

MEDICAL TERMINOLOGY

Castration (kas-TRĀ-shun = to prune) Removal, inactivation, or destruction of the gonads; commonly used in reference to removal of the testes only.

Colposcopy (kol-POS-kō-pē; *colpo-* = vagina; *-scopy* = to view) Visual inspection of the vagina and cervix of the uterus using a culposcope, an instrument that has a magnifying lens (between 5× and 50×) and a light. The procedure generally takes place after an unusual Pap smear.

Culdoscopy (kul-DOS-kō-pē; *-cul-* = cul-de-sac; *-scopy* = to examine) A procedure in which a culdoscope (endoscope) is inserted through the posterior wall of the vagina to view the rectouterine pouch in the pelvic cavity.

Dysmenorrhea (dis-men-ōr-Ē-a; *dys-* = difficult or painful) Pain associated with menstruation; the term is usually reserved to describe menstrual symptoms that are severe enough to prevent a woman from functioning normally for one or more days each month. Some cases are caused by uterine tumors, ovarian cysts, pelvic inflammatory disease, or intrauterine devices.

Dyspareunia (dis-pa-ROO-nē-a; *dys-* = difficult; *-para-* = beside; *-enue* = bed) Pain during sexual intercourse. It may occur in the genital area or in the pelvic cavity, and may be due to inadequate lubrication, inflammation, infection, an improperly fitting diaphragm or cervical cap, endometriosis, pelvic inflammatory disease, pelvic tumors, or weakened uterine ligaments.

Endocervical curettage (kū-re-TAHZH; *curette* = scraper) A procedure in which the cervix is dilated and the endometrium of the uterus is scraped with a spoon-shaped instrument called a curette; commonly called a *D and C* (dilation and curettage).

Fibroids (FĪ-broyds; *fibro-* = fiber; *-eidos* = resemblance) Noncancerous tumors in the myometrium of the uterus composed of muscular and fibrous tissue. Their growth appears to be related to high levels of estrogens. They do not occur before puberty and usually stop growing after menopause. Symptoms include abnormal menstrual bleeding and pain or pressure in the pelvic area.

Hermaphroditism (her-MAF-rō-dīt-izm) The presence of both ovarian and testicular tissue in one individual.

Hypospadias (hī'-pō-SPĀ-dē-as; *hypo-* = below) A common congenital abnormality in which the urethral opening is displaced. In males, the displaced opening may be on the underside of the penis, at the penoscrotal junction, between the scrotal folds, or in the perineum; in females, the urethra opens into the vagina. The problem can be corrected surgically.

Leukorrhea (loo'-kō-RĒ-a; *leuko-* = white) A whitish (nonbloody) vaginal discharge containing mucus and pus cells that may occur at any age and affects most women at some time.

Menorrhagia (men-ō-RA-jē-a; *meno-* = menstruation; *-rhage* = to burst forth) Excessively prolonged or profuse menstrual period. May be due to a disturbance in hormonal regulation of the menstrual cycle, pelvic infection, medications (anticoagulants), fibroids (noncancerous uterine tumors composed of muscle and fibrous tissue), endometriosis, or intrauterine devices.

Oophorectomy (ō'-of-ō-REK-tō-mē; *oophor-* = bearing eggs) Removal of the ovaries.

Orchitis (or-KĪ-tis; *orchi-* = testes; *-itis* = inflammation) Inflammation of the testes, for example, as a result of the mumps virus or a bacterial infection.

Ovarian cyst The most common form of ovarian tumor, in which a fluid-filled follicle or corpus luteum persists and continues growing.

Pelvic inflammatory disease (PID) A collective term for any extensive bacterial infection of the pelvic organs, especially the uterus, uterine tubes, or ovaries, which is characterized by pelvic soreness, lower back pain, abdominal pain, and urethritis. Often the early symptoms of PID occur just after menstruation. As infection spreads, fever may develop, along with painful abscesses of the reproductive organs.

Salpingectomy (sal'-pin-JEK-tō-mē; *salpingo* = tube) Removal of a uterine (fallopian) tube.

Smegma (SMEG-ma) the secretion, consisting principally of desquamated epithelial cells, found chiefly around the external genitals and especially under the foreskin of the male.

SELF-QUIZ QUESTIONS

Fill in the blanks in the following statement.

1. The period of time when secondary sexual characteristics begin to develop and the potential for sexual reproduction is reached is called _____. The first menses is called _____, and the permanent cessation of menses is called _____.

Indicate whether the following statements are true or false.

2. Spermatogenesis does not occur at normal core body temperature.

3. The route of sperm from the production in the testes to the exterior of the body is seminiferous tubules, straight tubules, rete testes, epididymis, ductus (vas) deferens, ejaculatory duct, prostatic urethra, membranous urethra, spongy urethra, external urethral orifice.

Choose the one best answer to the following questions.

4. Which of the following are functions of Sertoli cells? (1) protection of developing spermatogenic cells, (2) nourishment of spermatocytes, spermatids, and sperm, (3) phagocytosis of excess sperm cytoplasm as development proceeds, (4) mediation of the effects of testosterone and FSH, (5) control of movements of spermatogenic cells and release of sperm into the lumen of seminiferous tubules.
   (a) 1, 2, 4, and 5   (b) 1, 2, 3, and 5   (c) 2, 3, 4, and 5
   (d) 1, 2, 3, and 4   (e) 1, 2, 3, 4, and 5

5. Which of the following are *true*? (1) An erection is a sympathetic response initiated by sexual stimulation. (2) Dilation of blood vessels supplying

erectile tissue results in erection. (3) Nitric oxide causes smooth muscle within erectile tissue to relax, which results in widening of blood sinuses. (4) Ejaculation is a sympathetic reflex coordinated by the sacral region of the spinal cord. (5) The purpose of the corpus cavernosa penis is to keep the spongy urethra open during ejaculation.

(a) 1, 2, and 3   (b) 1, 2, 3, 4, and 5   (c) 2 and 3

(d) 2, 4, and 5   (e) 1, 2, 3, and 4

6. Which of the following are *true* concerning estrogens? (1) They promote development and maintenance of female reproductive structures and secondary sex characteristics. (2) They help control fluid and electrolyte balance. (3) They increase protein catabolism. (4) They lower blood cholesterol. (5) In moderate levels, they inhibit the release of GnRH and the secretion of LH and FSH.

(a) 1, 4, and 5   (b) 1, 3, 4, and 5   (c) 1, 2, 3, and 5

(d) 1, 2, 3, and 4   (e) 1, 2, 3, 4, and 5

7. Which of the following statements are *correct*? (1) A sperm head contains DNA and an acrosome. (2) An acrosome is a specialized lysosome that contains enzymes that enable sperm to produce the ATP needed to propel themselves out of the male reproductive tract. (3) Mitochondria in the midpiece of a sperm produce ATP for sperm motility. (4) A sperm's tail, a flagellum, propels it along its way. (5) Once ejaculated, sperm are viable and normally are able to fertilize a secondary oocyte for 5 days.

(a) 1, 2, 3, and 4   (b) 2, 3, 4, and 5   (c) 1, 3, and 4

(d) 2, 4, and 5   (e) 2, 3, and 4

8. Which of the following statements are *correct*? (1) Spermatogonia are stem cells because when they undergo mitosis, some of the daughter cells remain to serve as a reservoir of cells for future mitosis. (2) Meiosis I is a division of pairs of chromosomes resulting in daughter cells with only one member of each chromosome pair. (3) Meiosis II separates the chromatids of each chromosome. (4) Spermiogenesis involves the maturation of spermatids into sperm. (5) The process by which the seminiferous tubules produce haploid sperm is called spermatogenesis.

(a) 1, 2, 3, and 5   (b) 1, 2, 3, 4, and 5   (c) 1, 3, 4, and 5

(d) 1, 2, 3, and 4   (e) 1, 3, and 5

9. Which of the following statements are *correct*? (1) Cells from the yolk sac give rise to oogonia. (2) Ova arise from the germinal epithelium of the ovary. (3) Primary oocytes enter prophase of meiosis I during fetal development but do not complete it until after puberty. (4) Once a secondary oocyte is formed, it proceeds to metaphase of meiosis II and stops at this stage. (5) The secondary oocyte resumes meiosis II and forms the ovum and a polar body only if fertilization occurs. (6) A primary oocyte gives rise to an ovum and four polar bodies.

(a) 1, 3, 4, and 5   (b) 1, 3, 4, and 6   (c) 1, 2, 4, and 6

(d) 1, 2, 4, and 5   (e) 1, 2, 5, and 6

10. Which of the following statements are *correct*? (1) The female reproductive cycle consists of a menstrual phase, a preovulatory phase, ovulation, and a postovulatory phase. (2) During the menstrual phase, small secondary follicles in the ovary begin to enlarge while the uterus is shedding its lining. (3) During the preovulatory phase, a dominant follicle continues to grow and begins to secrete estrogens and inhibin while the uterine lining begins to rebuild. (4) Ovulation results in the release of an ovum and the shedding of the uterine lining to nourish and support the released ovum. (5) After ovulation, a corpus luteum forms from the ruptured follicle and begins to secrete progesterone and estrogens, which it will continue to do throughout pregnancy if the egg is fertilized. (6) If pregnancy does not occur, then the corpus luteum degenerates into a scar called the corpus albicans, and the uterine lining is prepared to be shed again.

(a) 1, 2, 4, and 5   (b) 2, 4, 5, and 6   (c) 1, 4, 5, and 6

(d) 1, 3, 4, and 6   (e) 1, 2, 3, and 6

11. Oral contraceptives work by (1) causing a thickening of the cervical mucus, (2) blocking the uterine tubes, (3) inhibiting the release of FSH and LH, (4) preventing ovulation, (5) disrupting the plasma membranes of sperm, (6) irritating the endometrial lining so that it is inhospitable for fetal development.

(a) 3 only   (b) 3 and 4   (c) 1, 2, and 5

(d) 1, 3, and 4   (e) 1, 2, 3, 4, and 5

12. Match the following:

_____ (a) the process during meiosis when portions of homologous chromosomes may be exchanged with each other

_____ (b) refers to cells containing one-half the chromosome number

_____ (c) the cell produced by the union of an egg and a sperm

_____ (d) the degeneration of oogonia before and after birth

_____ (e) a packet of discarded nuclear material from the first or second meiotic division of the egg

_____ (f) refers to cells containing the full chromosome number

(1) zygote
(2) haploid
(3) diploid
(4) crossing-over
(5) polar body
(6) atresia

13. Match the following:

_____ (a) modified sudoriferous glands involved in lactation

_____ (b) a small, cylindrical mass of erectile tissue and nerves in the female; homologue of the male glans penis

_____ (c) produce mucus in the female during sexual arousal and intercourse; homologous to the male bulbourethral glands

_____ (d) the group of cells that nourish the developing oocyte and begin to secrete estrogens

_____ (e) a pathway for sperm to reach the uterine tubes; the site of menstruation; the site of implantation of a fertilized ovum; the womb

_____ (f) produces progesterone, estrogens, relaxin, and inhibin

_____ (g) draw the ovum into the uterine tube

_____ (h) the opening between the uterus and vagina

_____ (i) muscular layer of uterus; responsible for expulsion of fetus from uterus

_____ (j) mucus-secreting glands in the female that are homologous to the prostate gland

_____ (k) the female copulatory organ; the birth canal

_____ (l) passageway for the ovum to the uterus; usual site of fertilization; site of tubal ligation

_____ (m) refers to the external genitals of the female

_____ (n) the layer of the uterine lining that is partially shed during each monthly cycle

(1) follicle
(2) corpus luteum
(3) uterine tube
(4) fimbriae
(5) uterus
(6) cervix
(7) endometrium
(8) vagina
(9) vulva
(10) clitoris
(11) paraurethral glands
(12) greater vestibular glands
(13) mammary glands
(14) myometrium

14. Match the following:

_____ (a) site of sperm maturation

_____ (b) the male copulatory organ; a passageway for ejaculation of sperm and excretion of urine

_____ (c) sperm-forming cells

_____ (d) produce an alkaline substance that protects sperm by neutralizing acids in the urethra

_____ (e) ejects sperm into the urethra just before ejaculation

_____ (f) the supporting structure for the testes

_____ (g) carries the sperm from the scrotum into the abdominopelvic cavity for release by ejaculation; is cut and tied as a means of sterilization

_____ (h) the shared terminal duct of the reproductive and urinary systems in the male

_____ (i) surrounds the urethra at the base of the bladder; produces secretions that contribute to sperm motility and viability

_____ (j) produce testosterone

_____ (k) supporting structure that consists of the ductus deferens, testicular artery, autonomic nerves, veins that drain the testes, lymphatic vessels, and cremaster muscle

_____ (l) support and protect developing spermatogenic cells; secrete inhibin; form the blood–testis barrier

_____ (m) secrete an alkaline fluid to help neutralize acids in the female reproductive tract; secrete fructose for use in ATP production by sperm

_____ (n) contraction and relaxation move testes near to or away from pelvic cavity

_____ (o) site of spermatogenesis

(1) spermatogenic cells

(2) Sertoli cells

(3) Leydig cells

(4) penis

(5) scrotum

(6) epididymis

(7) ductus deferens

(8) ejaculatory duct

(9) seminiferous tubules

(10) seminal vesicles

(11) prostate gland

(12) bulbourethral glands

(13) urethra

(14) spermatic cord

(15) cremaster muscle

29 | DEVELOPMENT AND INHERITANCE

29.1 Overview of Development

1. Pregnancy is a sequence of events that begins with fertilization, and proceeds to implantation, embryonic development, and fetal development. It normally ends in birth.
2. During the embryonic period (fertilization through the eighth week of development), the developing human is called an embryo.
3. During the fetal period (the ninth week of development until birth), the developing human is known as a fetus.

29.2 The First Two Weeks of the Embryonic Period

1. During fertilization a sperm cell penetrates a secondary oocyte and their pronuclei unite. Penetration of the zona pellucida is facilitated by enzymes in the sperm's acrosome. The resulting cell is a zygote.
2. Normally, only one sperm cell fertilizes a secondary oocyte because of the fast and slow blocks to polyspermy.
3. Early rapid cell division of zygote is called cleavage, and the cells produced by cleavage are called blastomeres. The solid sphere of cells produced by cleavage is a morula. The morula develops into a blastocyst, a hollow ball of cells differentiated into a tropoblast and an inner cell mass. The attachment of a blastocyst to the endometrium is termed implantation; it occurs as a result of enzymatic degradation of the endometrium. After implantation, the endometrium becomes modified and is known as the decidua. The trophoblast develops into the synctiotrophoblast and cytotrophoblast, both of which become part of the chorion. The inner cell mass differentiates into hypoblast and epiblast, the bilaminar (two-layered) embryonic disc.
4. The amnion is a thin protective membrane that develops from the cytotrophoblast.
5. The exocoelomic membrane and hypoblast form the yolk sac, which transfers nutrients to the embryo, forms blood cells, produces primordial germ cells, and forms part of the gut.
6. Erosion of sinusoids and endometrial glands provides blood and secretions, which enter lacunar networks to supply nutrition to and remove wastes from the embryo.
7. The extraembryonic coelom forms within extraembryonic mesoderm.
8. The extraembryonic mesoderm and trophoblast form the chorion, the principal embryonic part of the placenta.

29.3 The Remaining Weeks of the Embryonic Period

1. The third week of development is characterized by gastrulation, the conversion of the bilaminar disc into a trilaminar (three-layered) embryo consisting of ectoderm, mesoderm, and endoderm. The first evidence of gastrulation is formation of the primitive streak, after which the primitive node, notochordal process, and notochord develop. The three primary germ layers form all tissues and organs of the developing organism. **Table 29.1** summarizes the structures that develop from the primary germ layers. Also during the third week, the oropharyngeal and cloacal membranes form. The wall of the yolk sac forms a small vascularized outpouching called the allantois, which functions in blood formation and development of the urinary bladder.
2. The process by which the neural plate, neural folds, and neural tube form is called neurulation. The brain and spinal cord develop from the neural tube.
3. Paraxial mesoderm segments to form somites from which skeletal muscles of the neck, trunk, and limbs develop. Somites also form connective tissues and vertebrae.
4. Blood vessel formation, called angiogenesis, begins in mesodermal cells called angioblasts.
5. The heart forms from mesodermal cells called the cardiogenic area. By the end of the third week, the primitive heart beats and circulates blood. Chorionic villi, projections of the chorion, connect to the embryonic heart so that maternal and fetal blood vessels are brought into close proximity, allowing the exchange of nutrients and wastes between maternal and fetal blood.
6. Placentation refers to formation of the placenta, the site of exchange of nutrients and wastes between the mother and fetus. The placenta also functions as a protective barrier, stores nutrients, and produces several hormones to maintain pregnancy. The actual connection between the placenta and embryo (and later the fetus) is the umbilical cord.
7. Organogenesis refers to the formation of body organs and systems and occurs during the fourth week of development.
8. Conversion of the flat, two-dimensional trilaminar embryonic disc to a three-dimensional cylinder occurs by a process called embryonic folding. Embryonic folding brings various organs into their final adult positions and helps form the gastrointestinal tract.
9. Pharyngeal arches, clefts, and pouches give rise to the structures of the head and neck.
10. By the end of the fourth week, upper and lower limb buds develop, and by the end of the eighth week the embryo has clearly human features.

29.4 Fetal Period

1. The fetal period is primarily concerned with the growth and differentiation of tissues and organs that developed during the embryonic period.
2. The rate of body growth is remarkable, especially during the ninth and sixteenth weeks.
3. The principal changes associated with embryonic and fetal growth are summarized in **Table 29.2**.

29.5 Teratogens

1. Teratogens are agents that cause physical defects in developing embryos.
2. Among the more important teratogens are alcohol, pesticides, industrial chemicals, some prescription drugs, cocaine, LSD, nicotine, and ionizing radiation.

29.6 Prenatal Diagnostic Tests

1. Several prenatal diagnostic tests are used to detect genetic disorders and to assess fetal well-being. These include fetal ultrasonography, in which an image of a fetus is displayed on a screen; amniocentesis, the withdrawal and analysis of amniotic fluid and the fetal cells within it; and chorionic villi sampling (CVS), which involves withdrawal of chorionic villi tissue for chromosomal analysis.
2. CVS can be done earlier than amniocentesis, and the results are available more quickly, but it is slightly riskier than amniocentesis.
3. Noninvasive prenatal tests include the maternal alpha-fetoprotein (AFP) test to detect neural tube defects and the Quad AFP Plus test to detect Down syndrome, trisomy 18, and neural tube defects.

29.7 Maternal Changes during Pregnancy

1. Pregnancy is maintained by human chorionic gonadotropin (hCG), estrogens, and progesterone.
2. Human chorionic somatomammotropin (hCS) contributes to breast development, protein anabolism, and catabolism of glucose and fatty acids.
3. Relaxin increases flexibility of the pubic symphysis and helps dilate the uterine cervix near the end of pregnancy.
4. Corticotropin-releasing hormone, produced by the placenta, is thought to establish the timing of birth, and stimulates the secretion of cortisol by the fetal adrenal gland.
5. During pregnancy, several anatomical and physiological changes occur in the mother.

29.8 Exercise and Pregnancy

1. During pregnancy, some joints become less stable, and certain physical activities are more difficult to execute.
2. Moderate physical activity does not endanger the fetus in a normal pregnancy.

29.9 Labor

1. Labor is the process by which the fetus is expelled from the uterus through the vagina to the outside. True labor involves dilation of the cervix, expulsion of the fetus, and delivery of the placenta.
2. Oxytocin stimulates uterine contractions via a positive feedback cycle.

29.10 Adjustments of the Infant at Birth

1. The fetus depends on the mother for oxygen and nutrients, the removal of wastes, and protection.
2. Following birth, an infant's respiratory and cardiovascular systems undergo changes to enable them to become self-supporting during postnatal life.

29.11 The Physiology of Lactation

1. Lactation refers to the production and ejection of milk by the mammary glands.
2. Milk production is influenced by prolactin (PRL), estrogens, and progesterone.
3. Milk ejection is stimulated by oxytocin.
4. A few of the many benefits of breastfeeding include ideal nutrition for the infant, protection from disease, and decreased likelihood of developing allergies.

29.12 Inheritance

1. Inheritance is the passage of hereditary traits from one generation to the next.
2. The genetic makeup of an organism is called its genotype; the traits expressed are called its phenotype.
3. Dominant genes control a particular trait; expression of recessive genes is masked by dominant genes.
4. Many patterns of inheritance do not conform to the simple dominant–recessive patterns. In incomplete dominance, neither member of an allelic pair dominates; phenotypically, the heterozygote is intermediate between the homozygous dominant and the homozygous recessive. In multiple-allele inheritance, genes have more than two alternate forms. An example is the inheritance of ABO blood groups. In complex inheritance, a trait such as skin or eye color is controlled by the combined effects of two or more genes and may be influenced by environmental factors.
5. Each somatic cell has 46 chromosomes—22 pairs of autosomes and 1 pair of sex chromosomes.
6. In females, the sex chromosomes are two X chromosomes; in males, they are one X chromosome and a much smaller Y chromosome, which normally includes the prime male-determining gene, called *SRY*.
7. If the *SRY* gene is present and functional in a fertilized ovum, the fetus will develop testes and differentiate into a male. In the absence of *SRY*, the fetus will develop ovaries and differentiate into a female.
8. Red–green color blindness and hemophilia result from recessive genes located on the X chromosome. These sex-linked traits occur primarily in males because of the absence of any counterbalancing dominant genes on the Y chromosome.
9. A mechanism termed X-chromosome inactivation (lyonization) balances the difference in number of X chromosomes between males (one X) and females (two Xs). In each cell of a female's body, one X chromosome is randomly and permanently inactivated early in development and becomes a Barr body.
10. A given phenotype is the result of the interactions of genotype and the environment.

CLINICAL CONNECTIONS

Stem Cell Research and Therapeutic Cloning (Refer page 959 of Textbook)

Stem cells are unspecialized cells that have the ability to divide for indefinite periods and give rise to specialized cells. In the context of human development, a zygote (fertilized ovum) is a stem cell. Because it has the potential to form an entire organism, a zygote is known as a *totipotent stem cell* (tō-TIP-ō-tent; *totus-* = whole; *-potentia* = power). Inner cell mass cells, called *pluripotent stem cells* (ploo-RIP-ō-tent; *plur-* = several), can give rise to many (but not all) different types of cells. Later, pluripotent stem cells can undergo further specialization into *multipotent stem cells* (mul-TIP-ō-tent), stem cells with a specific function. Examples include keratinocytes that produce new skin cells, myeloid and lymphoid stem cells that develop into blood cells, and spermatogonia that give rise to sperm. Pluripotent stem cells currently used in research are derived from (1) the embryoblast of embryos in the blastocyst stage that were destined to be used for infertility treatments but were not needed and from (2) nonliving fetuses terminated during the first 3 months of pregnancy.

Scientists are also investigating the potential clinical applications of *adult stem cells*—stem cells that remain in the body throughout adulthood. Recent experiments suggest that the ovaries of adult mice contain stem cells that can develop into new ova (eggs). If these same types of stem cells are found in the ovaries of adult women, scientists could potentially harvest some of them from a woman about to undergo a sterilizing medical treatment (such as chemotherapy), store them, and then return the stem cells to her ovaries after the medical treatment is completed in order to restore fertility. Studies have also suggested that stem cells in human adult red bone marrow have the ability to differentiate into cells of the liver, kidney, heart, lung, skeletal muscle, skin, and organs of the gastrointestinal tract. In theory, adult stem cells from red bone marrow could be harvested from a patient and then used to repair other tissues and organs in that patient's body without having to use stem cells from embryos.

Ectopic Pregnancy (Refer page 960 of Textbook)

Ectopic pregnancy (ek-TOP-ik; *ec-* = out of; *-topic* = place) is the development of an embryo or fetus outside the uterine cavity. An ectopic pregnancy usually occurs when movement of the fertilized ovum through the uterine tube is impaired by scarring due to a prior tubal infection, decreased movement of the uterine tube smooth muscle, or abnormal tubal anatomy. Although the most common site of ectopic pregnancy is the uterine tube, ectopic pregnancies may also occur in the ovary, abdominal cavity, or uterine cervix. Women who smoke are twice as likely to have an ectopic pregnancy because nicotine in cigarette smoke paralyzes the cilia in the lining of the uterine tube (as it does those in the respiratory airways). Scars from pelvic inflammatory disease, previous uterine tube surgery, and previous ectopic pregnancy may also hinder movement of the fertilized ovum.

The signs and symptoms of ectopic pregnancy include one or two missed menstrual cycles followed by bleeding and acute abdominal and pelvic pain. Unless removed, the developing embryo can rupture the uterine tube, often resulting in death of the mother. Treatment options include surgery or the use of a cancer drug called methotrexate, which causes embryonic cells to stop dividing and eventually disappear.

Anencephaly (Refer page 965 of Textbook)

Neural tube defects (NTDs) are caused by arrest of the normal development and closure of the neural tube. These include spina bifida (discussed in Disorders: Homeostatic Imbalances in Chapter 7) and **anencephaly** (an-en-SEF-a-lē; *an-* = without; *-encephal* = brain). In anencephaly, the cranial bones fail to develop and certain parts of the brain remain in contact with amniotic fluid and degenerate. Usually, a part of the brain that controls vital functions such

as breathing and regulation of the heart is also affected. Infants with anencephaly are stillborn or die within a few days after birth. The condition occurs about once in every 1000 births and is two to four times more common in female infants than males.

Placenta Previa (Refer page 969 of Textbook)

In some cases, the entire placenta or part of it may become implanted in the inferior portion of the uterus, near or covering the internal os of the cervix. This condition is called **placenta previa** (PRĒ-vē-a = before or in front of). Although placenta previa may lead to spontaneous abortion, it also occurs in approximately 1 in 250 live births. It is dangerous to the fetus because it may cause premature birth and intrauterine hypoxia due to maternal bleeding. Maternal mortality is increased due to hemorrhage and infection. The most important symptom is sudden, painless, bright-red vaginal bleeding in the third trimester. Cesarean section is the preferred method of delivery in placenta previa.

Early Pregnancy Tests (Refer page 978 of Textbook)

Early pregnancy tests detect the tiny amounts of human chorionic gonadotropin (hCG) in the urine that begin to be excreted about 8 days after fertilization. The test kits can detect pregnancy as early as the first day of a missed menstrual period—that is, at about 14 days after fertilization. Chemicals in the kits produce a color change if a reaction occurs between hCG in the urine and hCG antibodies included in the kit.

Several of the test kits available at pharmacies are as sensitive and accurate as test methods used in many hospitals. Still, false-negative and false-positive results can occur. A false-negative result (the test is negative, but the woman is pregnant) may be due to testing too soon or to an ectopic pregnancy. A false-positive result (the test is positive, but the woman is not pregnant) may be due to excess protein or blood in the urine or to hCG production due to a rare type of uterine cancer. Thiazide diuretics, hormones, steroids, and thyroid drugs may also affect the outcome of an early pregnancy test.

Pregnancy-Induced Hypertension (Refer page 979 of Textbook)

About 10–15% of all pregnant women in the United States experience **pregnancy-induced hypertension (PIH)**, an elevated blood pressure that is associated with pregnancy. The major cause is **preeclampsia** (prē-ē-KLAMP-sē-a), an abnormal condition of pregnancy characterized by sudden hypertension, large amounts of protein in the urine, and generalized edema that typically appears after the 20th week of pregnancy. Other signs and symptoms are generalized edema, blurred vision, and headaches. Preeclampsia might be related to an autoimmune or allergic reaction resulting from the presence of a fetus. Treatment involves bed rest and various drugs. When the condition is also associated with convulsions and coma, it is termed **eclampsia**.

Dystocia and Cesarean Section (Refer page 981 of Textbook)

Dystocia (dis-TŌ-sē-a; *dys-* = painful or difficult; *-toc-* = birth), or difficult labor, may result either from an abnormal position (presentation) of the fetus or a birth canal of inadequate size to permit vaginal delivery. In a **breech presentation**, for example, the fetal buttocks or lower limbs, rather than the head, enter the birth canal first; this occurs most often in premature births. If fetal or maternal distress prevents a vaginal birth, the baby may be delivered surgically through an abdominal incision. A low, horizontal cut is made through the abdominal wall and lower portion of the uterus, through which the baby and placenta are removed. Even though it is popularly associated with the birth of Julius Caesar, the true reason this procedure is termed a **cesarean section** (*C-section*) is because it was described in Roman law, *lex cesarea*, about 600 years before Julius Caesar was born. Even a history of multiple C-sections need not exclude a pregnant woman from attempting a vaginal delivery.

Premature Infants (Refer page 981 of Textbook)

Delivery of a physiologically immature baby carries certain risks. A **premature infant** or "preemie" is generally considered a baby who weighs less than 2500 g (5.5 lb) at birth. Poor prenatal care, drug abuse, history of a previous premature delivery, and mother's age below 16 or above 35 increase the chance of premature delivery. The body of a premature infant is not yet ready to sustain some critical functions, and thus its survival is uncertain without medical intervention. The major problem after delivery of an infant under 36 weeks of gestation is respiratory distress syndrome (RDS) of the newborn due to insufficient surfactant. RDS can be eased by use of artificial surfactant and a ventilator that delivers oxygen until the lungs can operate on their own.

DISORDERS: HOMEOSTATIC IMBALANCES

Infertility

Female infertility, or the inability to conceive, occurs in about 10% of all women of reproductive age in the United States. Female infertility may be caused by ovarian disease, obstruction of the uterine tubes, or conditions in which the uterus is not adequately prepared to receive a fertilized ovum. **Male infertility** (*sterility*) is an inability to fertilize a secondary oocyte; it does not imply erectile dysfunction (impotence). Male fertility requires production of adequate quantities of viable, normal sperm by the testes, unobstructed transport of sperm though the ducts, and satisfactory deposition in the vagina. The seminiferous tubules of the testes are sensitive to many factors—x-rays, infections, toxins, malnutrition, and higher-than-normal scrotal temperatures—that may cause degenerative changes and produce male sterility.

One cause of infertility in females is inadequate body fat. To begin and maintain a normal reproductive cycle, a female must have a minimum amount of body fat. Even a moderate deficiency of fat—10% to 15% below normal weight for height—may delay the onset of menstruation (menarche), inhibit ovulation during the reproductive cycle, or cause amenorrhea (cessation of menstruation). Both dieting and intensive exercise may reduce body fat below the minimum amount and lead to infertility that is reversible, if weight gain or reduction of intensive exercise or both occur. Studies of very obese women indicate that they, like very lean ones, experience problems with amenorrhea and infertility. Males also experience reproductive problems in response to undernutrition and weight loss. For example, they produce less prostatic fluid and reduced numbers of sperm having decreased motility.

Many fertility-expanding techniques now exist for assisting infertile couples to have a baby. The birth of Louise Joy Brown on July 12, 1978, near Manchester, England, was the first recorded case of **in vitro fertilization (IVF)**—fertilization in a laboratory dish. In the IVF procedure, the mother-to-be is given follicle-stimulating hormone (FSH) soon after menstruation, so that several secondary oocytes, rather than the typical single oocyte, will be produced (superovulation). When several follicles have reached the appropriate size, a small incision is made near the umbilicus, and the secondary oocytes are aspirated from the stimulated follicles and transferred to a solution containing sperm, where the oocytes undergo fertilization. Alternatively, an oocyte may be fertilized in vitro by suctioning a sperm or even a spermatid obtained from the testis into a tiny pipette and then injecting it into the oocyte's cytoplasm. This procedure, termed **intracytoplasmic sperm injection (ICSI)** (in′-tra-sī-tō-PLAZ-mik), has been used when infertility is due to impairments in sperm motility or to the failure of spermatids to develop into spermatozoa. When the zygote achieved by IVF reaches the 8-cell or 16-cell stage, it is introduced into the uterus for implantation and subsequent growth.

In **embryo transfer**, a man's semen is used to artificially inseminate a fertile secondary oocyte donor. After fertilization in the donor's uterine tube, the morula or blastocyst is transferred from the donor to the infertile woman, who then carries it (and subsequently the fetus) to term. Embryo transfer is indicated for women who are infertile or who do not want to pass on their own genes because they are carriers of a serious genetic disorder.

In **gamete intrafallopian transfer (GIFT)** the goal is to mimic the normal process of conception by uniting sperm and secondary oocyte in the prospective mother's uterine tubes. It is an attempt to bypass conditions in the female reproductive tract that might prevent fertilization, such as high acidity or inappropriate mucus. In this procedure, a woman is given FSH and LH to stimulate the production of several secondary oocytes, which are aspirated from the mature follicles, mixed outside the body with a solution containing sperm, and then immediately inserted into the uterine tubes.

Congenital Defects

An abnormality that is present at birth, and usually before, is called a **congenital defect**. Such defects occur during the formation of structures that develop during the period of organogenesis, the fourth through eighth weeks of development, when all major organs appear. During organogenesis, stem cells are establishing the basic patterns of organ development, and it is during this time that developing structures are very susceptible to genetic and environmental influences.

Major structural defects occur in 2–3% of liveborn infants, and they are the leading cause of infant mortality, accounting for about 21% of infant deaths. Many congenital defects can be prevented by supplementation or avoidance of certain substances. For example, neural tube defects, such as spina bifida and anencephaly, can be prevented by having a pregnant female take folic acid. Iodine supplementation can prevent the mental retardation and bone deformation associated with cretinism. Avoidance of teratogens is also very important in preventing congenital defects.

Down Syndrome

Down syndrome (DS) is a disorder characterized by three, rather than two, copies of at least part of chromosome 21. Overall, one infant in 900 is born with Down syndrome. However, older women are more likely to have a DS baby. The chance of having a baby with this syndrome, which is less than 1 in 3000 for women under age 30, increases to 1 in 300 in the 35–39 age group and to 1 in 9 at age 48.

Down syndrome is characterized by mental retardation, retarded physical development (short stature and stubby fingers), distinctive facial structures (large tongue, flat profile, broad skull, slanting eyes, and round head), kidney defects, suppressed immune system, and malformations of the heart, ears, hands, and feet. Sexual maturity is rarely attained, and life expectancy is shorter.

MEDICAL TERMINOLOGY

Breech presentation A malpresentation in which the fetal buttocks or lower limbs present into the maternal pelvis; the most common cause is prematurity.

Conceptus (kon-SEP-tus) Includes all structures that develop from a zygote and includes an embryo plus the embryonic part of the placenta and associated membranes (chorion, amnion, yolk sac, and allantois).

Cryopreserved embryo (krī-ō-PRĒ-servd; *cryo-* = cold) An early embryo produced by in vitro fertilization (fertilization of a secondary oocyte in a laboratory dish) that is preserved for a long period by freezing it. After thawing, the early embryo is implanted into the uterine cavity. Also called a **frozen embryo**.

Deformation (dē-for-MĀ-shun; *de-* = without; *-forma* = form) A developmental abnormality due to mechanical forces that mold a part of the fetus over a prolonged period of time. Deformations usually involve the skeletal and/or muscular system and may be corrected after birth. An example is clubfeet.

Emesis gravidarum (EM-e-sis gra-VID-ar-um; *emeo* = to vomit; *gravida* = a pregnant woman) Episodes of nausea and possibly vomiting that are most likely to occur in the morning during the early weeks of pregnancy; also called **morning sickness**. Its cause is unknown, but the high levels of human chorionic gonadotropin (hCG) secreted by the placenta, and of progesterone secreted by the ovaries, have been implicated. If the severity of these symptoms requires hospitalization for intravenous feeding, the condition is known as **hyperemesis gravidarum**.

Epigenesis (ep-i-JEN-e-sis; *epi-* = upon; *-genesis* = creation) The development of an organism from an undifferentiated cell.

Fertilization age Two weeks less than the gestational age, since a secondary oocyte is not fertilized until about 2 weeks after the last normal menstrual period (LNMP).

Fetal alcohol syndrome (FAS) A specific pattern of fetal malformation due to intrauterine exposure to alcohol. FAS is one of the most common causes of mental retardation and the most common preventable cause of birth defects in the United States.

Fetal surgery A surgical procedure performed on a fetus; in some cases the uterus is opened and the fetus is operated on directly. Fetal surgery has been used to repair diaphragmatic hernias and remove lesions in the lungs.

Gestational age (jes-TĀ-shun-al; *gestatus* = to bear) The age of an embryo or fetus calculated from the presumed first day of the last normal menstrual period (LNMP).

Karyotype (KAR-ē-ō-tīp; *karyo-* = nucleus) The chromosomal characteristics of an individual presented as a systematic arrangement of pairs of metaphase chromosomes arrayed in descending order of size and according to the position of the centromere (see **Figure 29.24**); useful in judging whether chromosomes are normal in number and structure.

Klinefelter's syndrome A sex chromosome aneuploidy, usually due to trisomy XXY, that occurs once in every 500 births. Such individuals are somewhat mentally disadvantaged, sterile males with undeveloped testes, scant body hair, and enlarged breasts.

Lethal gene (LĒ-thal JĒN; *lethum* = death) A gene that, when expressed, results in death either in the embryonic state or shortly after birth.

Metafemale syndrome A sex chromosome aneuploidy characterized by at least three X chromosomes (XXX) that occurs about once in every 700 births. These females have underdeveloped genital organs and limited fertility, and most are mentally retarded.

Primordium (prī-MOR-dē-um; *primus-* = first; *-ordior* = to begin) The beginning or first discernible indication of the development of an organ or structure.

Puerperal fever (pū-ER-per-al; *puer* = child) An infectious disease of childbirth, also called puerperal sepsis and childbed fever. The disease, which results from an infection originating in the birth canal, affects the mother's endometrium. It may spread to other pelvic structures and lead to septicemia.

Turner syndrome A sex chromosome aneuploidy caused by the presence of a single X chromosome (designated XO); occurring about once in every 5000 births, it produces a sterile female with virtually no ovaries and limited development of secondary sex characteristics. Other features include short stature, webbed neck, underdeveloped breasts, and widely spaced nipples. Intelligence usually is normal.

SELF-QUIZ QUESTIONS

Fill in the blanks in the following statements.

1. The three stages of true labor, in order of occurrence, are _____, _____, and _____.

2. Hormones produced by the _____ are responsible for maintaining the pregnancy during the first 3–4 months. The hormone responsible for preventing degeneration of the corpus luteum is _____ produced by the trophoblast.

3. Indicate the germ layers responsible for development of the following structures: (a) muscle, bone, and peritoneum: _____; (b) nervous system and epidermis: _____; (c) epithelial linings of respiratory and gastrointestinal tracts: _____ .

Indicate whether the following statement is true or false.

4. Labor is an example of a negative feedback cycle that ends with the birth of the infant.

Choose the one best answer to the following questions.

5. Which of the following are *true*? (1) During implantation the outer cell mass of the blastocyst orients toward the endometrium. (2) The decidua basalis provides glycogen and lipids for the developing fetus. (3) The decidua parietalis becomes the maternal part of the placenta. (4) During implantation, the syncytiotrophoblast secretes enzymes that allow the blastocyst to penetrate the uterine lining. (5) After fetal delivery, the decidua separates from the endometrium and is released from the uterus.
   (a) 2, 4, and 5 (b) 1, 2, and 3 (c) 2, 3, 4, and 5
   (d) 1, 2, 3, 4, and 5 (e) 1, 3, and 5

6. Which of the following are maternal changes that occur during pregnancy? (1) altered pulmonary function; (2) increased stroke volume, cardiac output, and heart rate, and decreased blood volume; (3) weight gain; (4) increased gastric motility, causing a delay in gastric emptying time; (5) edema and possible varicose veins.
   (a) 1, 2, 3, and 4 (b) 2, 3, 4, and 5 (c) 1, 3, 4, and 5
   (d) 1, 3, and 5 (e) 2, 4, and 5

7. Which of the following statements is *correct*? (a) Normal traits always dominate over abnormal traits. (b) Occasionally an error in meiosis called nondisjunction results in an abnormal number of chromosomes. (c) The mother always determines the sex of the child because she has either an X or Y gene in her oocytes. (d) Most patterns of inheritance are simple dominant–recessive inheritances. (e) Genes are expressed normally regardless of any outside influence such as chemicals or radiation.

8. Which of the following are *true* concerning fertilization? (1) The sperm first penetrate the zona pellucida and then the corona radiata. (2) The binding of specific membrane proteins in the sperm head to ZP3 causes the release of acrosomal contents. (3) Sperm are able to fertilize the oocyte within minutes after ejaculation. (4) Depolarization of the cell membrane of the secondary oocyte inhibits fertilization by more than one sperm. (5) The oocyte completes meiosis II after fertilization.

   (a) 1, 2, 4, and 5   (b) 1, 3, and 5   (c) 1, 2, 3, and 4

   (d) 1, 4, and 5   (e) 2, 4, and 5

9. Amniotic fluid (1) is derived entirely from a filtrate of maternal blood, (2) acts as a fetal shock absorber, (3) provides nutrients to the fetus, (4) helps regulate fetal body temperature, (5) prevents adhesions between the skin of the fetus and surrounding tissues.

   (a) 1, 2, 3, 4, and 5   (b) 2, 4, and 5   (c) 2, 3, 4, and 5

   (d) 1, 4, and 5   (e) 1, 2, 4, and 5

10. Which of the following structures develop during the fourth week after fertilization? (1) embryonic folding, (2) the neural tube, (3) otic placode (beginning of the ear), (4) beginning of the eyes, (5) upper and lower limb buds.

   (a) 1 and 2   (b) 1, 2, and 5   (c) 1, 2, 3, 4, and 5

   (d) 2, 3, and 5   (e) 1, 3, 4, and 5

11. Match the following:

   _____ (a) a fluid-filled sphere of cells that enters the uterine cavity

   _____ (b) cells produced by cleavage

   _____ (c) the developing individual from week nine of pregnancy until birth

   _____ (d) the outer covering of cells of the blastocyst

   _____ (e) membrane derived from trophoblast

   _____ (f) early divisions of the zygote

   _____ (g) a solid sphere of cells still surrounded by the zona pellucida

   _____ (h) event in which differentiation into the three primary germ layers occurs

   _____ (i) embryonic development of structures that will become the nervous system

   _____ (j) the formation of blood vessels to support the developing embryo

   _____ (k) result of the fusion of female and male pronuclei

   (1) cleavage
   (2) blastomeres
   (3) morula
   (4) angiogenesis
   (5) trophoblast
   (6) blastocyst
   (7) zygote
   (8) gastrulation
   (9) neurulation
   (10) chorion
   (11) fetus

12. Match the following:

   _____ (a) stimulates the corpus luteum to continue production of progesterone and estrogens

   _____ (b) increases the flexibility of the pubic symphysis and helps dilate the uterine cervix during labor

   _____ (c) secreted by the placenta; helps establish the timing of birth and increases the secretion of cortisol for fetal lung maturation

   _____ (d) helps prepare mammary glands for lactation; regulates certain aspects of maternal and fetal metabolism

   _____ (e) stimulates uterine contractions; responsible for the milk ejection reflex

   _____ (f) promotes milk synthesis and secretion; inhibited by progesterone during pregnancy

   (1) oxytocin
   (2) human chorionic somatomammo-tropin
   (3) human chorionic gonadotropin
   (4) prolactin
   (5) corticotropin-releasing hormone
   (6) relaxin

13. Match the following:

   _____ (a) the penetration of a secondary oocyte by a single sperm cell

   _____ (b) fertilization of a secondary oocyte by more than one sperm

   _____ (c) the attachment of a blastocyst to the endometrium

   _____ (d) the fusion of the genetic material from a haploid sperm and a haploid secondary oocyte into a single diploid nucleus

   _____ (e) the induction by the female reproductive tract of functional changes in sperm that allow them to fertilize a secondary oocyte

   _____ (f) the examination of embryonic or fetal cells sloughed off into the amniotic fluid

   _____ (g) an abnormal condition of pregnancy characterized by sudden hypertension, large amounts of protein in urine, and generalized edema

   _____ (h) noninvasive test that can detect fetal neural tube defects

   _____ (i) the process of giving birth

   _____ (j) the period of time (about 6 weeks) during which the maternal reproductive organs and physiology return to the prepregnancy state

   (1) fertilization
   (2) capacitation
   (3) syngamy
   (4) polyspermy
   (5) implantation
   (6) amniocentesis
   (7) preeclampsia
   (8) parturition
   (9) puerperium
   (10) maternal AFP test

14. Match the following:

_____ (a) the control of inherited traits by the combined effects of many genes

_____ (b) the two alternative forms of a gene that code for the same trait and are at the same location on homologous chromosomes

_____ (c) abnormal number of chromosomes due to failure of homologous chromosomes or chromatids to separate

_____ (d) inheritance based on genes that have more than two alternative forms; an example is the inheritance of blood type

_____ (e) a cell in which one or more chromosomes of a set is added or deleted

_____ (f) refers to an individual with different alleles on homologous chromosomes

_____ (g) traits linked to the X chromosome

_____ (h) permanent inheritable change in an allele that produces a different variant of the same trait

_____ (i) neither member of the allelic pair is dominant over the other, and the heterozygote has a phenotype intermediate between the homozygous dominant and the homozygous recessive

_____ (j) refers to how the genetic makeup is expressed in the body; the physical or outward expression of a gene

_____ (k) a homozygous dominant, homozygous recessive, or heterozygous genetic makeup; the actual gene arrangement

_____ (l) refers to a person with the same alleles on homologous chromosomes

_____ (m) inactivated X chromosome in females

_____ (n) heterozygous individuals who possess a recessive gene (but do not express it) and can pass the gene on to their offspring

_____ (o) interchange of portions of nonhomologous chromosomes

_____ (p) an allele that masks the presence of another allele and is fully expressed

(1) genotype
(2) phenotype
(3) alleles
(4) aneuploid
(5) incomplete dominance
(6) multiple-allele inheritance
(7) polygenic inheritance
(8) sex-linked inheritance
(9) homozygous
(10) heterozygous
(11) carriers
(12) dominant trait
(13) mutation
(14) nondisjunction
(15) translocation
(16) Barr body

15. Match the following:

_____ (a) the embryonic membrane that entirely surrounds the embryo

_____ (b) functions as an early site of blood formation; contains cells that migrate into the gonads and differentiate into the primitive germ cells

_____ (c) becomes the principal part of the embryonic placenta; produces human chorionic gonadotropin

_____ (d) modified endometrium after implantation has occurred; separates from the endometrium after the fetus is delivered

_____ (e) contains the vascular connections between mother and fetus

_____ (f) the fetal portion is formed by the chorionic villi and the maternal portion is formed by the decidua basalis of the endometrium; allows oxygen and nutrients to diffuse from maternal blood into fetal blood

_____ (g) serves as an early site of blood vessel formation

_____ (h) fingerlike projections of the chorion that bring maternal and fetal blood vessels into close proximity

_____ (i) plays an important role in induction whereby an inducing tissue stimulates the development of an unspecialized responding tissue into a specialized tissue

(1) decidua
(2) placenta
(3) amnion
(4) chorion
(5) allantois
(6) yolk sac
(7) notochord
(8) chorionic villi
(9) umbilical cord

ANSWERS

Answers to Self-Quiz Questions

Chapter 1   1. tissue 2. metabolism, anabolism, catabolism 3. intracellular fluid (ICF), extracellular fluid (ECF) 4. true 5. false 6. false 7. e 8. d 9. a 10. c 11. c 12. (a) 4, (b) 6, (c) 8, (d) 1, (e) 9, (f) 5, (g) 2, (h) 7, (i) 3, (j) 10 13. (a) 1, (b) 12, (c) 1, 6, (d) 6, (e) , (f) 8, (g) 7 (h) 3, (i) 2, (j) 10 14. (a) 4, (b) 1, (c) 3, (d) 6, (e) 5, (f) 7, (g) 2 15. (a) 6, (b) 1, (c) 11, (d) 5, (e) 10, (f) 8, (g) 7, (h) 9, (i) 4, (j) 3, (k) 2

Chapter 2   1. 8 2. solid, liquid, gas 3. monosaccharides, amino acids 4. true 5. false 6. true 7. c 8. a 9. d 10. b 11. e 12. a 13. e 14. (a) 1, (b) 2, (c) 1, (d) 4, (e) 3 15. (a) 11, (b) 1, (c) 8, (d) 3, (e) 7, (f) 4, (g) 5, (h) 9, (i) 10, (j) 12, (k) 6, (l) 2

Chapter 3   1. plasma membrane, cytoplasm, nucleus 2. apoptosis, necrosis 3. Telomeres 4. UAG 5. false 6. true 7. true 8. e 9. c 10. c, g, i, b, d, k, f, j, a, e, h 11. a 12. c 13. (a) 2, (b) 3, (c) 5, (d) 7, (e) 6, (f) 8, (g) 1, (h) 4 14. (a) 2, (b) 9, (c) 3, (d) 5, (e) 11, (f) 8, (g) 1, (h) 6, (i) 10, (j) 7, (k) 13, (l) 4, (m) 12 15. (a) 3, (b) 9, (c) 1, (d) 5, (e) 11, (f) 4, (g) 8, (h) 7, (i) 2, (j) 10, (k) 6

Chapter 4   1. epithelial, connective, muscle, nervous 2. arrangement of cells in layers, cell shape 3. true 4. true 5. e 6. b 7. a 8. c 9. e 10. b 11. d 12. c 13. (a) 4, (b) 8, (c) 5, (d) 2, (e) 6, (f) 3, (g) 1, (h) 7 14. (a) C, (b) M, (c) N, (d) E, (e) C, (f) E, (g) M, (h) E, (i) C, (j) M, (k) N, (l) E, (m) C, (n) E, (o) M and N 15. (a) 3, (b) 5, (c) 8, (d) 13, (e) 9, (f) 7, (g) 11, (h) 6, (i) 2, (j) 4, (k) 10, (l) 12, (m) 1

Chapter 5   1. stratum lucidum 2. eccrine, ceruminous, apocrine 3. false 4. true 5. c 6. e 7. a 8. c 9. b 10. e 11. a 12. c 13. (a) 3, (b) 5, (c) 4, (d) 1, (e) 6, (f) 11, (g) 2, (h) 8, (i) 9, (j) 10, (k) 7 14. (a) 3, (b) 4, (c) 1, (d) 2 15. (a) 4, (b) 3, (c) 2, (d) 1, inflammatory, migratory, proliferative, maturation

Chapter 6   1. interstitial, appositional 2. hardness, tensile strength 3. true 4. true 5. true 6. d 7. a 8. e 9. c 10. a 11. (a) 1, (b) 4, (c) 3, (d) 2 12. (a) 2, (b) 6, (c) 4, (d) 5, (e) 7, (f) 3, (g) 1 13. (a) 3, (b) 9, (c) 8, (d) 1, (e) 5, (f) 4, (g) 6, (h) 7, (i) 12, (j) 2, (k) 11 (l) 10 14. (a) 3, (b) 7, (c) 6, (d) 1, (e) 4, (f) 2, (g) 5, (h) 9, (i) 8, (j) 10 15. (a) 12, (b) 4, (c) 8, (d) 6, (e) 3, (f) 9, (g) 13, (h) 10, (i) 7, (j) 5, (k) 2, (l) 11, (m) 1

Chapter 7   1. fontanels 2. pituitary gland 3. sacrum, coccyx 4. false 5. false 6. c 7. c 8. a 9. e 10. d 11. e 12. (a) 4, (b) 9, (c) 7, (d) 5, (e) 3, (f) 1, (g) 2, (h) 8, (i) 6 13. (a) 7, (b) 5, (c) 1, (d) 6, (e) 2, (f) 4, (g) 8, (h) 9, (i) 3, (j) 10, (k) 11, (l) 13, (m) 12 14. (a) 2, (b) 3, (c) 5, (d) 6, (e) 4, (f) 1, (g) 5, (h) 4, (i) 2, (j) 4, (k) 3 15. (a) 3, (b) 1, (c) 6, (d) 9, (e) 13, (f) 12, (g) 2, (h) 4, (i) 5, (j) 7, (k) 10, (l) 15, (m) 8, (n) 11, (o) 14

Chapter 8   1. metacarpals 2. ilium, ischium, pubis 3. true (lesser), false (greater) 4. false 5. true 6. b 7. c 8. e 9. c 10. a 11. b and e 12. a 13. (a) 2, (b) 6, (c) 9, (d) 7, (e) 4, (f) 5, (g) 8, (h) 10, (i) 1, (j) 3 14. (a) 3, (b) 8, (c) 4, (d) 11, (e) 9, (f) 13, (g) 5, (h) 6, (i) 10, (j) 14, (k) 2, (l) 1, (m) 7, (n) 12 15. (a) 4, (b) 3, (c) 3, (d) 6, (e) 7, (f) 1, (g) 3, (h) 2, (i) 5, (j) 9, (k) 8, (l) 2, (m) 4, (n) 6, (o) 7, (p) 9, (q) 6, (r) 3, (s) 4, (t) 9, (u) 4 and 5

Chapter 9   1. joint, articulation or arthrosis 2. arthroplasty 3. false 4. false 5. false 6. e 7. d 8. b 9. c 10. a 11. c 12. e 13. (a) 5, (b) 3, (c) 7, (d) 2, (e) 6, (f) 4, (g) 1 14. (a) 6, (b) 4, (c) 5, (d) 1, (e) 3, (f) 2 15. (a) 8, (b) 11, (c) 10, (d) 13, (e) 15,

(f) 9, (g) 6, (h) 12, (i) 3, (j) 4, (k) 16, (l) 2, (m) 18, (n) 1, (o) 7, (p) 14, (q) 17, (r) 5

Chapter 10   1. motor unit 2. muscular atrophy, fibrosis 3. acetylcholine 4. true 5. true 6. e 7. a 8. c 9. e 10. d 11. b 12. (a) 5, (b) 6, (c) 9, (d) 7, (e) 2, (f) 4, (g) 10, (h) 3, (i) 1, (j) 8 13. (a) 7, (b) 10, (c) 9, (d) 12, (e) 8, (f) 11, (g) 6, (h) 1, (i) 2, (j) 3, (k) 4, (l) 13, (m) 5 14. (a) 2, (b) 3, (c) 1, (d) 1 and 2, (e) 3, (f) 2, (g) 1, (h) 3, (i) 1 and 2, (j) 3, (k) 2 and 3, (l) 3 15. (a) 10, (b) 2, (c) 4, (d) 3, (e) 6, (f) 5, (g) 1, (h) 12, (i) 7, (j) 9, (k) 11, (l) 8

Chapter 11   1. buccinator 2. gastrocnemius, soleus, plantaris, calcaneus 3. true 4. true 5. b 6. c 7. c 8. a 9. e 10. e 11. (a) 6, (b) 2, (c) 8, (d) 5, (e) 3, (f) 1, (g) 7, (h) 4 12. (a) 13, (b) 9, (c) 8, (d) 6, (e) 3, (f) 11, (g) 10, (h) 1, (i) 2, (j) 7, (k) 12, (l) 4, (m) 5 13. (a) 6, (b) 3, (c) 7, (d) 4, (e) 2, (f) 9, (g) 5, (h) 1, (i) 8 14. (a) 10, (b) 1, (c) 9, (d) 8, (e) 12, (f) 17, (g) 2, (h) 6, (i) 8, (j) 14, (k) 5, (l) 4, (m) 2, (n) 15, (o) 1, (p) 11, (q) 13, (r) 12, (s) 7, (t) 16, (u) 11, (v) 17, (w) 16, (x) 15, (y) 3, (z) 10 15. (a) 3, (b) 1, (c) 2, (d) 1, (e) 2, (f) 3, (g) 3

Chapter 12   1. somatic, autonomic, enteric 2. sympathetic, parasympathetic 3. false 4. false 5. c 6. d 7. c 8. e 9. e 10. b 11. e 12. b 13. (a) 6, (b) 12, (c) 1, (d) 2, (e) 9, (f) 14, (g) 4, (h) 8, (i) 7, (j) 13, (k) 5, (l) 3, (m) 10, (n) 15, (o) 11 14. (a) 2, (b) 1, (c) 10, (d) 9, (e) 6, (f) 3, (g) 4, (h) 5, (i) 12, (j) 8, (k) 7, (l) 13, (m) 11 15. (a) 4, (b) 5, (c) 16, (d) 8, (e) 7, (f) 1, (g) 2, (h) 10, (i) 15, (j) 6, (k) 3, (l) 13, (m) 9, (n) 11, (o) 14, (p) 12

Chapter 13   1. mixed 2. sensory receptor, sensory neuron, integrating center, motor neuron, effector 3. true 4. false 5. c 6. c 7. a 8. c 9. d 10. e 11. a 12. d 13. (a) 1, (b) 8, (c) 4, (d) 2, (e) 11, (f) 1, (g) 6, (h) 5, (i) 3, (j) 9, (k) 1, (l) 12, (m) 7, (n) 2, (o) 10 14. (a) 14, (b) 12, (c) 13, (d) 1, (e) 2, (f) 5, (g) 11, (h) 8, (i) 10, (j) 9, (k) 15, (l) 4, (m) 7, (n) 3, (o) 6 15. (a) 2, (b) 1, (c) 3, (d) 4, (e) 1, (f) 5, (g) 3, (h) 2, (i) 4, (j) 1, (k) 2, (l) 4, (m) 3, (n) 5, (o) 1

Chapter 14   1. corpus callosum 2. frontal, temporal, parietal, occipital, insula 3. longitudinal fissure 4. false 5. true 6. d 7. c 8. d 9. e 10. d 11. e 12. (a) 3, (b) 5, (c) 6, (d) 8, (e) 11, (f) 10, (g) 7, (h) 9, (i) 1, (j) 4, (k) 2, (l) 12, (m) 1, (n) 8, (o) 5, (p) 7, (q) 12, (r) 10, (s) 9, (t) 1 and 2, (u) 3, 4, and 6, (v) 11 13. (a) 9, (b) 2, (c) 6, (d) 10, (e) 4, (f) 11, (g) 1, (h) 2, (i) 5, (j) 8, (k) 12, (l) 7, (m) 3, (n) 6 and 8, (o) 13, (p) 7, (q) 1 14. (a) 5, (b) 9, (c) 11, (d) 6, (e) 3, (f) 1, (g) 10, (h) 8, (i) 2, (j) 4, (k) 7 15. (a) 10, (b) 2, (c) 6, (d) 8, (e) 7, (f) 5, (g) 3, (h) 11, (i) 14, (j) 13, (k) 4, (l) 1, (m) 12, (n) 9, (o) 15

Chapter 15   1. acetylcholine, epinephrine or norepinephrine 2. thoracolumbar, craniosacral 3. true 4. true 5. d 6. e 7. b 8. c 9. e 10. a 11. a 12. c 13. e, b, g, f, d, a, c 14. (a) 3, (b) 2, (c) 1, (d) 1, (e) 2, (f) 3, (g) 3, (h) 1, (i) 4, (j) 2, (k) 5 15. (a) 2, (b) 1, (c) 1, (d) 2, (e) 1, (f) 1, (g) 2, (h) 2

Chapter 16   1. sensation, perception 2. decussation 3. false 4. true 5. c 6. 1 7. d 8. b 9. d 10. e 11. e 12. d 13. (a) 3, (b) 2, (c) 5, (d) 7, (e) 1, (f) 9, (g) 3, (h) 11, (i) 8, (j) 4, (k) 6, (l) 10 14. (a) 10, (b) 8, (c) 7, (d) 1, (e) 4, (f) 3, (g) 5, (h) 6, (i) 9, (j) 2 15. (a) 9, (b) 8, (c) 4, (d) 7, (e) 10, (f) 2, (g) 3, (h) 1, (i) 5, (j) 6, (k) 11, (l) 12

Chapter 17   1. sweet, sour, salty, bitter, umami 2. static, dynamic 3. true 4. false 5. d 6. a 7. d 8. b 9. b 10. c, j, k, d, h, l, e, b, f, i, a, m, g 11. c 12. a 13. (a) 1, (b) 5, (c) 7, (d) 6, (e) 8, (f) 2, (g) 4, (h) 3 14. (a) 3, (b) 5, (c) 8, (d) 2, (e) 7, (f) 4, (g) 1, (h) 6, (i) 9 15. (a) 2, (b) 11, (c) 14, (d) 13, (e) 3, (f) 10, (g) 6, (h) 12, (i) 4, (j) 5, (k) 9, (l) 1, (m) 7, (n) 8

Chapter 18 1. fight-or-flight response, resistance reaction, exhaustion 2. hypothalamus 3. less, more 4. false 5. true 6. b 7. e 8. d 9. a 10. e 11. c 12. a 13. (a) 8, (b) 2 (and 13, 18, 22), (c) 7, (d) 1, (e) 12, (f) 20, (g) 5, (h) 18, (i) 22, (j) 15 (and 11), (k) 3, (l) 17, (m) 21, (n) 6, (o) 13, (p) 11 (and 15), (q) 4, (r) 10, (s) 14, (t) 9, (u) 16, (v) 19 14. (a) 10, (b) 8, (c) 2, (d) 12, (e) 15, (f) 4, (g) 1, (h) 16, (i) 6, (j) 9, (k) 13, (l) 7, (m) 5, (n) 14, (o) 3, (p) 11 15. (a) 12, (b) 1, (c) 11, (d) 7, (e) 3, (f) 10, (g) 2, (h) 9, (i) 4, (j) 8, (k) 5, (l) 6

Chapter 19 1. serum 2. clot retraction 3. true 4. true 5. e 6. a 7. b 8. c 9. d 10. a 11. d 12. e 13. (a) 4, (b) 7, (c) 6, (d) 1, (e) 3, (f) 5, (g) 2 14. (a) 4, (b) 6, (c) 2, (d) 7, (e) 1, (f) 5, (g) 3 15. (a) 4, (b) 6, (c) 1, (d) 3, (e) 5, (f) 2

Chapter 20 1. left ventricle 2. systole, diastole 3. false 4. true 5. a 6. c 7. d 8. b 9. b 10. e 11. c 12. (a) 3, (b) 6, (c) 1, (d) 5, (e) 2, (f) 4 13. (a) 8, (b) 4, (c) 11, (d) 5, (e) 1, (f) 9, (g) 7, (h) 2, (i) 10, (j) 6, (k) 3 14. (a) 3, (b) 2, (c) 9, (d) 14, (e) 8, (f) 7, (g) 11, (h) 12, (i) 15, (j) 4, (k) 5, (l) 1, (m) 6, (n) 21, (o) 22, (p) 19, (q) 17, (r) 18, (s) 20, (t) 16, (u) 13, (v) 10 15. (a) 3, (b) 7, (c) 2, (d) 5, (e) 1, (f) 6, (g) 4 and 7

Chapter 21 1. carotid sinus, aortic 2. skeletal muscle pump, respiratory pump 3. true 4. true 5. b 6. a 7. c 8. e 9. a 10. d 11. (a) D, (b) C, (c) C, (d) D, (e) D, (f) C, (g) C, (h) C, (i) D, (j) D, (k) C 12. (a) 2, (b) 5, (c) 1, (d) 4, (e) 3 13. (a) 11, (b) 1, (c) 4, (d) 9, (e) 3, (f) 8, (g) 6, (h) 2, (i) 7, (j) 5, (k) 10, (l) 12, (m) 13 14. (a) 2, (b) 6, (c) 4, (d) 1, (e) 3, (f) 5 15. (a) 5, (b) 3, (c) 1, (d) 4, (e) 2, (f) 4, (g) 1, (h) 5, (i) 3

Chapter 22 1. skin, mucous membranes, antimicrobial proteins, natural killer cells, phagocytes 2. antigens 3. true 4. true 5. c 6. d 7. e 8. d 9. e 10. b 11. c 12. e, h, b, f, a, d, g, i, c 13. (a) 3, (b) 1, (c) 7, (d) 4, (e) 2, (f) 5, (g) 6 14. (a) 2, (b) 3, (c) 4, (d) 7, (e) 1, (f) 6, (g) 5

Chapter 23 1. oxyhemoglobin; dissolved CO_2, carbamino compounds (primarily carbaminohemoglobin), and bicarbonate ion 2. $CO_2 + H_2O \rightarrow H_2CO_3 \rightarrow H^+ + HCO_3^{-3}$. 3. false 4. true 5. c 6. e 7. b 8. d 9. a 10. e 11. e, g, b, h, a, d, f, c 12. (a) 3, (b) 8, (c) 5, (d) 9, (e) 2, (f) 7, (g) 10, (h) 1, (i) 4, (j) 6 13. (a) 2, (b) 11, (c) 3, (d) 9, (e) 1, (f) 12, (g) 10, (h) 5, (i) 13, (j) 8, (k) 4, (l) 6, (m) 7 14. (a) 7, (b) 8, (c) 1, (d) 5, (e) 6, (f) 9, (g) 2, (h) 3, (i) 4 15. (a) 9, (b) 11, (c) 3, (d) 4, (e) 6, (f) 1, (g) 5, (h) 10, (i) 7, (j) 8, (k) 2

Chapter 24 1. monosaccharides; amino acids; monoglycerides, fatty acids; pentoses, phosphates, nitrogenous bases 2. diffusion, facilitated diffusion, osmosis, active transport 3. true 4. true 5. b 6. d 7. e 8. c 9. a 10. b 11. d 12. a 13. (a) 5, (b) 13, (c) 8, (d) 11, (e) 9, (f) 1, (g) 7, (h) 12, (i) 14, (j) 2, (k) 4, (l) 10, (m) 6, (n) 3 14. (a) 4, (b) 6, (c) 7, (d) 1, (e) 5, (f) 3, (g) 9, (h) 2, (i) 11, (j) 10, (k) 8

Chapter 25 1. hypothalamus 2. glucose 6-phosphate, pyruvic acid, acetyl coenzyme A 3. true 4. false 5. e 6. c 7. b 8. d 9. a 10. b 11. e 12. a 13. (a) 2 and 3, (b) 1, (c) 3 and 5, (d) 2, 4, 5, and 6, (e) 1, (f) 2, (g) 1, 4, and 6 14. (a) 9, (b) 12, (c) 11, (d) 10, (e) 4, (f) 13, (g) 3, (h) 5, (i) 8, (j) 1, (k) 6, (l) 2, (m) 7 15. (a) 17, (b) 15, (c) 8, (d) 19, (e) 1, (f) 7, (g) 4, (h) 10, (i) 16, (j) 14, (k) 11, (l) 2, (m) 13, (n) 6, (o) 20, (p) 9, (q) 5, (r) 18, (s) 3, (t) 12

Chapter 26 1. glomerulus, glomerular (Bowman's) capsule 2. micturition 3. false 4. true 5. d 6. e 7. c 8. b 9. e 10. a 11. g, a, h, c, o, n, l, j, i, b, d, e, m, k, f 12. g, f, b, d, h, c, a, e, i 13. (a) 8, (b) 2, (c) 10, (d) 5, (e) 3, (f) 1, (g) 7, (h) 4, (i) 11, (j) 9, (k) 6 14. (a) 4, (b) 3, (c) 7, (d) 6, (e) 2, (f) 8, (g) 1, (h) 5 15. (a) 5, (b) 4, (c) 6, (d) 8, (e) 1, (f) 2, (g) 7, (h) 3, (i) 9

Chapter 27 1. metabolic 2. bicarbonate ion, carbonic acid 3. true 4. false 5. d 6. e 7. a 8. b 9. a 10. (a) 8, (b) 9, (c) 7, (d) 1, (e) 6, (f) 2, (g) 4, (h) 5, (i) 3 10. (a) 8, (b) 12, (c) 7, (d) 5, (e) 10, (f) 6, (g) 9, (h) 13, (i) 1, (j) 11, (k) 3, (l) 4, (m) 2

Chapter 28 1. puberty, menarche, menopause 2. true 3. true 4. e 5. c 6. a 7. c 8. b 9. a 10. e 11. d 12. (a) 4, (b) 2, (c) 1, (d) 6, (e) 5, (f) 3 13. (a) 13, (b) 10, (c) 12, (d) 1, (e) 5, (f) 2, (g) 4, (h) 6, (i) 14, (j) 11, (k) 8, (l) 3, (m) 9, (n) 7 14. (a) 6, (b) 4, (c) 1, (d) 12, (e) 8, (f) 5, (g) 7, (h) 13, (i) 11, (j) 3, (k) 14, (l) 2, (m) 10, (n) 15, (o) 9

Chapter 29 1. dilation, expulsion, placental 2. corpus luteum, human chorionic gonadotropin 3. mesoderm, ectoderm, endoderm 4. false 5. a 6. d 7. b 8. e 9. b 10. c 11. (a) 6, (b) 2, (c) 11, (d) 5, (e) 10, (f) 1, (g) 3, (h) 8, (i) 9, (j) 4, (k) 7 12. (a) 3, (b) 6, (c) 5, (d) 2, (e) 1, (f) 4 13. (a) 3, (b) 4, (c) 5, (d) 1, (e) 2, (f) 6, (g) 7, (h) 10, (i) 8, (j) 9 14. (a) 7, (b) 3, (c) 14, (d) 6, (e) 4, (f) 10, (g) 8, (h) 13, (i) 5, (j) 2, (k) 1, (l) 9, (m) 16, (n) 11, (o) 15, (p) 12 15. (a) 3, (b) 6, (c) 4, (d) 1, (e) 9, (f) 2, (g) 5, (h) 8, (i) 7

Answers to Figure Questions

Chapter 5

SG5.1 Basal cell carcinoma is the most common type of skin cancer.

SG5.2 The seriousness of a burn is determined by the depth and extent of the area involved, the individual's age, and general health.

SG5.3 About 22.5% of the body would be involved (4.5% [anterior arm] + 18% [anterior trunk]).

SG5.4 Pressure ulcers typically develop in tissues that overlie bony projections subjected to pressure, such as the shoulders, hips, buttocks, heels, and ankles.

Chapter 6

SG6.1 A drug that inhibits the activity of osteoclasts might lessen the effects of osteoporosis, because osteoclasts are responsible for bone resorption.

Chapter 7

SG7.1 Most herniated discs occur in the lumbar region because it bears most of the body weight and most flexing and bending occur there.

SG7.2 Kyphosis is common in individuals with advanced osteoporosis.

SG7.3 Deficiency of folic acid is associated with spina bifida.

Chapter 18

SG18.1 Antibodies that mimic the action of TSH are produced in Graves disease.

Chapter 19

SG19.1 Agglutination refers to clumping of red blood cells.

SG19.2 Some symptoms of sickle-cell disease are anemia, jaundice, bone pain, shortness of breath, rapid heart rate, abdominal pain, fever, and fatigue.

Chapter 20

SG20.1 HDL removes excess cholesterol from body cells and transports it to the liver for elimination.

SG20.2 Coronary angiography is used to visualize many blood vessels.

SG20.3 Tetralogy of Fallot involves an interventricular septal defect, an aorta that emerges from both ventricles, a stenosed pulmonary valve, and an enlarged right ventricle.

SG20.4 In ventricular fibrillation, ventricular pumping stops, blood ejection ceases, and circulatory failure and death can occur without immediate medical intervention.

Chapter 22

SG22.1 HIV attacks helper T cells.

BMA LIBRARY
BRITISH MEDICAL ASSOCIATION
WITHDRAWN FROM LIBRARY